Future City

Volume 22

As of 2008, for the first time in human history, half of the world's population now live in cities. And with concerns about issues such as climate change, energy supply and environmental health receiving increasing political attention, interest in the sustainable development of our future cities has grown dramatically.

Yet despite a wealth of literature on green architecture, evidence-based design and sustainable planning, only a fraction of the current literature successfully integrates the necessary theory and practice from across the full range of relevant disciplines.

Springer's *Future City* series combines expertise from designers, and from natural and social scientists, to discuss the wide range of issues facing the architects, planners, developers and inhabitants of the world's future cities. Its aim is to encourage the integration of ecological theory into the aesthetic, social and practical realities of contemporary urban development.

Giulia Caneva • Flavia Bartoli
Nicoletta Cannone
Carmelo Maria Musarella
Fabio Orlandi
Editors

Urban Botany

Ecological Principles for Green Cities

Springer

Editors
Giulia Caneva
Department of Sciences, University Roma Tre
Rome, Italy

Flavia Bartoli
Institute Heritage Science (ISPC)
National Research Council (CNR)
Montelibretti, Rome, Italy

Nicoletta Cannone
Department of Theoretical and Applied Sciences
University of Insubria
Varese, Italy

Carmelo Maria Musarella
Department of AGRARIA
Mediterranean University of Reggio Calabria
Reggio Calabria, Italy

Fabio Orlandi
Department of Civil and Environmental Engineering
University of Perugia
Perugia, Italy

ISSN 1876-0899 ISSN 1876-0880 (electronic)
Future City
ISBN 978-3-032-28985-8 ISBN 978-3-032-28986-5 (eBook)
https://doi.org/10.1007/978-3-032-28986-5

This Springer imprint is published by the registered company Springer Nature Switzerland AG
The registered company address is: Gewerbestrasse 11, 6330 Cham, Switzerland

Cities and Urban Green Areas

Nowadays, human beings mainly live in metropolitan areas: almost 60% of the world's population resides in urban contexts, and this share is expected to increase further in the coming years. Cities have become the main centers of development, innovation, and economic growth, with the consequent emergence of important environmental issues such as biodiversity loss, noise, poor air quality, pollution, and rising temperatures that contribute to the formation of urban heat islands. It is therefore clear that the most accessible form of nature in the future will ever more be urban. Consequently, biodiversity and the expansion of urban green areas will represent an increasingly strategic priority for ensuring a healthier coexistence within cities, which will need to accommodate further population growth while remaining livable environments.

Humans, as organisms, have evolved and adapted their functioning to natural environmental stimuli. Today, however, the urban anthropogenic environment influences ecosystems to the point of damaging them and compromising the physiological, psychological, sociological, cultural, and economic relationships linked to the surrounding environment.

Urban life, characterized by traffic, noise, artificial light, technology, screens, frenetic rhythms, time pressure, and crowding, produces stress and continuous stimuli that lead to fatigue. As a result, mental health is significantly affected, with increased problems related to concentration, irritability, anxiety, sleep disorders, depression, aggression, and discomfort across different age groups. Children spend most of their time indoors and many hours in front of electronic devices. Their daily lives in cities leave less and less time for leisure and spontaneous play in contact with nature, while more time is spent on indoor activities with limited or no exposure to natural environments. This deprives them of essential stimuli and experiences needed for healthy growth. At the same time, parents, pediatricians, and educators lose valuable tools for preventing and treating a wide range of disorders, from sedentary lifestyles and obesity to video addiction and attention deficit hyperactivity disorder.

There are also several physiological issues related to the impact on the immune system, which defends the body against external pathogens and internal "abnormalities" and is strongly influenced by environmental conditions, stress, and pollution.

Human activities also contribute to climate change (rising temperatures and extreme weather events such as floods and droughts), which can directly or indirectly increase the disease burden. Two phenomena make cities particularly inhospitable during the summer months: the "urban heat island," measured as the temperature difference between urban and rural areas and most intense in the evening, and "heat waves," which are independent of urban systems, but whose effects overlap with those of heat islands when they occur. During heat waves, concrete and asphalt absorb large amounts of heat and release it at night, causing prolonged physiological stress over consecutive days. Numerous studies have highlighted the profound relationship between nature and well-being. For example, the presence of greenery near residential areas has been associated with reduced medical expenses and a lower incidence of psychological and psychiatric disorders in adulthood among people who grew up in areas rich in urban vegetation.

There are three mechanisms through which nature exerts positive effects on health and well-being in urban contexts: mitigation/protection, through the reduction of environmental pollution (air, noise, water, etc.); restoration, through mechanisms by which contact with nature regenerates and restores proper physiological, cognitive, and psychological functioning; generativity (instoration), providing access to fundamental psychological and social experiences through physical activity, social cohesion, and opportunities for interaction.

The benefits produced by trees and plants derive from oxygen production, the removal of atmospheric pollutants, and the mitigation of heat compared with asphalted and cemented areas. Greenery and biodiversity also play fundamental aesthetic, architectural, and cultural roles, contributing to well-being and reducing stress in the urban environment. Contact with nature reduces psychiatric risk and improves physical and emotional health, social and personal abilities, sensory skills, and the capacity for focused attention, concentration, and creative, healthy, and participatory interests. Experiencing nature supports the harmonious development of the musculoskeletal system and helps prevent cardiorespiratory, metabolic, and cancer-related diseases. Certain molecules released by plants also appear to have direct effects on the immune system.

Many studies show that spending at least 80 minutes a day outdoors reduces the risk of developing myopia in children, while playing and engaging in sensory experiences in green spaces increase creativity and collaboration, contributing to higher cognitive development. Spending at least 20 minutes a day in green areas engaging in activities is also an effective and inexpensive way to manage symptoms of attention-deficit/hyperactivity disorder. The World Health Organization has set specific targets for cities to improve the health and well-being of urban populations, including:

- Promoting safe, inclusive and accessible urban environments for all, with a minimum of 9 m^2 of green space per inhabitant
- Improving green infrastructure within cities

- Protecting urban ecosystems to ensure sustainability and guaranteeing universal access to green and public spaces

One of the UN Sustainable Development Goals (SDG 11.7) aims precisely to make cities inclusive, safe, resilient, and sustainable, including by creating and making green areas more accessible. To achieve this, we must restore space to nature, conceived as a genuine strategic "infrastructure" for building healthy and resilient communities and territories. Humans are an integral part of the ecosystems in which they live: restoring and conserving biodiversity and ecosystems is the most concrete long-term investment in the health and well-being of the entire population. Since 2022, these principles have also been enshrined in the Constitution, with the reform of Article 9, which states that "the environment, biodiversity and ecosystems shall be protected, also in the interests of future generations."

To change lifestyles and effectively address environmental, social, and cultural challenges, such as increasing urbanization, an ageing population, and children's detachment from nature, it is necessary to design new public green spaces, protect existing ones, ensure fair access to green areas near homes, and implement nature-based solutions. These actions are now urgent and essential for growing well, feeling well, and living well. Green spaces are therefore not urban accessories but fundamental structures for well-being, health, and sustainability. They help improve air quality, mitigate urban heat islands, regulate the urban microclimate, and promote natural drainage, reducing flood risk. They also protect biodiversity, restore degraded areas, and provide spaces for socialization, recreation, and environmental education.

Cities, with their high population density and intense human activity, are among the areas most exposed to the effects of climate change. Green Management Plans must go beyond the simple maintenance of existing green areas, enhancing cities' natural capital, integrating green and blue solutions into the urban fabric, and supporting climate and social justice policies. They offer solutions to:

- Mitigate climate impacts through nature-based interventions
- Increase green and blue infrastructure for greater resilience
- Reconnect urban populations with nature through accessible, high-quality green areas

Urban greenery, integrated with blue infrastructure such as waterways and artificial lakes, forms a complex and multifunctional system that provides environmental benefits (improved air quality, noise reduction, hydrogeological protection, and biodiversity conservation) and socio-cultural benefits (meeting places, educational opportunities, landscape beauty, and tourism attraction). Investing in green spaces means promoting community well-being and responding to global challenges with local, nature-based solutions. Through careful, participatory management, green spaces can become the beating heart of cities of the future.

University of Perugia
Perugia, Italy

Bruno Romano

Contents

Part I
Introduction to Urban Botany

Chapter 1
The Urban Ecosystem: Features and Needs

Nicoletta Cannone, Francesco Malfasi, and Mauro Guglielmin

1.1 Characteristics of the Urban Ecosystem

The origin of cities dates back several millennia to the Near East and the Indus Valley (Gates and Goldman 2024). The name "urban" derives from Latin "*urbs/urbis*", meaning "city", and originally referred par excellence to Rome, since it was considered the *Urbe*.

The structure and size of cities have changed significantly over the centuries, and today the phenomenon has reached its highest level of relevance (Figs. 1.1 and 1.2). In fact, urbanization is one of the most significant social transformations of modern times, with urbanized areas becoming the most rapidly expanding habitat type worldwide (Grimm et al. 2008). According to modern classifications (e.g., the US Bureau of the Census), urban areas are defined as those with a human population density reaching or exceeding 186 people per km^2 (Pickett et al. 2001).

Urbanization constitutes a near-permanent alteration of land use that eliminates locally dominant natural ecosystems (Güneralp and Seto 2013). It is characterized by high population density and by a large proportion of land surface occupied by built infrastructure (Pickett et al. 2001).

It has been estimated that approximately 55% of the world's population currently lives in cities, encompassing most people, goods, and infrastructure (Masson et al. 2020). Moreover, current and future patterns of urbanization involve large increases

N. Cannone (✉)
Department of Theoretical and Applied Sciences, University of Insubria, Varese, Italy
e-mail: nicoletta.cannone@uninsubria.it

F. Malfasi
Department of Science and High Technology, University of Insubria, Como, Italy

M. Guglielmin
Climate Change Research Center (CCRC), University of Insubria, Varese, Italy

G. Caneva et al. (eds.), *Urban Botany*, Future City 22,
https://doi.org/10.1007/978-3-032-28986-5_1

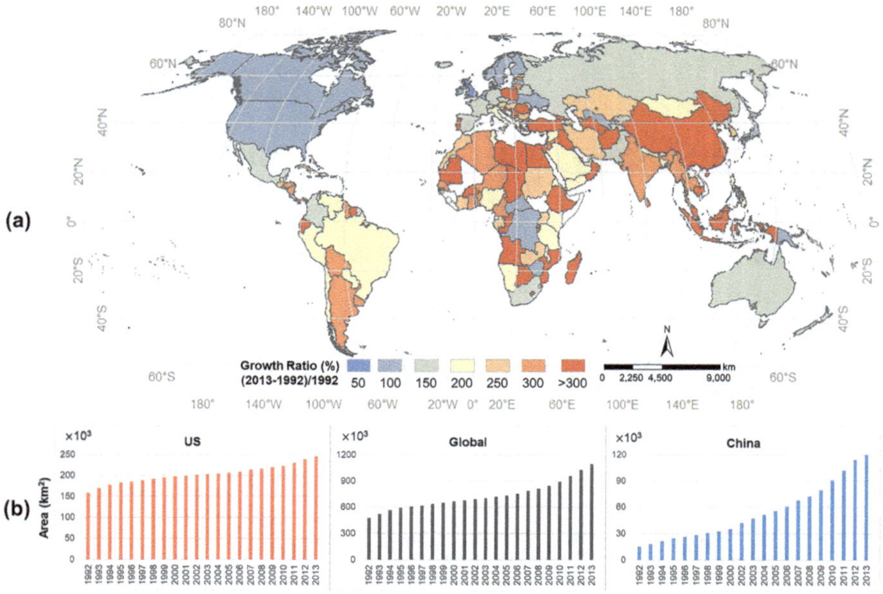

Fig. 1.1 Growth ratios of urban areas at the global scale from 1992 to 2013 (**a**) and annual dynamics of urban areas for representative regions (**b**) World map shows the growth ratio of land area from 1992 to 2013, with countries colored from blue to red indicating low to very high growth percentages. Blue shades represent growth ratios between 50% and 100%, green to yellow shades indicate moderate growth from 100% to 200%, orange to red shades show high growth from 250% to over 300%. Most of North America, parts of Europe, and Australia have lower growth ratios in blue and green, while many countries in Africa, Asia, and South America show higher growth in orange and red. Below the map, three bar charts compare land area growth over time from 1992 to 2013 for the US, globally, and China. The US shows steady moderate growth, the global chart shows a consistent increase, and China's chart shows rapid growth accelerating especially after 2000. This image highlights differences in land area growth rates worldwide and emphasizes China's significant increase compared to global and US trends (From Li et al. 2019)

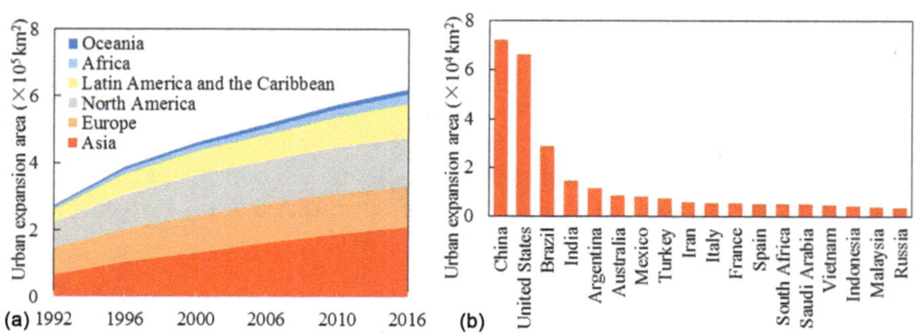

Fig. 1.2 The resulting global urban expansion, 1992 to 2016. The area chart on the left shows the growth of urban expansion from 1992 to 2016 across six regions: Asia, Europe, North America, Latin America and the Caribbean, Africa, and Oceania. (**a**) Asia has the largest increase in urban area, followed by Europe and North America, with smaller contributions from Latin America and the Caribbean, Africa, and Oceania. The stacked colors represent each region's share of total urban growth, highlighting Asia's dominant role in global urban expansion. (**b**) The bar chart on the right ranks countries by their urban expansion area in the same period. China leads by a wide margin, followed by the United States and Brazil. Other countries like India, Argentina, Australia, and Mexico show moderate urban growth, while countries such as Malaysia and Russia have the smallest increases. This comparison emphasizes the significant urban growth in Asia and the Americas, with China and the U.S. as the top contributors. (Modified from He et al. 2019)

in both the size and human density of cities, especially in megacities (i.e., cities with more than ten million inhabitants).

Urban ecosystems are highly modified and complex environments that combine artificial infrastructures and natural elements, characterized by high human density and multiple activities. This combination makes the urban environment a mixture of artificial, semi-natural, and/or natural **habitats** (Fig. 1.3) (Deutschewitz et al. 2003).

All urban areas worldwide have undergone major expansion in the last century and have given rise to dramatic changes in landscapes, with a progressive loss of natural areas, forests, and agricultural fields. The case of Rome, shown in Fig. 1.3, is an emblematic example of a widespread phenomenon of transformation (Salvati 2013).

Urban systems share some **common features**: they are complex and open systems, characterized by a mix of artificial and natural elements, with constant exchanges of energy, resources, products, services, waste, people; they are self-organizing, complex and in constant adaptation and change; and they are part of broader ecological, economic, technical, institutional, legal, and governance structures (Bai et al. 2016, 2017).

All urban ecosystems are characterized by high levels of environmental stress due to development, pollution, and direct human use (Elmqvist et al. 2013; Bai et al. 2017). Moreover, urbanization triggers several major environmental changes exacerbating abiotic stresses, including changes in local and regional climate, habitat fragmentation, and anthropogenic disturbance, with biotic impacts such as losses of native biodiversity and increases in non-native species, and the suppression of natural disturbances (which hampers the natural patterns of succession, especially in earlier stages), hence affecting ecosystem regeneration and evolution (Müller et al. 2010; Bai et al. 2017).

Urban ecosystems include cities but also the surrounding socio-ecological systems where most people live. Given the heterotrophic nature of cities and their dependence on other territories for energy and material supply, the environmental impacts of urbanization are not limited to the land occupied by cities but extend to local, regional, and global scales (Pickett et al. 2001; Bai et al. 2017). Indeed, the ecological footprint of cities extends well beyond their boundaries and may involve tens to hundreds of times the area occupied by a city (Folke et al. 1997; Kaye et al. 2006), making urbanization one of the major threats to global biodiversity (Grimm et al. 2008). Therefore, compared to the area it occupies, the impact of urban land-use change is disproportionately large (Bai et al. 2017).

Cities are artificial mosaics of patches of various sizes and land-use types (ranging from remnants of natural environments to totally artificial surfaces), resulting in **highly fragmented environments** (Faeth et al. 2011) and including other ecosystem types such as forests, lakes, rivers, and agricultural areas. The structural physical base of cities is composed of different types of substrates, including natural substrates (geological characteristics of the site), material imported from other geological regions (e.g., railway ballast, mine spoil), artificial substrates (e.g., slag), and mixtures of artificial and natural substrates (e.g., slag/clay) (Rebele 1994).

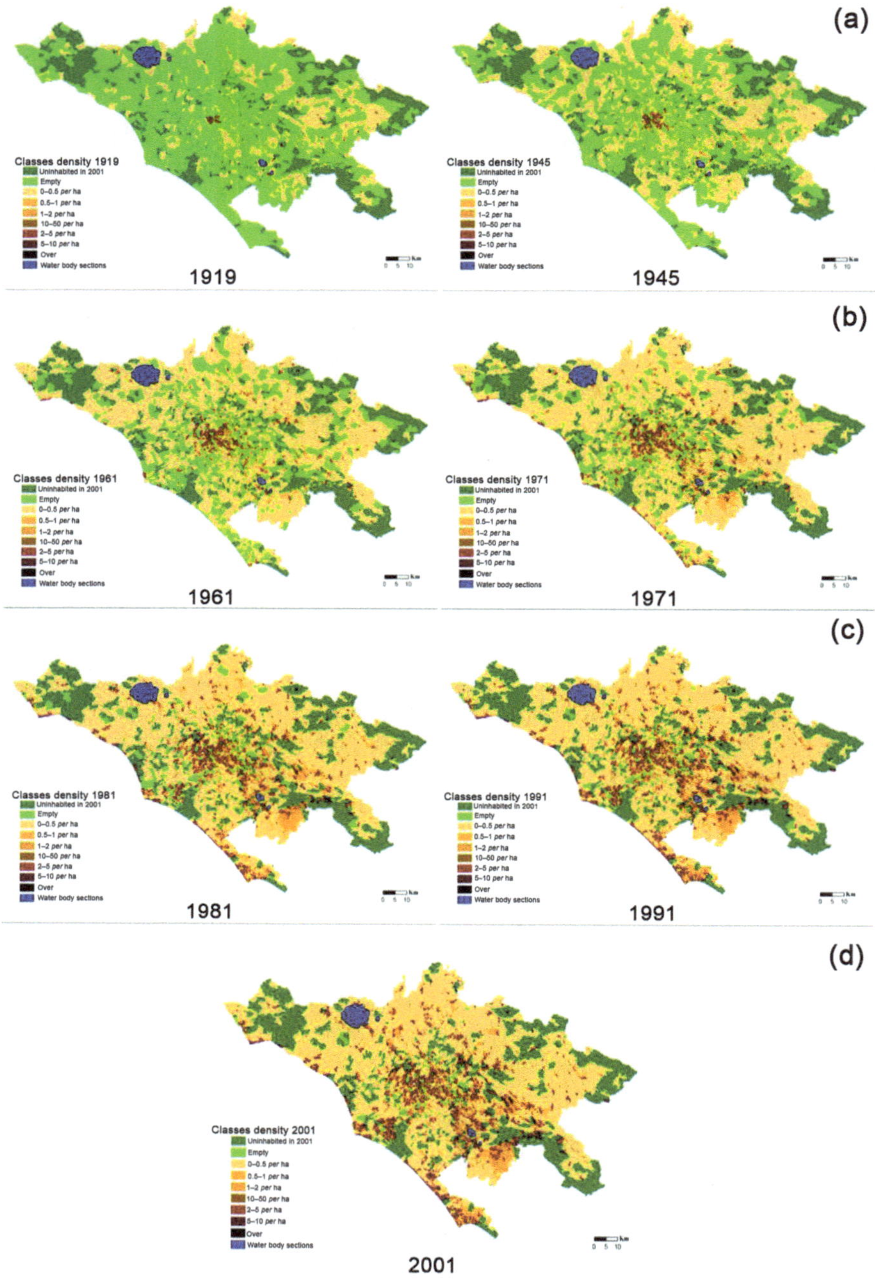

Fig. 1.3 Urban expansion and land consumption in Rome during the last century. Seven maps show changes in population density in a region from 1919 to 2001. (**a–d**) Each map uses colors from green to dark red to represent increasing population density, with green for very low density and dark red for the highest. Blue areas mark water bodies. Over time, green areas shrink while yellow, orange, and red areas expand, especially near the center and southern parts, indicating growing urbanization and population concentration. The maps highlight how the region became more densely populated and urbanized over the 82 years. (Courtesy of Salvati 2013)

The urban matrix is characterized by both **spatial and temporal habitat heterogeneity** (Fig. 1.2). Spatial heterogeneity is associated with the types and densities of buildings, vegetation, and infrastructure, and these features may vary over short distances (Cadenasso et al. 2007). Temporal heterogeneity is due to historical patterns of urban development, which result in habitats of different ages and successional stages, with distinct vegetation, surrounding buildings, ground surfaces, and soil legacies (Hruska 2006; Cadenasso et al. 2007).

Therefore, urban ecosystems are **complex mosaics of habitats** characterized by different levels of fragmentation (leading to smaller sizes and increased edge effects), connectivity, age and successional stages, and are subject to specific climatic, topographic, historical and economic factors trigger selective forces capable of modifying physiology, phenology and behavior of city dwelling organisms (Grimm et al. 2008; Faeth et al. 2011). These selective factors may act both directly (e.g., the presence of buildings and impervious surfaces, habitat modification and fragmentation, wildlife feeding, human management of vegetation) and indirectly (through changes in climate, biogeochemical cycles, light, acoustic emissions, and pollution), as well as over short-term (e.g. phenotypic changes) or over long-terms periods through changes in the population genetics of urban species (Grimm et al. 2008).

The **drivers of urban biodiversity** include abiotic stresses due to habitat fragmentation and degradation, the suppression of natural disturbance, which is replaced by anthropogenic disturbances, and alterations and gradients in productivity, with implications for food-web and trophic interactions and dynamics. They also include shifts in habitat preference, immigration, and emigration, and interspecific interactions (such as competition, predation, and mutualism) affecting survival and reproduction rates, as well as increased risks and impacts of natural hazards and climate change (Shochat et al. 2006, 2010; Grimm et al. 2008; Faeth et al. 2011; Bai et al. 2017).

Usually, the **biodiversity** of urban areas is characterized by decreased species richness and evenness, and/or low species richness with higher population densities (and/or abundance) for most taxonomic groups, with a loss of native biodiversity and a gain of non-native species compared to the nonurban areas, as well as altered species composition and dominance within communities (Fig. 1.3) (Grimm et al. 2008; Faeth et al. 2011; Bai et al. 2017).

Important exceptions are related to the observed increase of abundance and biomass of birds and arthropods, as well as increases in plant species richness and evenness (Faeth et al. 2011; Bai et al. 2017). Bird species richness often peaks at intermediate levels of urbanization, driven by greater habitat heterogeneity.

Birds often exhibit peaks in species richness at intermediate levels of urbanization due to greater heterogeneity of edge habitats. In urban environments, the availability and nature of nutrients drive shifts in species dominance, favoring granivorous over insectivorous birds, medium-sized and generalist predators among mammals, and generalist over specialist among arthropods. Some species may become more efficient competitors in urban areas, becoming dominant or increasing in abundance, also due to the lack of natural enemies.

Plant species richness, composition, and abundance in urban areas reflect both the remnants of natural vegetation and the impacts of land-use change, landscape transformation, and habitat fragmentation due to human management. These factors shape the structure, diversity, dynamics, and floristic composition of plant communities, with cascading effects on food-web and ecosystem functioning (Shochat et al. 2006, 2010). Indeed, the original landscape and vegetation have been deeply transformed and reconstructed to create gardens, lawns, urban forests, parks, and recreational areas, mostly using non-native species (Faeth et al. 2011). Therefore, plants are the biotic component of urban areas most subject to human control, with few exceptions related to some pests or domesticated species (Grimm et al. 2008).

Urban floras are ecologically rich and diverse, including both native and non-native (or alien) species, with origins in distinct geographical areas, arriving spontaneously and/or being accidentally or intentionally introduced (Lososová et al. 2012). The pool of native species is mainly determined by regional climate, reflecting the remnants of the area's original natural vegetation. The link to climate is well established for trees, with urban tree species showing similar taxonomic diversity in cities with similar climates (Jenerette et al. 2016). However, urban parks and forests are mainly characterized by lower stem density, with the exception of old-growth trees, while urban and suburban forests often show increased species richness due to the larger occurrence of non-native species, and to differences in the composition of the pool of herbaceous species (Pickett et al. 2001).

Non-native species have been classified into two categories: archaeophytes and neophytes, the former introduced in Europe before, and the latter after, c. AD 1500 (Pyšek et al. 2004). Usually, the number of non-native species increases from the suburbs toward the city center (Pickett et al. 2001), also because non-native species in some cases occupy their urban niche better than native species and thus rapidly expand. Urbanization also leads to the homogenization of plant assemblages, especially among non-native species able to withstand stresses and pressures associated with urban conditions (Lokatis and Jeschke 2022).

However, urban ecosystems may also host rare and even threatened species, especially when human management is aimed at conserving and/or reconstructing native habitats. Interestingly, in many cases, native species account for more than 50% of urban flora, especially in extra-European cities (La Sorte et al. 2014). It has also been noted that within similar bioclimatic envelopes, plant species richness tends to increase with city size as well as with its age, with more species occurring in older than in younger cities, and similarly in larger than in smaller ones (Pickett et al. 2001). However, this pattern shows large variability among and within cities, depending on the availability of green spaces, which usually range from less than 5% to approximately 45% (Beninde et al. 2015). Given the heterogeneity of urban ecosystems, urban vegetation is characterized by communities with novel and urban-specific species combinations. Urban vegetation has also been classified based on the combination of prevailing CSR (Competitive, Stress-tolerant, and Ruderal) strategies and management types into ruderal, managed, and residual categories (Pickett et al. 2001), with ruderal species also called "urban specialists."

Urban ecosystems are characterized by specific and, in some cases, **unique physical/abiotic features** that influence their structure, composition, and functioning. Such unique urban features include: the widespread occurrence of impervious surfaces (e.g., concrete, asphalt, and other non-porous materials that disrupt natural water flow and infiltration); the urban heat island (UHI) effect and associated climate alterations; altered and artificial soils; pollution of air, soil, and water; light pollution (due to large amounts of artificial and polarized light); and recurrent, strong acoustic emissions caused by engines, alarms, and radio systems. Due to the presence of these unique features, urban habitats are often considered novel environments, distinct from natural habitats (Rebele 1994; Hobbs et al. 2006; Seastedt et al. 2008). However, while some features are unique, others **resemble specific natural habitats** from a biological and ecological perspective. Severe and frequent disturbances in cities resemble those in natural environments, subject to seasonal hurricanes, storms, fires, and landslides. Fragmentation also occurs in natural habitats at various spatial scales (Fahrig 2003), while rock outcrops, cliffs, and rocky deserts impose environmental constraints similar to those of impervious surfaces such as buildings and paved areas. Likewise, the surface energy balance and thermal regime of bare rocky surfaces share characteristics with those of the urban heat island, and the polarized light reflection from asphalt resembles that from water.

Despite their analogs in the natural environment, these **urban features are unique in terms of the intensity, scale, extent, and combination of selective environmental pressures** that they impose and, consequently, the impacts they have on living organisms and ecosystem processes (Faeth et al. 2011). These selective pressures have already influenced both non-adaptive and adaptive evolution, promoting new biotic patterns and **driving novel evolutionary dynamics in urban systems** (Johnson and Munshi-South 2017).

Knowledge of the environmental similarities between unique urban abiotic features and their natural analogs suggests that species colonizing and inhabiting urban environments share functional and ecological traits with their counterparts in natural habitats (Richardson et al. 2010). This may be a key factor in improving the management of urban ecosystems to make them more resilient and better adapted to current and future challenges, and in enhancing the well-being and safety of urban environments (Faeth et al. 2011).

1.2 Energy and Resources in the Cities

Cities are heterotrophic ecosystems that depend on external energy and material sources and have a large ecological footprint, requiring an energy input 10–100 times greater than that of unmanaged ecosystems (Collins et al. 2000; Kaye et al. 2006). These complex, dynamic, spatially and temporally heterogeneous systems share several major problems (Grimm et al. 2008): land-use and land-cover change, alterations of biogeochemical cycles (particularly C, N, S, and P), changes in hydrological cycles and systems, soil alteration, climate alteration, and loss of biodiversity. Cities have historically developed in areas of high productivity (e.g., near lakeshores, coastlines, rivers, river deltas, and estuaries) (Faeth et al. 2011), where human activities profoundly affect the major **biogeochemical cycles**. They generate elevated emissions of greenhouse gases per unit of land area, particularly CO_2, CH_4, and N_2O, as well as NO, O_3, SO_2, HNO_3, and various organic acids (Kaye et al. 2005; Grimm et al. 2008; Bai et al. 2017).

Urban systems are hotspots for the emission and accumulation of N, P, metals, and micropollutants, including both organic contaminants (such as pharmaceuticals—especially antibiotics and hormones—and personal care products) and industrial chemicals. They are also characterized by acid rain and N deposition, which affect biogeochemical cycles, ecosystem processes, and organism physiology (Grimm et al. 2008; Bai et al. 2017; Du et al. 2022). These nutrients and pollutants accumulate in urban soils and hydro systems, and their impact on biogeochemical cycles and biodiversity varies depending on the structure and functioning of the urban fabric, particularly the extent of impervious surfaces and the residual capacity of urban areas to provide key ecosystem services such as water purification and pollution attenuation (Bai et al. 2017). An emerging issue concerns the elevated concentrations of pollutants and anthropogenic substances undergoing **human-accelerated weathering process** and corrosion in the artificial built environment, with both acidification and alkalinization processes, enhanced rates of geochemical transformations triggered by changes in temperature, pH, redox potentials and ionic strength, thus promoting the combination of these pollutants into novel combinations named "**chemical cocktails**" (Kaushal et al. 2020). These mixtures accumulate in urban soils and watersheds, where they may exert interactive effects on urban ecosystems, producing impacts greater than the sum of individual chemical stressors (Kaushal et al. 2020). Urban green areas and hydro systems can play a key role in counterbalancing the adverse effects of urban emissions on biogeochemical cycles by enhancing ecosystem services such as water purification and pollution attenuation (Du et al. 2022). Urban greening contributes to C and N sequestration, improves groundwater and soil quality and functioning, and helps remove air and soil pollutants (Pickett et al. 2001; Bai et al. 2017; Du et al. 2022).

Water cycles and hydrological systems are drastically modified in urban environments compared to agricultural and wild lands (Pickett et al. 2001) due to human water use (e.g., domestic and industrial consumption), alterations in water quantity and quality, changes in drainage and infiltration patterns, and modifications of natural hydro systems (Grimm et al. 2008). These modifications include designed and altered streams, canals, flood channels, and natural or artificial lakes and ponds that serve urban areas by providing flood conveyance and water delivery. Urban hydro systems neither preserve the characteristics of the original hydrological networks nor fully replace them. Spatial constraints often prevent water from following natural flow patterns during peak precipitation events, generally increasing water velocity. This is one of the main reasons for the **high hydrogeological risk affecting nearly all cities**, especially during extreme climatic events such as floods, hurricanes, and, in rare cases, tsunamis (Walsh et al. 2012). The water cycle is deeply altered in urban systems because the increase in impervious and impermeable ground surfaces, combined with limited vegetation cover, reduces water drainage and infiltration, prevents groundwater aquifer recharge, increases surface runoff, and decreases evapotranspiration (Carlson and Arthur 2000; Qiu et al. 2023; Xu et al. 2025).

Urban water management must therefore address key issues such as adequate sanitation, and the conservation and restoration of efficient hydrological cycles that enhance groundwater drainage and recharge, increase evapotranspiration, and provide protection from natural hazards, particularly hydrogeological risks. A large portion of the global urban population is already facing water shortages, and it has been estimated that by 2050, almost half of the world's urban residents will experience **water scarcity** (He et al. 2021). Urban water scarcity is becoming a global concern and a potential future emergency, as both urban populations and water demand continue to increase. One cause of water scarcity in urban systems is an imbalance in the water budget compared to non-urban areas. Under an ideal water balance from precipitation, urban areas are characterized by increased surface runoff and decreased groundwater recharge and evapotranspiration.

Another issue closely linked to water consumption and conservation in urban environments is urban greening. **Urban greening** is an important tool for mitigating the negative impacts of altered urban climates, particularly the urban heat island effect (Meili et al. 2021a, 2021b; Wong et al. 2021; Zhang et al. 2023). However, urban vegetation requires substantial water, either from rainfall or irrigation, especially in cities located in dry regions or experiencing hot, dry seasons. This demand may intensify existing water scarcity challenges (Reyes-Paecke et al. 2019; Cheung et al. 2022, 2024; Zhang et al. 2023). Proper management of urban green spaces can partially compensate for impairments in the urban water cycle by enhancing groundwater infiltration and reducing surface runoff through canopy rainfall interception. As a cascade effect, urban greening can reduce downstream material transport, increase nutrient uptake from stormwater runoff, lower hydrological risks, and improve water quality (Kaye et al. 2006; Bai et al. 2017). In urban systems water management needs to be finalized both to water supply (for human population, activities and services), stormwater drainage, and wastewater disposal, but also to provide mandatory ecosystem services associated to hydro systems, such as protection from floods and drought, habitat availability for the biota through the active conservation and/or the restoration of natural, semi-natural and artificial hydro systems (Pickett et al. 2001; Bai et al. 2017).

Soils in urban systems include natural soils altered by human activities and **urban anthropogenic soils** (UAS, also known as *made land*), which derive from substrates created through the deposition of debris, soil, and rock in urban sites (Pickett et al. 2001). Three main processes have been identified in the formation of UAS (Howard 2021): (1) *Metapedogenesis, where pre-existing natural soils are transformed by human additions of organic matter with changes of the soil profile;* (2) *Technopedogenesis, with the formation of a new soil due to mechanical mixture of earth materials with sediments deposited by humans and developed in artifact-bearing;* (3) *Ekranopedogenesis, where impervious materials and structures seal pre-existing soils. UAS are classified into two main categories, Anthrosols (i.e.,* *human-altered soils) and Technosoils (i.e., human-transported soils). Anthrosols include* Hortic (garden) and Nekric (cemetery) soils, while Technosols include Urbic (residential) and Ekranic (sealed), which are the most widespread, as well as Industric (industrial) and Spolic (mine) soils (Howard 2021).

Urban soils exhibit altered physical and chemical characteristics and functions, including nutrient accumulation or depletion and contamination by heavy metals and other pollutants (e.g., from air and water pollution, fertilization, irrigation, and acid rain). They are also subject to physical disturbance due to reworking, the addition of fill material, sealing by impervious surfaces, and compaction that may lead to anoxic conditions (Howard 2021). Compared to natural soils, urban soils often exhibit altered thermal regimes (due to the urban heat island effect), increased hydrophobicity resulting from changes in water retention capacity, and reduced infiltration potential due to compaction. Alterations in soil chemistry can profoundly affect soil biota, particularly by accumulating toxic substances that induce physiological stress, including sublethal or lethal effects. These conditions strongly select for soil decomposers and primary producers (Pickett et al. 2001; Howard 2021). Such features have broad impacts on both abiotic and biotic soil environments, with cascading consequences for soil development. The ecological role of urban soils is critical, as they provide key ecosystem services, including water drainage, substrate and physical support for green infrastructure, carbon sequestration, and habitat for biodiversity (Pickett et al. 2001; Morel et al. 2015; Vazenev and Kuzyakov 2018; Howard 2021).

Urbanization is characterized by landscape transformations that alter both the physical properties and the morphology of surfaces (Bai et al. 2017), with significant impacts on the climatic conditions of cities and their surrounding areas. The energy associated with human activities and the physical structure of cities strongly influence urban energy budgets (Oke 1988; Pickett et al. 2001). Therefore, the climate of urban areas differs markedly from that of non-urban regions, showing altered patterns of temperature, wind, and precipitation (Masson et al. 2020).

Urban climate is strictly linked to its **energy balance** (Oke 1988; Phelan et al. 2015; Masson et al. 2020), with the net radiation (Q*) being the balance between the incoming and outcoming short (K) and long-wave (L) radiation (Fig. 1.4). The urban energy balance involves the energy from incoming short (K↓) and long-wave radiation (L↓) (i.e., the solar radiation and the re-irradiated energy from the upper-atmosphere, clouds and gray infrastructure), the energy reflected (K↑) and re-irradiated (L↑) back to the atmosphere, and the energy released by human activities (Qf) represented by the heat sources associated to human activities (such as heating and cooling of urban buildings and structures, thermal energy from transportation, energy from industrial activities).

$$\mathrm{Q}* = \mathrm{K}\downarrow + \mathrm{L}\downarrow - \mathrm{K}\uparrow - \mathrm{L}\uparrow + Qf$$

The energy of the net radiation Q* is partitioned into stored energy (ΔQs), transformed energy (i.e., latent heat, Qe, and sensible heat, Qh), and anthropogenic heat (Qf).

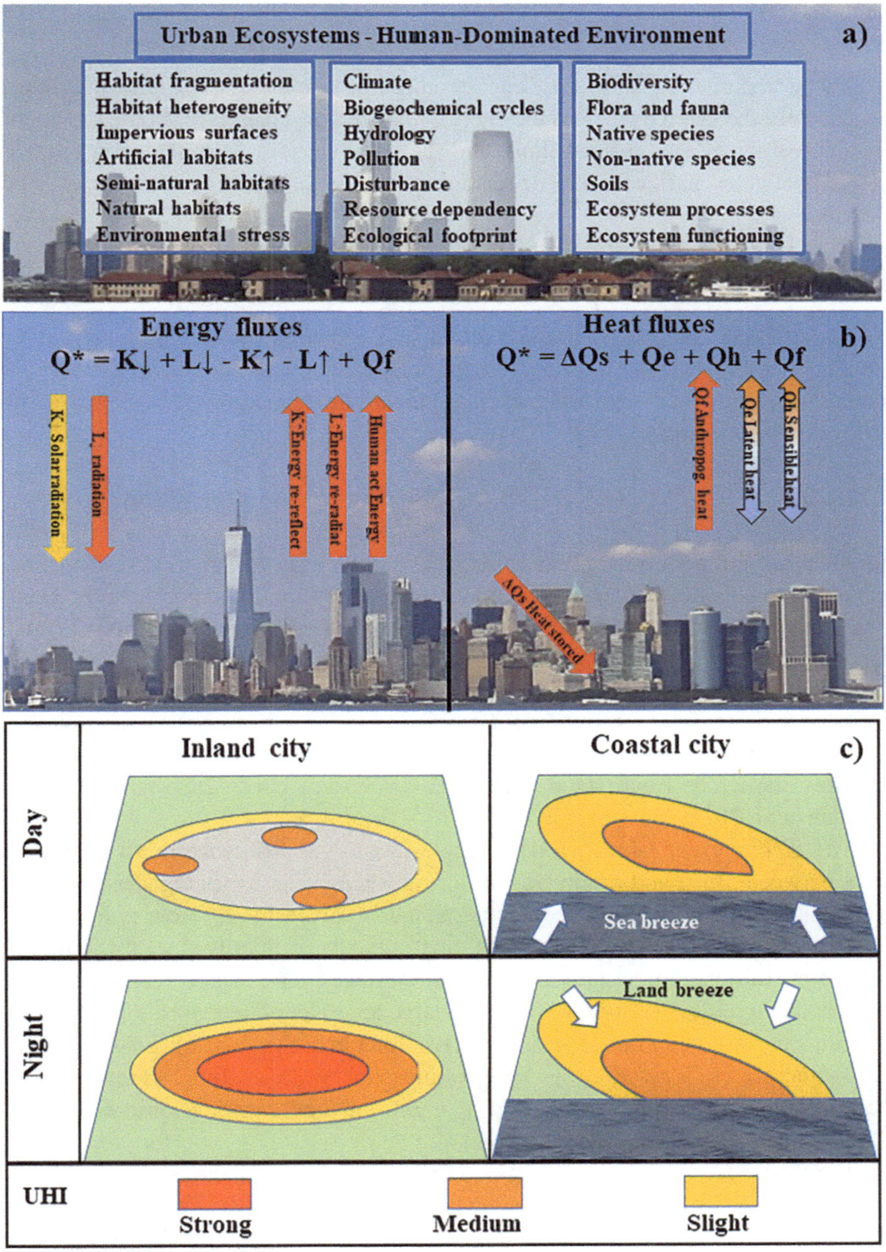

Fig. 1.4 The image explains how urban ecosystems in human-dominated environments affect energy and heat flows, and how these influence urban heat islands (UHI) in inland and coastal cities. The top section lists key factors shaping urban ecosystems, including habitat types, climate and pollution, and biodiversity elements. The middle section shows two formulas for energy fluxes and heat fluxes with arrows indicating different energy components like solar radiation, human energy use, and anthropogenic heat, highlighting how cities absorb and release energy differently. The bottom section compares UHI intensity patterns during day and night in inland versus coastal cities, using colored zones from strong (red) to slight (yellow) heat effects. Coastal cities show sea and land breezes that modify heat distribution, while inland cities have more uniform heat zones. This image helps understand how urban features and location influence city temperatures and energy balance. (Elaboration by the Authors)

The stored energy ΔQs is the energy stored in urban materials (like buildings, roads, soil), the latent heat Qe is the energy adsorbed or released for liquid or vapour changes of state (e.g., evaporation), and the sensible heat Qs is the energy transferred between surfaces and the atmosphere (e.g., through convection and conduction).

$$Q* = \Delta Qs + Qe + Qh + Qf$$

These relations could be expressed also using a thermal model, as proposed by Phelan et al. (2015) and referred to the thermal exchange of built environment with both below-ground and atmosphere, where the urban temperature T is expressed by the following equation:

$$\left(mc / \mathrm{A}\right) * \left(dT / dt\right) = \left(1 - \alpha\right) \mathrm{Qsol} + \mathrm{Qantro} - \mathrm{Qcond} - \mathrm{Qevap} - \mathrm{Qconv} - \mathrm{Q\,rad}$$

In which: m = mass of the urban fabric, c = specific heat, A = normal surface area, t = time, α = albedo, Qsol = is the the incoming solar flux (including both the beam and diffuse components) (W/m^2), Qanthro = anthropogenic heat addition (W/m^2), Qcond = conduction to the deep ground (W/m^2), Qevap = heat loss via evapotranspiration (W/m^2), Qconv = heat loss via convection to the surrounding air (W/m^2), Qrad = radiative heat loss to the sky (W/m^2).

The alteration of temperature in urban systems is characterized by two specific features: the Urban Heat Island (UHI) and the Surface Urban Heat Island (SUHI) (Phelan et al. 2015; Masson et al. 2020).

The Urban Heat Island (UHI) effect refers to the difference in **air temperature** between urban and rural areas. It occurs when urban zones register higher air temperatures than their rural counterparts and is driven by differences in net all-wave radiation (Q*) between the two environments, primarily due to the higher proportion of impervious and non-permeable surfaces (e.g., concrete, asphalt, brick, stonework) in cities. These surfaces store large amounts of heat during the day and release it as long-wave radiation at night, resulting in higher sensible heat flux (Qh). Consequently, in large cities, air temperatures can be up to 10 °C warmer than in surrounding rural areas, particularly at night and under anticyclonic conditions (Masson et al. 2020) (Fig. 1.5).

The temperature differential between urban and non-urban areas increases with the percentage of impervious surfaces. The intensity of the UHI effect also increases with the size and population density of urban areas. These thermal differences vary seasonally, with greater differences during the summer. The amount of energy received and emitted from urban areas is further influenced by geographic features—particularly latitude, orographic conditions, maritime effects, and continentality—which in turn affect wind and precipitation patterns. These factors impact the urban energy balance and shape the diurnal and seasonal expression of UHI in inland versus coastal cities (Figs. 1.4c and 1.5).

Due to their greater capacity to store heat, impervious surfaces are hotter than those in rural areas during the day, an effect known as the **Surface Urban Heat**

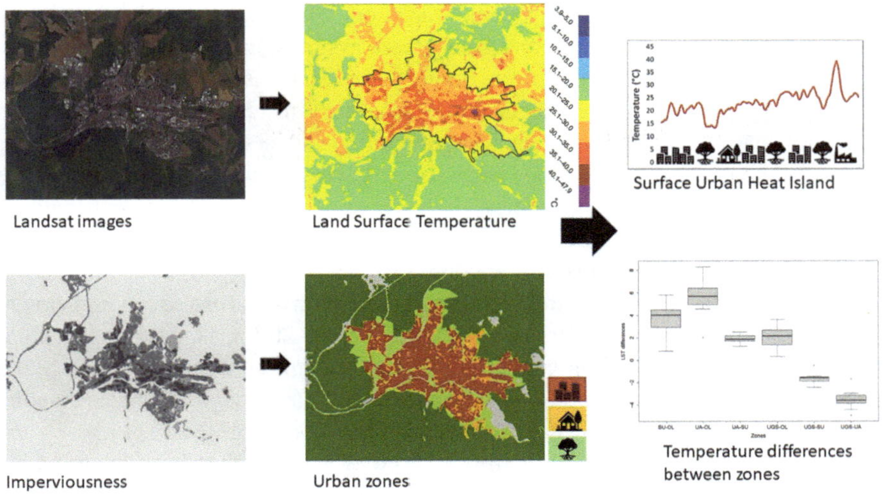

Fig. 1.5 Land Surface Temperature (LST) and Urban Heat Island (UHI) intensity between different urban zones: example of a medium-sized town in Europe. OL—open land, UA—dense urban area, SU—discontinuous urban area, UGS—urban green spaces. This image explains how satellite data is used to study urban heat. It starts with a clear city photo from Landsat images, showing the area's layout. Next, a colorful map shows land surface temperatures, with red areas being hottest and green areas coolest, highlighting how heat spreads in the city. A line graph connects temperature to different urban features, showing that built-up areas are warmer than natural ones. Below, a black-and-white map shows impervious surfaces like roads and buildings, which don't absorb water. Next to it, a map divides the city into urban zones marked by colors: red for dense buildings, yellow for residential areas, and green for vegetation. Finally, a box plot compares temperature differences across these zones, showing that dense urban areas are significantly hotter than green spaces. The image highlights how urban design affects heat, important for planning cooler, healthier cities. (Courtesy of Murtinová et al. 2022)

Island (SUHI) (Moffett et al. 2019). SUHI is the difference in the temperature of the surfaces (e.g., roofs, roads, grass, etc.) between urban and non-urban areas. Typically, SUHI is characterized by higher temperatures, especially during daytime, although exceptions occur in cities located in desert or arid climates (Masson et al. 2020).

Importantly, the thermal influence of cities extends vertically from the top of urban surfaces up to approximately 1 km into the atmosphere, forming the **Urban Boundary Layer** (UBL). This vertical layer, which corresponds to the upper limits of urban pollution, is characterized by atmospheric temperature inversions and is transported downwind in a "plume effect," contributing to increased warming and enhanced precipitation in rural areas located downwind of cities (Oke 1988; Song and Wang 2016; Masson et al. 2020).

The aerodynamically rough urban surface, together with its structure and architecture, interacts with the UHI to modify **wind speed and direction**, either increasing or decreasing airflow and altering circulation patterns. These changes influence air quality, pollution dispersion, and the precipitation regime. In coastal and mountain cities, winds exhibit specific behaviors. High-rise urban structures can disrupt the complex sea-breeze and land-breeze systems, and interactions between sea breezes and the UHI may even trigger thunderstorms (Masson et al. 2020). Mountainous terrain also influences urban wind patterns, depending on relief: low relief can lead to air stagnation, reduced air quality, and an enhanced UHI, whereas high relief can generate downslope katabatic or drainage flows that help mitigate the UHI. At the local scale within cities, the configuration of buildings—particularly the height-to-width aspect ratio of streets—can modify wind speed and direction, promoting the formation of street canyon winds.

Increased precipitation and cloudiness characterize urban areas and downwind areas compared to non-urban areas (Pickett et al. 2001). This is due to a combination of factors, including convective fluxes of warm air associated with SUHI and UHI, and higher concentrations of aerosols, pollutants, and particulates creating condensation nuclei in urban atmospheres. The result of these processes are increased precipitations both over the city core as well as extended downwind the city, with the amount of rainfall being proportional with the size of the city and the extent of the could condensation nuclei, as well as increase of thunderstorms, in particular in big cities located in regions characterized by warm and humid climates (Sarangi et al. 2018). The strongest contrasts in precipitation between urban and non-urban areas occur where background aerosol concentrations are low relative to urban levels, such as in coastal cities or those embedded in tropical forests. In such settings, high aerosol concentrations can increase city-edge precipitation by up to 100 mm/hr (Schmid and Niyogi 2017). Importantly, increased precipitation is not spatially uniform across urban and downwind areas; instead, rainfall tends to occur in distinct, localized, and intense clusters.

The characteristics of urban climate increase cities' vulnerability to **extreme climatic events** and high-impact weather hazards, as well as to broader **climatic change** (Masson et al. 2020). The impacts of UHI extend beyond climatic alterations and have a wide range of biological and ecological implications, affecting human health, air quality, water resources, biodiversity, and ecosystem functioning (Grimm et al. 2008). UHI contributes to heat stress in humans and urban biota, further modifies the surface energy balance by altering heat and moisture fluxes, increases ozone and photochemical smog formation, may promote air stagnation and worsen pollution episodes, and elevates energy demand for cooling (Pickett et al. 2001; Grimm et al. 2008).

Quantitative assessments of urban impacts on climate, and in particular of the UHI effect, are challenging because of the structural, climatic, and topographic heterogeneity of cities. This complexity underscores the need for consistent, broadly applicable classification systems. One of the most effective systems is the **Local Climate Zone (LCZ)** classification (Stewart and Oke 2012), a universal scheme for characterizing urban and non-urban land cover. LCZ includes ten urban classes, defined at spatial scales between 0.1 and 1 km^2, based on a combination of factors such as building height (high-rise, mid-rise, low-rise), building density (dense, open, sparse), urban fabric (informal settlements, light industry, heavy industry), construction materials, and the presence of vegetated land cover. LCZ allows relating the urban microclimate to specific, visible (and suitable for assessment using remote sensing techniques) and categorical spatial structures. Areas within the same LCZ class are expected to have a similar local effect on UHI, characterized by spatially homogeneous temperatures that differ from those of their neighborhoods in different LCZ classes. The LCZ classification has also been widely adopted because the assessment of UHI intensity at the local (microscale) scale in urban and nonurban areas provides key information for management and risk prevention, as UHI is one of the most significant climate hazards in this complex system. The use of LCZ highlighted the importance of vegetation as a major driver of heat reduction through shading and evapotranspiration. However, the physical layout of built structures may also provide an "oasis effect" because compact residential areas, such as high-rise buildings in compact streets, provide better shading than open landscapes, with a cooling effect especially during the peak of the day.

1.3 The Cities and the Equilibrium Need Between Humans and Nature

Urban areas are increasingly **vulnerable to the impacts of climate and environmental change**, including associated disasters (e.g., extreme heat events, flooding, pollution). This vulnerability stems from their limited recovery capacity and resilience, high exposure to disaster risks, and frequent lack of minimum conditions for well-being, safety, health, and vitality among human populations, animals, and vegetation. The growing need to achieve a balance between humans and nature within cities has promoted the concept of **urban sustainability** (Fig. 1.6), which is recognized among the **Sustainable Development Goals** (SDGs), specifically **SDG 11:** Make cities and human settlements inclusive, safe, resilient and sustainable (UN 2015). Because cities account for approximately 80% of global economic production and around 70% of global carbon emissions, progress toward urban sustainability can contribute significantly to decarbonization, supporting **SDG 12** on responsible consumption and production. Achieving urban sustainability requires that urban system management addresses several key challenges, including the major societal challenges identified by the IUCN (Fig. 1.7): climate change mitigation and adaptation, food security, water security, disaster risk reduction, human health, economic and social development, environmental degradation, and biodiversity loss (IUCN 2020).

Nature-based solutions (NbS) are actions that protect, sustainably manage, and restore natural and/or modified ecosystems, addressing societal challenges in effective and adaptive ways while simultaneously delivering benefits for human well-being and biodiversity (IUCN 2020). NbS recognize and respond to the interactions among the economy, society, and ecosystems, and aim to support biodiversity conservation and ecosystem integrity (Bona et al. 2023). They contribute to achieving urban sustainability by addressing specific urban challenges such as flooding, heat islands, and air pollution.

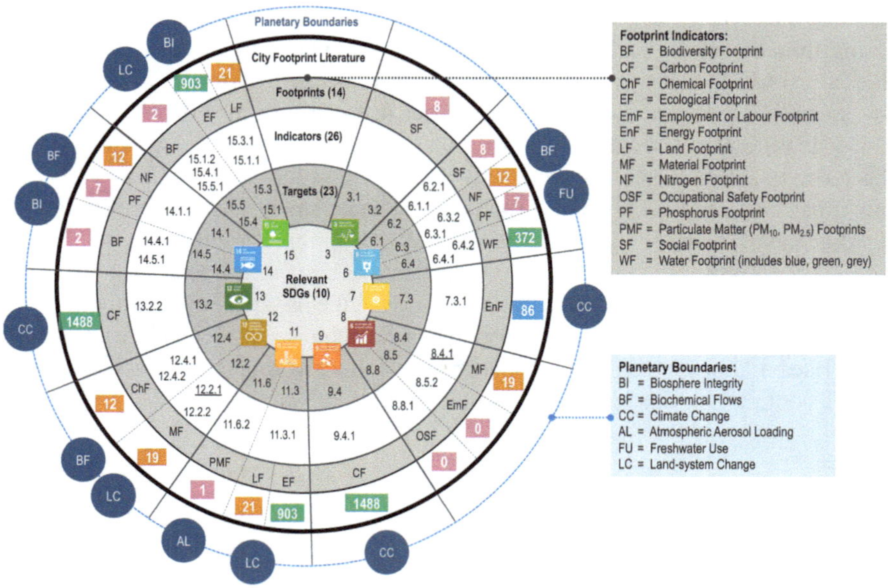

Fig. 1.6 A radar chart displays the relationship between 10 relevant Sustainable Development Goals (SDGs) and their associated environmental footprints and planetary boundaries. The chart is divided into four concentric rings labeled Targets, Indicators, Footprints, and Planetary Boundaries, moving outward from the center. Each SDG is represented by a colored icon in the Targets ring, linked to specific indicators and footprints in the outer rings. Footprint indicators include measures like biodiversity, carbon, chemical, ecological, energy, land, material, nitrogen, occupational safety, phosphorus, particulate matter, social, and water footprints, each abbreviated with two or three letters. Planetary boundaries are shown as blue circles around the outer edge, representing biosphere integrity, biochemical flows, climate change, atmospheric aerosol loading, freshwater use, and land-system change. Numbers in colored boxes near the planetary boundaries indicate quantitative values for each footprint. The chart highlights how different SDGs connect to environmental impacts and planetary limits, emphasizing the importance of balancing development goals with ecological sustainability. (Courtesy of Wiedmann and Allen 2021)

Managing urban ecosystems—and particularly **urban greening**—is an effective way to implement NbS. Relevant actions include the creation of regenerated soils, community-based urban farms and gardens, green walls and roofs, accessible green corridors, local environmental compensation processes, and the enhancement of pollinator biodiversity (Fig. 1.7).

NbS contribute to **mitigating the adverse impacts of urban climate and urban climate change**, reducing urban energy demand, and providing multiple beneficial **ecosystem services,** including climate regulation, water regulation, air-quality improvement, carbon sequestration, habitat provision, water purification, recreation and ecotourism opportunities, food and fiber production, and aesthetic value (Fang et al. 2024). In addition, the beneficial effects of NbS support several Sustainable Development Goals (SDGs), both directly and indirectly—especially SDG 2 (zero hunger), SDG 3 (good health and well-being), SDG 6 (clean water and sanitation), SDG 8 (decent work and economic growth), SDG 10 (reduced inequalities), SDG

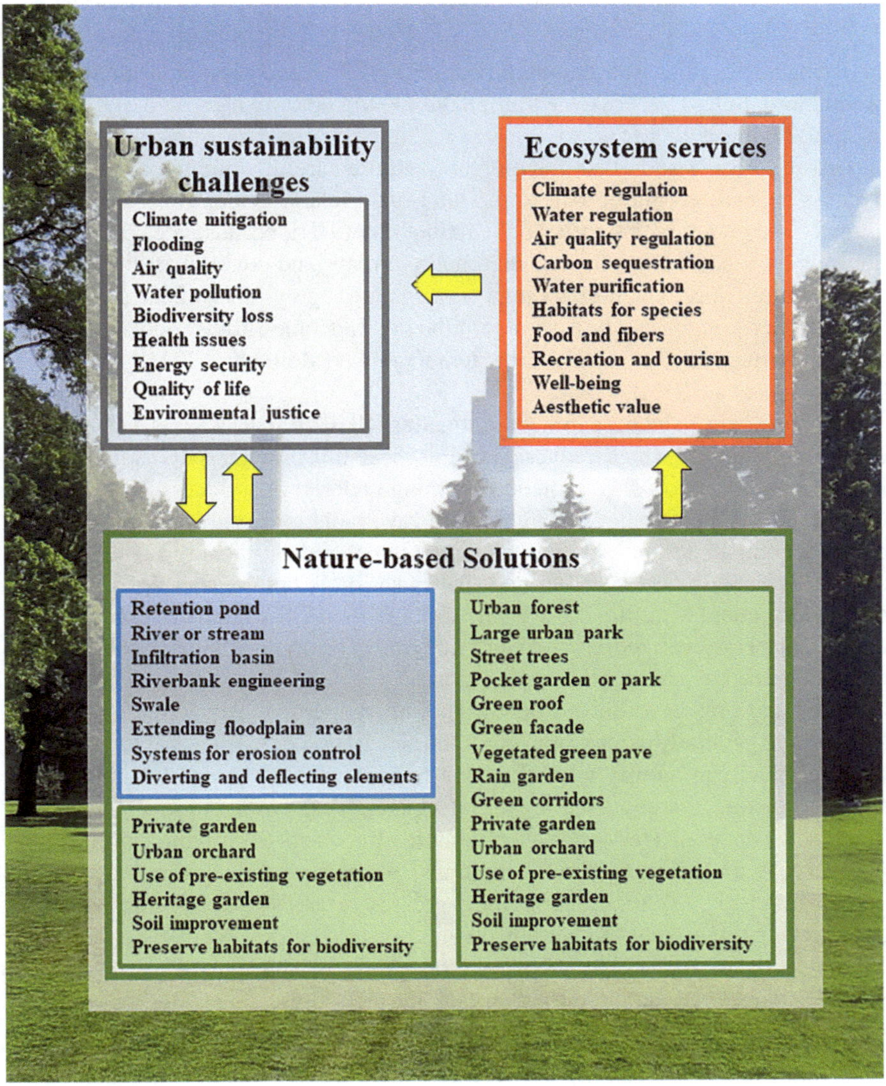

Fig. 1.7 Conceptual framework linking urban sustainability challenges with Nature-based Solutions and ecosystem services provided by NbS. (Original elaboration). This diagram explains how nature-based solutions can address urban sustainability challenges and provide ecosystem services. It is divided into three main sections connected by arrows. The top left section lists urban sustainability challenges such as climate mitigation, flooding, air quality, water pollution, biodiversity loss, health issues, energy security, quality of life, and environmental justice. The top right section shows ecosystem services like climate regulation, water regulation, air quality regulation, carbon sequestration, water purification, habitats for species, food and fibers, recreation and tourism, well-being, and aesthetic value. The bottom section details nature-based solutions grouped into three colored boxes: blue for water-related solutions like retention ponds, rivers or streams, infiltration basins, riverbank engineering, swales, floodplain extension, erosion control, and diverting elements; and two green boxes listing vegetation-based solutions such as urban forests, large urban parks, street trees, pocket gardens or parks, green roofs and facades, vegetated green pavements, rain gardens, green corridors, private gardens, urban orchards, use of pre-existing vegetation, heritage gardens, soil improvement, and habitat preservation for biodiversity. Arrows indicate that nature-based solutions help reduce urban challenges and enhance ecosystem services, highlighting the role of natural elements in creating healthier, more sustainable cities

11 (sustainable cities and communities), SDG 12 (responsible consumption and production), SDG 13 (climate action), SDG 14 (life below water), and SDG 15 (life on land) (Bona et al. 2023).

Urban greening and other "green NbS" require careful planning, as their effectiveness depends on multiple factors. This is particularly evident for NbS aimed at climate-change mitigation through the reduction of UHI, because the cooling effect depends on topography, climate, seasonality, urban land-cover characteristics and configuration (e.g., patch shape and aggregation), grey-infrastructure properties (e.g., surface roughness, heat storage, albedo), vegetation traits, and whether the aim is to reduce daytime or nighttime temperatures (Meili et al. 2021b; Ibsen et al. 2024; Beele et al. 2024).

Greening is a valuable NbS for mitigating UHI through several mechanisms: evapotranspiration (which reduces sensible heat flux), increased shading (which lowers incoming shortwave radiation reaching underlying surfaces), higher albedo (which reflects a portion of incoming shortwave radiation), and lower specific heat capacity compared to non-natural materials (which reduces heat re-radiation to the atmosphere and lowers heat storage). If vegetation is not irrigated, greening may also reduce latent heat flux (Oke 1988; Meili et al. 2021b; Ibsen et al. 2024; Beele et al. 2024). However, the extent and timing of these cooling effects depend on vegetation type, cover, and spatial aggregation.

Trees and grasses exhibit different cooling properties. Trees provide stronger shading and generally exhibit higher evapotranspiration rates than grasses; however, their structure can modify wind flow due to differences in surface roughness and sky-view factors, thereby affecting turbulent energy exchange and potentially limiting heat dissipation. Trees are typically more effective at improving daytime cooling—except under water-stress conditions, when stomatal closure may reduce cooling efficiency (Meili et al. 2021b). In contrast, grasses generally enhance nighttime cooling due to their higher albedo and greater ventilation (Beele et al. 2024). The selection of appropriate greening strategies and NbS must also consider the energy and resources required for their maintenance. Grasses and lawns may require substantial inputs of water, fertilizers, and herbicides, as well as intensive management (e.g., weeding, pruning, mulching, plant replacement), particularly because they often consist of monocultures or include non-native species poorly adapted to local climatic conditions (Faeth et al. 2011). Moreover, urban greening and NbS should address habitat conservation and restoration and strengthen the educational and political role of urban ecosystems by fostering public awareness of biodiversity and nature conservation (Faeth et al. 2011).

These interventions often contribute to the development of **Green Infrastructure (GI)**, formally defined by the European Commission as *"a strategically planned network of natural and semi-natural areas...designed and managed to deliver a wide range of ecosystem services while enhancing biodiversity."* This concept underpins major EU strategies such as the EU Green Infrastructure Strategy (2013), the EU Biodiversity Strategy for 2030, and the recently adopted **Nature Restoration Law** (Regulation 2024/1991), which mandates the restoration of 20% of EU land and sea areas by 2030 and the improvement of urban green spaces and tree-canopy cover. Additionally, the **Green City Accord**, launched under the European Green

Deal, commits cities to measurable targets for air and water quality, biodiversity enhancement, circular-economy practices, and noise reduction by 2030.

References

Gates C, Goldman A (2024) Ancient cities: the archaeology of urban life in the ancient near east and Egypt, Greece, and Rome. Routledge

Bai X, Surveyer A, Elmqvist T et al (2016) Defining and advancing a systems approach for sustainable cities. Curr Opin Environ Sustain 23:69–78

Bai X, McPhearson T, Cleugh H et al (2017) Linking Urbanization and the environment: conceptual and empirical advances. Annu Rev Environ Resour 42:215–240. https://doi.org/10.1146/annurev-environ-102016-061128

Beele E, Aerts R, Reyniers M, Somers B (2024) Spatial configuration of green space matters: associations between urban land cover and air temperature. Land Urb Plan 249:105121. https://doi.org/10.1016/j.landurbplan.2024.105121

Beninde J, Veith M, Hochkirch A (2015) Biodiversity in cities needs space: a meta-analysis of factors determining intra-urban biodiversity variation. Ecol Lett 18:581–592. https://doi.org/10.1111/ele.12427

Bona S, Silva-Afonso A, Gomes R et al (2023) Nature-based solutions in urban areas: a European analysis. Appl Sci 13:168. https://doi.org/10.3390/app13010168

Cadenasso ML, Pickett STA, Schwarz K (2007) Spatial heterogeneity in urban ecosystems: reconceptualizing land cover and a framework for classification. Front Ecol Environ 5:80–88

Carlson TN, Arthur ST (2000) The impact of land use—land cover changes due to urbanization on surface microclimate and hydrology: a satellite perspective. Glob Planet Chang 25(1–2):49–65. https://doi.org/10.1016/S0921-8181(00)00021-7

Cheung PK, Nice KA, Livesley SJ (2022) Irrigating urban green space for cooling benefits: the mechanisms and management considerations. Environ Res Clim 1(1):015001. https://doi.org/10.1088/2752-5295/ac6e7c

Cheung PK, Meili N, Nice KA, Livesley SJ (2024) Identifying the mechanisms by which irrigation can cool urban green spaces in summer. Urban Clim 55:101914. https://doi.org/10.1016/j.uclim.2024.101914

Collins JP, Kinzig A, Grimm NB et al (2000) A new urban ecology. Am Sci 88:416–425

Deutschewitz K, Lausch A, Kühn I, Klotz S (2003) Native and alien plant species richness in relation to spatial heterogeneity on a regional scale in Germany. Glob Ecol Biogeogr 12:299–311

Du E, Xia N, Guo Y et al (2022) Ecological effects of nitrogen deposition on urban forests: an overview. Front Agr Sci Eng 9(3):445–456. https://doi.org/10.15302/J-FASE-2021429

Elmqvist T, Goodness J, Marcotullio PJ et al (2013) Urbanization, biodiversity and ecosystem services: challenges and opportunities: a global assessment. Springer, Dordrecht, Neth

European Commission (2013) Green Infrastructure (GI) – Enhancing Europe's Natural Capital (COM(2013) 249 final). Brussels: European Commission. https://eur-lex.europa.eu/resource.html?uri=cellar:d41348f2-01d5-4abe-b817-4c73e6f1b2df.0014.03/DOC_1&format=PDF

Faeth SH, Bang C, Saari S (2011) Urban biodiversity: patterns and mechanisms. Ann N Y Acad Sci 1223:69–81

Fahrig L (2003) Effects of habitat fragmentation on biodiversity. Annu Rev Ecol Evol Syst 34:487–515

Fang X, Li J, Ma Q et al (2024) A quantitative review of nature-based solutions for urban sustainability (2016–2022): from science to implementation. Sci Tot Env 927:172219. https://doi.org/10.1016/j.scitotenv.2024.172219

Folke C, Jansson A, Larsson J, Costanza R (1997) Ecosystem appropriation by cities. Ambio 26:167

Grimm NB, Faeth SH, Golubiewski NE et al (2008) Global change and the ecology of cities. Science 319:756–760

Güneralp B, Seto KC (2013) Futures of global urban expansion: uncertainties and implications for biodiversity conservation. Environ Res Lett 8:014025

He C, Liu Z, Gou S et al (2019) Detecting global urban expansion over the last three decades using a fully convolutional network. Environ Res Lett 14:34008. https://doi.org/10.1088/1748-9326/aaf936

He C, Liu Z, Wu J et al (2021) Future global urban water scarcity and potential solutions. Nat Commun 12:4667. https://doi.org/10.1038/s41467-021-25026-3

Hobbs RJ et al (2006) Novel ecosystems: theoretical and management aspects of the new ecological world order. Glob Ecol Biogeogr 15:1–7

Howard JL (2021) Urban anthropogenic soils—a review. Adv Agronomy 165:1–57. https://doi.org/10.1016/bs.agron.2020.08.001

Hruska K (2006) Notes on the evolution and organization of the urban ecosystem. Urban Ecosyst 9:291–298

Ibsen PC, Crawford BR, Corro LM et al (2024) Urban tree cover provides consistent mitigation of extreme heat in arid but not humid cities. Sustain Cities Soc 113:105677. https://doi.org/10.1016/j.scs.2024.105677

IUCN (International Union for Conservation of Nature) (2020) Global standard for nature-based solutions. 30 pp. https://doi.org/10.2305/IUCN.CH.2020.08.en

Jenerette GD, Clarke LW, Avolio ML et al (2016) Climate tolerances and trait choices shape continental patterns of urban tree biodiversity. Glob Ecol Biogeogr 25:1367–1376

Johnson MTJ, Munshi-South J (2017) Evolution of life in urban environments. Science 358:eaam8327. https://doi.org/10.1126/science.aam8327

Kaushal SS, Wood KL, Galella JG et al (2020) Making 'chemical cocktails'—evolution of urban geochemical processes across the periodic table of elements. Appl Geochem 119:104632. https://doi.org/10.1016/j.apgeochem.2020.104632

Kaye JP, McCulley RL, Burke IC (2005) Carbon fluxes, nitrogen cycling, and soil microbial communities in adjacent urban, native, and agricultural ecosystems. Glob Change Biol 11:575–587

Kaye JP, Groffman PM, Grimm NB et al (2006) A distinct urban biogeochemistry? Trends Ecol Evol 21:192

La Sorte FA, Aronson MFJ, Williams NSG et al (2014) Beta diversity of urban floras among European and non-European cities. Glob Ecol Biogeogr 23(7):769–779. https://doi.org/10.1111/geb.12159

Li X, Zhou Y, Eom J et al (2019) Projecting global urban area growth through 2100 based on historical time series data and future shared socioeconomic pathways. Earth's Futur 7:351–362. https://doi.org/10.1029/2019EF001152

Lokatis S, Jeschke JM (2022) Urban biotic homogenization: approaches and knowledge gaps. Ecol Appl 32(8):e2703. https://doi.org/10.1002/eap.2703

Lososová Z, Chytrý M, Tichý L et al (2012) Native and Alien Floras in urban habitats: a comparison across 32 cities of Central Europe. Global Ecol Biogeogr 21(5):545–555. https://doi.org/10.1111/j.1466-8238.2011.00704.x

Masson V, Lemonsu A, Hidalgo J, Voogt J (2020) Urban climates and climate change. Annu Rev Environ Resour 45:411–444. https://doi.org/10.1146/annurev-environ-012320-083623

Meili N, Acero JA, Peleg N et al (2021a) Vegetation cover and plant-trait effects on outdoor thermal comfort in a tropical city. Build Environ 195:107733. https://doi.org/10.1016/j.buildenv.2021.107733

Meili N, Manoli G, Burlando P et al (2021b) Tree effects on urban microclimate: diurnal, seasonal, and climatic temperature differences explained by separating radiation, evapotranspiration, and roughness effects. Urban For Urban Green 58:126970. https://doi.org/10.1016/j.ufug.2020.126970

Moffett KB, Makido Y, Shandas V (2019) Urban-rural surface temperature deviation and intra-urban variations contained by an urban growth boundary. Remote Sens 11(22):2683

Morel JL, Chenu C, Lorenz K (2015) Ecosystem services provided by soils of urban, industrial, traffic, mining, and military areas (SUITMAs). J Soils Sediments 15(8):1659–1666

Murtinová V, Gallay I, Olah B (2022) Mitigating effect of urban green spaces on surface urban Heat Island during summer period on an example of a medium size town of Zvolen, Slovakia. Remote Sens 14. https://doi.org/10.3390/rs14184492

Müller F, de Groot R, Willemen L (2010) Ecosystem services at the landscape scale: the need for integrative approaches. Landscape 23:1–11
Oke TR (1988) The urban energy balance. Prog Phys Geogr 12:471–508
Phelan PE, Kaloush K, Miner M et al (2015) Urban heat Island: mechanisms, implications, and possible remedies. Annu Rev Environ Resour 40:285–307. https://doi.org/10.1146/annurev-environ-102014-021155
Pickett STA, Cadenasso ML, Grove JM et al (2001) Urban ecological systems: linking terrestrial ecological, physical, and socioeconomic components of metropolitan areas. Annu Rev Ecol Syst 32:127–157
Pyšek P, Chocholousková Z, Pyšek A et al (2004) Trends in species diversity and composition of urban vegetation over three decades. J Veg Sci 15:781–788
Qiu GY, Yan C, Liu Y (2023) Urban evapotranspiration and its effects on water budget and energy balance: review and perspectives. Earth Sci Rev:104577. https://doi.org/10.1016/j.earscirev.2023.104577
Rebele F (1994) Urban ecology and special features of urban ecosystems. Global Ecol Biogeogr Lett 4:173–187
Richardson PJ, Lundholm JT, Larson DW (2010) Natural analogues of degraded ecosystems enhance conservation and reconstruction in extreme environments. Ecol Appl 20:728–740
Reyes-Paecke S, Gironás J, Melo O et al (2019) Irrigation of green spaces and residential gardens in a Mediterranean metropolis: gaps and opportunities for climate change adaptation. Landsc Urban Plan 182:34–43. https://doi.org/10.1016/j.landurbplan.2018.10.006
Salvati L (2013) A chronicle of a death foretold: urban expansion and land consumption in Rome, Italy. Eur Plan Stud 21(8):1176–1188
Sarangi C, Tripathi SN, Qian Y et al (2018) Aerosol and urban land-use effects on rainfall around cities in the Indo-Gangetic basin from observations and cloud-resolving model simulations. J Geophys Res Atm 123:3645–3667
Schmid PE, Niyogi D (2017) Modeling urban precipitation modification by spatially heterogeneous aerosols. J Appl Meteorol Climatol 56:2141–2153
Seastedt TR, Hobbs RJ, Suding KN (2008) Management of novel ecosystems: are novel approaches required? Front Ecol Environ 6:547–553
Shochat E, Warren PS, Faeth SH et al (2006) From patterns to emerging processes in mechanistic urban ecology. Trends Ecol Evol 21:186–191. https://doi.org/10.1016/j.tree.2005.11.019
Shochat E, Lerman SB, Anderies JM et al (2010) Invasion, competition, and biodiversity loss in urban ecosystems. Bioscience 60:199–208. https://doi.org/10.1525/bio.2010.60.3.6
Song J, Wang Z (2016) Diurnal changes in urban boundary layer environment induced by urban greening. Environ Res Lett 11:114018
Stewart ID, Oke TR (2012) Local climate zones for urban temperature studies. Bull Am Meteorol Soc 93:1879–1900
United Nations (2015) Transforming our world: the 2030 agenda for sustainable development. Division for Sustainable Development Goals, New York, NY
Vazenev V, Kuzyakov Y (2018) Urban soils as hot spots of anthropogenic carbon accumulation: review of stocks, mechanisms and driving factors. Land Degrad Dev 29:1607–1622. https://doi.org/10.1002/ldr.2944
Walsh CJ, Fletcher TD, Burns MJ (2012) Urban stormwater runoff: a new class of environmental flow problem. PLoS One 7(9):e45814. https://doi.org/10.1371/journal.pone.0045814
Wiedmann T, Allen C (2021) City footprints and SDGs provide untapped potential for assessing city sustainability. Nat Commun 12:3758. https://doi.org/10.1038/s41467-021-23968-2
Wong NH, Tan CL, Kolokotsa DD et al (2021) Greenery as a mitigation and adaptation strategy to urban heat. Nat Rev Earth Environ 2:166–181. https://doi.org/10.1038/s43017-020-00129-5
Xu YL, Meili L, Fatichi S (2025) Long-term hydrological budget over urban areas: approaches and challenges. J Hydrol 662:134000. https://doi.org/10.1016/j.jhydrol.2025.134000
Zhang Z, Dobson B, Moustakis Y et al (2023) Assessing the co-benefits of urban greening coupled with rainwater harvesting management under current and future climates across USA cities. Environ Res Lett 18(3):034036. https://doi.org/10.1088/1748-9326/acbc90

Chapter 2
The Wild Plants in Urban Ecosystems: Characteristics, Evolutionary Trends and Conservation Needs

Flavia Bartoli, Francesca Bretzel, Ricardo Quinto Canas, Giulia Albani Rocchetti, and Enrica Roccotiello

Abstract Urban ecosystems host diverse spontaneous plant communities shaped by intense habitat heterogeneity, disturbance, and human influence. Wild urban plants display high ecological plasticity and traits, colonizing marginal spaces such as pavements, walls, brownfields, and railways, that enable survival under urban stressors, like pollution, soil compaction, heat islands, and altered hydrology. The flora often is dominated by ruderal, nitrophilous, and non-native species, and its composition reflects a complex interplay between ecological filters, landscape fragmentation, and anthropogenic factors. However, native biodiversity can persist where semi-natural substrates or low-management regimes remain.

Urban succession differs from rural paths, often beginning with pioneer therophytes and trampling-tolerant species before shifting, when disturbance decreases, toward perennial grasslands, tall-herb stands, and in some cases, pre-wood and woodland communities. Dynamic-catenal phytosociology, linking vegetation patterns to potential natural vegetation, edaphic gradients, and disturbance regimes, provides a framework to understand these processes.

F. Bartoli (✉)
Institute of Heritage Science (ISPC), National Research Council (CNR), Montelibretti, Rome, Italy
e-mail: flavia.bartoli@cnr.it

F. Bretzel
National Research Council (CNR), Research Institute on Terrestrial Ecosystems (IRET), Pisa, Italy

R. Q. Canas
Faculty of Sciences and Technology, University of the Algarve, Centre Marine Sciences (CCMAR), Faro, Portugal

G. A. Rocchetti
Department of Sciences, University Roma Tre, Rome, Italy

E. Roccotiello
Department of Earth, Environmental and Life Sciences, University of Genoa, Genoa, Italy

G. Caneva et al. (eds.), *Urban Botany*, Future City 22,
https://doi.org/10.1007/978-3-032-28986-5_2

Cities result as laboratories to study plant species turnover, functional convergence, and the emergence of novel ecosystems. The evaluation of functional trait reveals reliable selection for stress-tolerant and fast-reproducing taxa, while phylogenetic findings show non-random clustering among successful urban plants. Long-term monitoring demonstrates that urban floras display high temporal dynamism, with recognised or increasing richness despite significant compositional change determined by climate warming, land-use transformation, and constant non-native species introductions.

Conservation strategies in urban contexts emphasize the need of habitat connectivity, nature-based solutions, as green infrastructure, and adaptive restoration approaches focusing on soil remediation, native species reintroduction, and the careful use of non-natives. Corridors and stepping stones—such as riverbanks, railway verges, pocket parks, and green roofs—enhance ecological resilience by promoting dispersal and maintaining gene flow across fragmented landscapes.

Keywords Urban succession · Spontaneous vegetation · Novel ecosystems · Habitat connectivity · Ecological resilience · Phytosociological frameworks

2.1 Ecological Features of Urban Wild Plants (F.B., F.B.)

Wild urban plants are species that establish spontaneously in cities without deliberate human planting or maintenance. They colonize cracks in pavements, abandoned lots, walls, railways, and other marginal spaces, are dispersed by wind, water, animals, and human activity, and are sometimes not correctly called *emergent vegetation or wasteland flora* (Del Tredici 2010). These assemblages include native, non-native, and sometimes invasive species, unified by their ability to persist in disturbed habitats.

Urbanization profoundly transforms natural environments, causing habitat loss, fragmentation, and environmental homogenization. Abiotic vectors, such as wind and water, further shape plant dispersal. Spontaneous vegetation occurs in rarely managed spaces—brownfields, commons, industrial peripheries, and historical monuments—where successional dynamics can reveal vegetation age. Cities act as crossroads for transport and biotic vectors, facilitating the arrival of generalist and alien species (Fig. 2.1).

However, frequent disturbance often interrupts succession, leaving communities in early stages dominated by herbaceous swards and shrubs, which provide habitat for insects, reptiles, and birds.

Urban soils, often derived from fertile alluvial substrates, host rich seed banks of crops and weeds, contributing to heterogeneous flora and novel ecosystems not found in natural sites. Stressors driving urban flora include habitat transformation,

Fig. 2.1 Wild urban vegetation on disturbed soils composed of therophytes (*Echium plantagineum*, *Papaver rhoeas*, *Coleostephus myconis*). (Photo by F. Bretzel)

fragmentation, pollution, soil sealing, and human preferences for ornamentals (Williams et al. 2015). Urban plants tend to be thermophilic, tolerant of alkaline, nitrogen-rich soils, and dominated by functional types such as therophytes and trees, while geophytes are less favoured. Traits like heavy seeds and taller growth forms are advantageous. Typical herbaceous vegetation includes ruderal-nitrophilous assemblages on fertile soils, with perennials or biennials in stable sites and therophytes in disturbed areas. Dispersal limitation also influences species composition in vacant lots (Maehara et al. 2024).

These dynamics raise questions about whether spontaneous vegetation is dominated by natives or exotics and which species regenerate best after disturbance. Evidence suggests that disturbance can limit the regeneration of invasive species (Latzel et al. 2008), but this remains debated. Approaches such as habitat analogues—matching pedoclimatic conditions to introduce germplasm from similar habitats—offer strategies to enhance biodiversity in anthropized sites. Remarkably, some rare species now prefer urban habitats, highlighting cities' potential role in conservation (Itani et al. 2020).

2.1.1 Urban Ecological Filters and Plant Traits

Urban habitats are heterogeneous, ranging from cracks in pavements and rooftops to abandoned lots and railways. Wild urban plants display remarkable morphological and physiological plasticity to exploit these niches (Fig. 2.2).

Common adaptations include *Morphological traits:* Reduced leaf size and waxy cuticles minimize water loss and resist pollutants (Del Tredici 2010). *Root systems:* Deep or fibrous roots penetrate compacted soils or exploit surface moisture. *Microhabitat colonization:* Ability to grow in minimal substrates such as wall crevices and asphalt fissures. These adaptations allow species to persist in environments with limited resources and frequent disturbance.

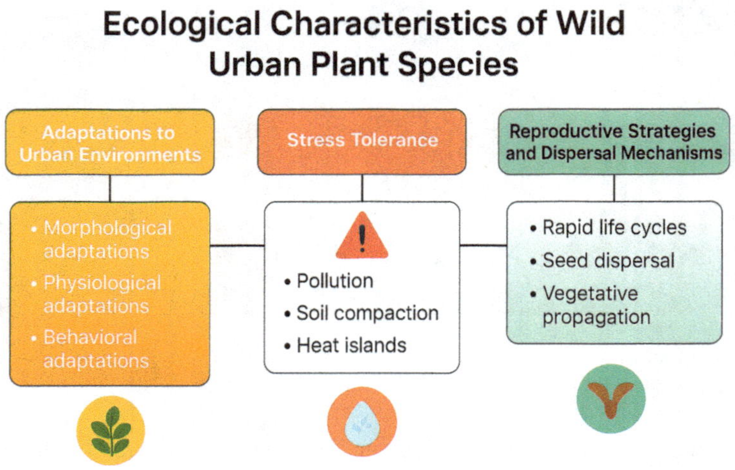

Fig. 2.2 The diagram illustrates the three main categories. The Colour coding and icons highlight each category for clarity and visual appeal. (Image by F. Bartoli)

Urban ecosystems are subject to multiple stressors, including pollution, soil compaction, and heat islands. *Pollution resistance:* Plants develop mechanisms to tolerate heavy metals and nitrogen enrichment (Williams et al. 2015). *Soil compaction:* Structural root adaptations enable anchorage and nutrient uptake in degraded soils. *Heat island adaptation:* Thermophilic traits and efficient water-use strategies support survival under elevated temperatures and reduced humidity. Such tolerance reflects evolutionary pressures favouring species with high ecological plasticity.

Successful colonization of urban habitats depends on effective reproduction and dispersal: *Rapid life cycles*: Annual species dominate disturbed sites due to quick germination and reproduction (Maehara et al. 2024). *Seed dispersal:* Wind, water runoff, and anthropogenic vectors (vehicles, footwear) facilitate spread. *Vegetative propagation:* Rhizomes, stolons, and fragmentation ensure persistence in harsh conditions. These strategies enable plants to exploit transient habitats and maintain populations despite frequent disturbance.

2.1.2 *Species Composition in Urban Settings*

Urban floras emerge from the interplay of **ecological filters**—such as environmental conditions and disturbance regimes—**landscape processes** like fragmentation and connectivity, and **human actions** including planting preferences, transport, and trade. Together, these drivers shape the balance between native and non-native taxa, bias species pools toward certain families and genera, and determine the role and impact of invasive species across cities. This dynamic results in urban plant communities dominated by ruderal species, generalists, and alien taxa adapted to anthropogenic conditions, while stress-tolerant species associated with nutrient-poor or extreme habitats are strongly underrepresented (Pyšek 1991; Godefroid and Koedam 2007; Kalusová et al. 2016). Functional traits such as small seed size, short-term persistent seed banks, and human-mediated dispersal (*hemerochory*) confer advantages in highly disturbed environments, whereas taller perennials and woody ornamentals prevail in less disturbed habitats, such as parks and residential gardens (Williams et al. 2005; Vega et al. 2021). Urban heterogeneity further modulates these patterns: densely built-up areas favour low-stature therophytes tolerant of trampling, industrial zones support species adapted to alkaline rubble-rich soils, and open residential areas harbour taller perennials and garden escapes. Ultimately, the composition of urban floras reflects a complex socio-ecological system where abiotic filters, successional trajectories, and cultural practices converge to shape biodiversity in cities.

2.1.2.1 Native Versus Non-native Species

Urban environments typically host mixed assemblages of **native species**, often persisting in remnant habitats or lightly managed green spaces, and **non-native species** introduced through horticulture, commerce, and transport corridors that subsequently naturalize in disturbed, resource-rich microsites (Williams et al. 2005). Across many cities, non-native richness and frequency tend to increase along urbanization gradients, driven by propagule pressure, ruderal opportunities, and environmental filtering toward warmer, more alkaline, and nutrient-enriched conditions associated with the urban heat island, atmospheric deposition, and substrate alkalinity (Kalusová et al. 2016; Godefroid and Koedam 2007). Conversely, native species losses are concentrated among stress-tolerant taxa from nutrient-poor or moisture-dependent habitats—such as wetlands and fens—that are disproportionately eliminated or degraded in urban landscapes (Williams et al. 2005). Trait-based syntheses reveal that urbanization favours thermophilic, nutrient-demanding, and often taller species, patterns that promote widespread generalists and alien taxa while disfavouring short-statured specialists, such as hemicryptophytes and geophytes (Pyšek 1991; Kalusová et al. 2016). Evidence from applied contexts reinforces this dynamic: roadside floras functioning as dispersal corridors frequently exhibit high proportions of aliens and disturbance-adapted natives, with surveys in South Asia reporting >50% alien species on highly disturbed verges dominated by hypertolerant grasses such as *Axonopus*, *Eleusine*, and *Cynodon*, illustrating how disturbance and propagule pressure tip composition toward cosmopolitan ruderals (Godefroid and Koedam 2007). Nonetheless, recent reviews of spontaneous urban vegetation emphasize that cities can retain substantial native diversity—particularly where small habitat islands persist, and mowing regimes are relaxed—and that both native and alien richness may peak at intermediate disturbance levels depending on context (Vega et al. 2021). Ultimately, whether natives or non-natives dominate a given site depends on local habitat history, the four filters of urban floras (habitat change, fragmentation, urban environmental conditions, and human preference), and the trait composition of the regional species pool (Williams et al. 2005; Kalusová et al. 2016).

2.1.2.2 Common Families and Genera in Urban Settings

Across regions, urban floras are consistently enriched in families that exhibit ruderal, nitrophilous, and disturbance-tolerant strategies, alongside woody ornamentals that frequently escape cultivation. Prominent among these are Asteraceae, which includes numerous ruderals and strong dispersers, such as *Taraxacum*, *Senecio*, and *Ambrosia*, that thrive on nutrient-rich substrates and open ground (Vega et al. 2021; Williams et al. 2005). *Poaceae* dominate lawns, verges, and vacant lots, with hyper-tolerant species such as *Cynodon dactylon*, *Eleusine indica*, and *Poa annua*, adapted to trampling, heat, and even heavy metal contamination (Capotorti et al. 2013). Brassicaceae, represented by genera such as *Capsella* and *Sisymbrium*, specialize in cracks and compacted soils, benefiting from high nutrient availability and frequent disturbance (Pyšek 1991). Fabaceae, including *Trifolium* and *Medicago*, contribute nitrogen fixation in urban grasslands and edges, while woody Fabaceae occasionally escape cultivation, adding to the ornamental component of spontaneous floras (Kalusová et al. 2016).

Globally, these families recur across continents (Fig. 2.3), reflecting the convergence of urban floras toward cosmopolitan ruderal dominance under shared

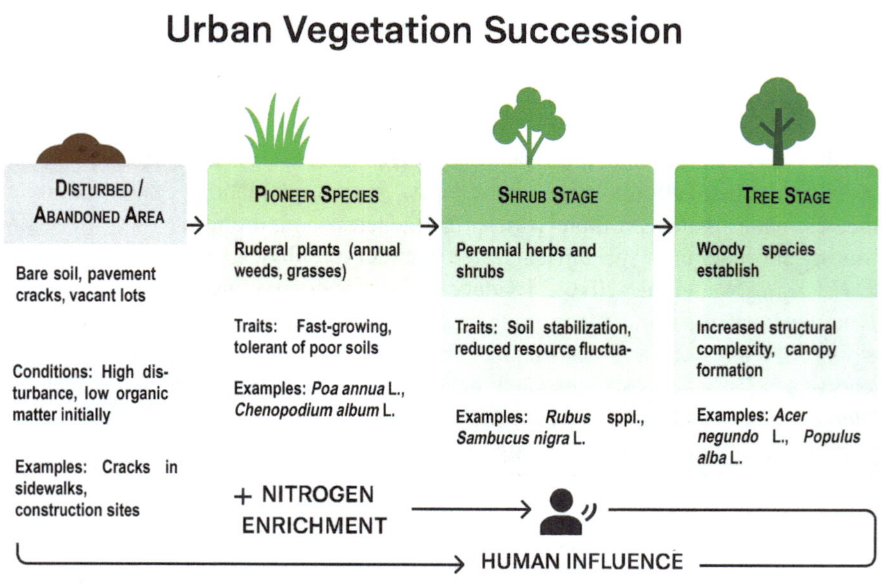

Fig. 2.3 Urban vegetation succession from disturbed or abandoned areas through pioneer, shrub, and tree stages which also favour alien species. (Image by F. Bartoli)

ecological filters. Surveys in Europe, North America, and Asia consistently identify Asteraceae and Poaceae as the most frequent families in spontaneous urban vegetation, while Brassicaceae and Fabaceae are common in verges and semi-natural patches (Godefroid and Koedam 2007; Kalusová et al. 2016). In South Asian cities, roadside floras often contain >50% alien species, dominated by hyper-tolerant grasses such as *Axonopus*, *Eleusine*, and *Cynodon*, illustrating how disturbance and propagule pressure tip composition toward cosmopolitan ruderals (Godefroid and Koedam 2007). Similar patterns occur in Latin America and Africa, where Asteraceae and Poaceae dominate vacant lots and transport corridors, while Fabaceae and Brassicaceae maintain strong representation in managed green spaces. Trait syntheses indicate that urban filters favour therophytes (annuals) and taller growth forms, with increasing representation of woody phanerophytes—partly due to ornamental escapes—and declines in geophytes sensitive to competition and nutrient enrichment. These shifts align with Ellenberg-type indicator patterns toward higher nutrients and alkalinity, reflecting the strong influence of anthropogenic substrates and management regimes (Williams et al. 2005; Pyšek 1991).

2.1.2.3 Role of Invasive Species

Invasive plants, along with other alien species (see Chap. 3), can reconfigure urban floras by shifting species composition, trait distributions, and successional pathways as they capitalize on resource pulses, repeated disturbance, and dense dispersal networks created by roads, railways, and human movement. Many high-impact invaders originate in the ornamental trade, establishing first in managed green spaces and then spreading into spontaneous habitats such as walls, rail verges, and brownfields. In Rome, for example, *Ailanthus altissima* and *Robinia pseudoacacia* reshape vegetation dynamics and outcompete natives, aided by tall stature, clonal strategies, and relatively heavy propagules that secure establishment in compacted or impervious microsites (Capotorti et al. 2013). Comparative trait syntheses across European cities show that urban filters tend to favour taller, longer-lived, clonal taxa and species with heavier seeds, while human-mediated dispersal (hemerochory) provides a steady flow of propagules from gardens and supply chains into the urban matrix (Kalusová et al. 2016; Godefroid and Koedam 2007). Yet disturbance does not uniformly promote invasion: along intensely trafficked verges and other chemically or physically extreme sites, regeneration of some woody aliens is curtailed, with vegetation tipping toward a subset of hyper-tolerant graminoids and ruderal grasses, illustrating that the magnitude and type of stress (e.g., trampling, metals) can filter even aggressive invaders (Godefroid and Koedam 2007; Williams et al. 2005). Human connectivity further accelerates the spread of citizen science experiments in Zurich, documenting rapid, spontaneous colonization of seed-free trays by wind-dispersed taxa and ornamental escapes, underscoring the roles of urban networks and propagule pressure (Vega et al. 2021). From a management perspective, steering the spontaneous flora through adjusted mowing regimes, maintaining connectivity for natives (while interrupting vectors for high-risk ornamentals), and risk-aware planting can harness urban vegetation for cooling, filtration, and habitat provision while reducing the propagule supply of problem species (Capotorti et al. 2013; Godefroid and Koedam 2007).

2.1.3 Community Structure in Urban Settings

Urban plant communities exhibit distinctive structural features shaped by spatial heterogeneity, successional dynamics, and biotic interactions.

Spatial **distribution patterns** are strongly influenced by the urban matrix, with vegetation often fragmented into small habitat islands (parks, vacant lots, roadside verges) connected by dispersal corridors such as railways and streets (Godefroid and Koedam 2007). These patches vary in size and connectivity, creating mosaics where ruderal assemblages dominate highly disturbed microsites, while semi-natural remnants persist in less managed zones (Kalusová et al. 2016).

Interactions with **fauna and microorganisms** further shape community structure: pollinators and seed dispersers facilitate connectivity among patches, while soil microbial communities influence nutrient cycling and plant establishment, particularly in compacted or contaminated substrates (Vega et al. 2021). These multi-scale interactions underscore the complexity of urban ecosystems, where abiotic stress, human management, and biotic networks converge to define the composition and resilience of spontaneous vegetation.

Long-term observations from permanent plots show that ruderal succession in cities follows a repeatable trajectory: therophytes and annuals dominate freshly disturbed substrates, then give way to hemicryptophytes and geophytes, and, where disturbance abates, to woody phanerophytes in later stages (e.g., *Poa annua, Stellaria media* → perennial grasses/forbs → shrub/tree recruits).

Family representation shifts accordingly, with Chenopodiaceae and Brassicaceae prevalent in the earliest phases and Poaceae and Asteraceae increasing through mid—to late succession (Pyšek 1991). Functionally, communities exhibit (i) a decline of R-strategists and a rise of C-strategists, (ii) a retreat of *archaeophytes* and neophytes as native species increase in dominance, and (iii) a drift in Ellenberg-type indicators toward higher nitrogen and moisture, consistent with soil enrichment during succession—patterns documented across urban ruderal habitats and corroborated by quantitative analyses of plot persistence and species exchange over 6–15 years (Pyšek 1991). Contemporary Park studies extend this picture: across 96 plots in nine parks, spontaneous herb layers clustered into repeatable community types that span from managed lawn/ground-cover mosaics (e.g., *Cynodon dactylon–Lolium perenne* assemblages) to less managed, species-rich patches dominated by spontaneous herbs (e.g., *Stellaria media–Veronica persica* groups) (Xu et al. 2024). Community composition and alpha diversity are shaped by management frequency and light, with moderate management often increasing diversity, and by park-scale features such as area and time since establishment, which track species–area and secondary succession effects in the urban core versus newer outskirts (Xu et al. 2024). Taken together, these findings indicate that spontaneous succession in cities tends toward semi-natural vegetation mosaics where disturbance is episodic and management calibrated but is continually reset where disturbance is chronic producing the patchy, dynamic structure typical of urban landscapes.

2.2 Natural Dynamisms and Phytosociological Framework (R.Q.C., F.B.)

2.2.1 Vegetation Dynamics and Successions

Vegetation is the spatial assembly of plant communities that occupy a given area and constitute the structural and functional backbone of ecosystems (Rivas-Martínez 2011; Van der Maarel 2005). It sustains biodiversity and underpins soil formation, nutrient cycling, water regulation, and primary production, while mediating energy flow and the oxygen, carbon, nitrogen, and water cycles (Rivas-Martínez 2011; Quinto-Canas 2014). Because vegetation integrates climate, substrate, and biotic interactions, vegetation types are widely used to classify ecosystems, and vegetation science investigates composition, structure, habitat relations, environmental drivers, and temporal dynamics (Rivas-Martínez 2011; Van der Maarel 2005).

Rapid urbanization and land-use change is eroding and fragmenting natural vegetation, demanding planning that explicitly accounts for vegetation dynamics. Dynamic-catenal phytosociology, grounded in Potential Natural Vegetation (PNV) and articulated through vegetation series, geoseries, geominoriseries, and geopermaseries, provides a process-based template for land management and urban planning that can enhance biodiversity and ecosystem services. Clear, objective-driven management with stakeholder involvement yields durable ecological and social benefits (Rivas-Martínez 2011).

Succession is the ordered replacement of communities within an ecologically homogeneous tassel; it can be progressive (pioneers ➔ climax) or regressive under recurrent disturbance. PNV helps distinguish primeval, primary (potential), and secondary (replacement) vegetation and clarifies how present communities relate to regional climate–substrate templates and disturbance legacies. Observing natural recovery even in modified landscapes helps anticipate future trajectories (Biondi 2011).

In dynamic phytosociology, the vegetation series (sigmetum/synassociation) bundles the series head (PNV) and its subseral replacements (Rivas-Martínez 2011). Along geomorphological gradients (ridge ➔ slope ➔ piedmont ➔ valley bottom), series typically grade from edaphoxerophilous to climatophilous to edaphohygrophilous, with temporihygrophilous and relict series capturing additional hydrological and historical nuances (Rivas-Martínez 2011). At the landscape scale, geoseries assemble multiple series along edaphic/topographic or altitudinal (cliserial) gradients under a specific bioclimate; minoriseries/geominoriseries and permasigmeta capture persistent, stress-limited climaxes (Del Río et al. 2022).

2.2.2 *Successional Mechanisms in Vegetation: Competition, Facilitation, and Disturbance Regimes*

Successional change in vegetation arises from shifting species interactions along disturbance and resource gradients (Fig. 2.3). Early in succession, facilitation is common: pioneer species modify microclimate and soils in ways that ease the establishment of later colonizers, whereas in more advanced stages, competition for light, water, and nutrients intensifies. These interaction patterns correspond to broader life-history strategies, typically framed through Grime's CSR model—*Ruderal* (R) species dominating frequently disturbed habitats, *Competitive* (C/K) strategists prevailing in stable conditions, and *Stress-tolerant* (S) taxa persisting in harsh environments (Grime 1979)—as well as the r/K selection theory that distinguishes fast-growing, short-lived pioneers from slower, more competitive species in mature communities (MacArthur and Wilson 1967). Such frameworks help explain why targeted management—such as reducing mowing intensity or frequency—can shift vegetation from short-lived ruderals toward more structurally persistent, multilayered communities (Biondi 2011).

Superimposed on these interactions are disturbance regimes, which repeatedly reset successional trajectories. Natural events (fire, flood, storms) and human-driven disturbances (soil removal, construction, trampling, pollution) generate sub-seral communities and truncate development (Rivas-Martínez 2011). When disturbance frequency or intensity increases, succession tends toward regressive pathways, whereas a reduction in disturbance allows progressive development. Plant persistence in such contexts depends on two fundamental strategies: resprouting, where species recover from protected buds in lignotubers or rhizomes, and seedling recruitment, which draws on persistent, often disturbance-activated seed banks. This duality—resprouters vs seeders—structures post-disturbance recovery in fire-prone and frequently disturbed ecosystems (Pausas and Keeley 2014). Phytosociological analysis clarifies how these mechanisms give rise to recurrent community types along the urban disturbance–recovery gradient. In initial or pioneer stages, bare substrates, construction spoil, and nutrient-enriched soils are rapidly colonized by annual ruderal communities of the *Chenopodietea and Secalietea* classes (e.g., *Chenopodium album, Atriplex* spp., *Sisymbrium* spp.), and by trampling-tolerant turfs of the *Plantaginetea majoris/Polygono–Poetea annuae* alliance (e.g., *Poa annua, Plantago major*) typical of paths and squares. These communities mirror the early dominance of therophytes documented in permanent plots and decline rapidly as succession proceeds (Pyšek 1991).

Under mid-successional conditions, where soil stabilization increases and disturbance becomes intermittent, vegetation shifts toward perennial ruderal grasslands of the *Agropyretea (repentis)* class (e.g., *Elytrigia/Agropyron repens*). These often integrate with nitrophilous tall-herb stands of *Galio–Urticetea (e.g., Urtica dioica, Artemisia vulgaris, Tanacetum vulgare*).

Long-term urban plots confirm this turnover: declines in Chenopodiaceae and Brassicaceae ruderals, an increase of Poaceae (ruderal perennials) and Asteraceae (tall herbs), and a shift toward higher Ellenberg N and F values as soils accumulate organic matter and moisture (Pyšek 1991).

Where disturbance abates, **late-seral** or **woody** stages may develop. In Mediterranean–temperate cities such as Rome, this includes submesophilous pre-woods dominated by *Ulmus minor* within the *Carpino orientalis–Quercetum cerris* and *Roso sempervirentis–Quercetum pubescentis* series. These native pre-wood stages can counter the spread of non-native broadleaf stands such as *Robinia pseudoacacia* and *Ailanthus altissima*. Remnant or regenerating forest fragments—*Echinopo siculi–Quercetum frainetto, Cytiso villosi–Quercetum suberis, Carpino orientalis–Quercetum cerridis, Cyclamino hederifolii–Quercetum ilicis*—persist as high-value ecological patches that provide microclimate regulation and habitat functions inside the urban matrix (Attorre et al. 1997).

Phytosociological evidence thus links functional shifts (e.g., R/C strategists; declining archaeophytes/neophytes; rising native dominance; enrichment in N and moisture indicators) to **diagnostic community** types that recur across cities. Early *Chenopodietea/Secalietea* and trampling turfs transition toward *Agropyretea* and *Galio–Urticetea* mosaics, and—where human disturbance is moderated—toward ruderal pre-woods or native forest fragments that anchor semi-natural urban mosaics.

These patterns, consistently documented in permanent plots and city-scale vegetation assessments, demonstrate why spontaneous urban succession often reconstructs semi-natural vegetation complexes rather than a single predictable climax, and why targeted management (e.g., calibrated mowing, disturbance moderation, soil enrichment) can steer vegetation from short-lived ruderal classes toward *more structurally complex, ecologically functional plant communities* (Pyšek 1991; Attorre et al. 1997).

2.2.3 Translating the Framework to Cities

2.2.3.1 Urban Succession in Place

In cities, natural dynamisms persist but are repatterned by urban stressors and management. Succession typically initiates in *highly disturbed substrates* (vacant lots, verges, pavement cracks) and can progress toward perennial herbs/shrubs and woody layers as disturbance relaxes, though many sites remain stabilized in early- to mid-seral mosaics (Pickett et al. 2011; Kowarik and Lippe 2018). Compared with

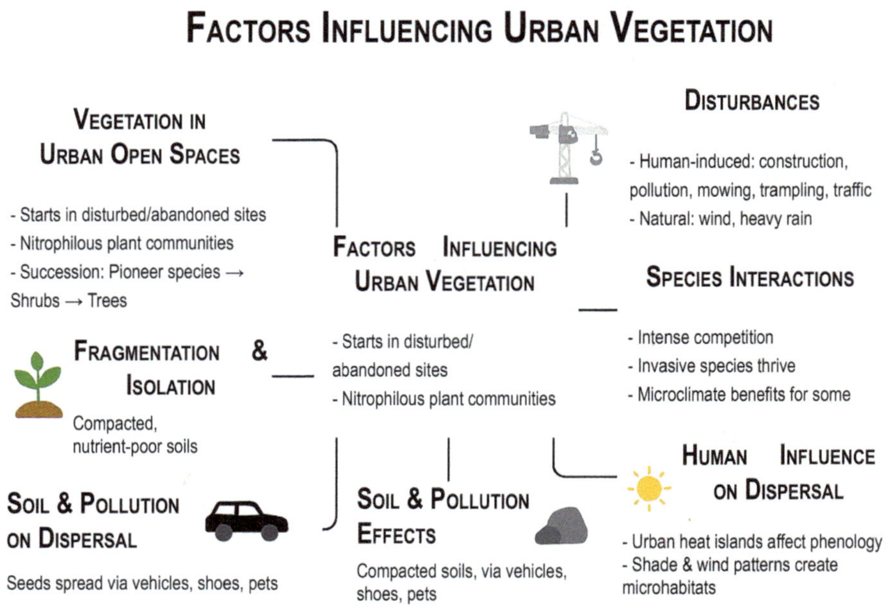

Fig. 2.4 The key drivers shaping plant communities in urban environments. (Image by F. Bartoli)

rural systems, urban trajectories are shaped by soil compaction, pollution, and microclimate, producing novel assemblages often dominated by nitrophilous and non-native taxa (Sukopp 2004) (Fig. 2.4).

2.2.3.2 The Urban Disturbance Templates

Urban green spaces experience high-frequency, patchy disturbance: soil removal/sealing and redevelopment alter hydrology and nutrient cycles; depositions of metals and nitrogen and particulates shift growth and composition; maintenance (mowing/pruning) interrupts succession; and trampling/traffic reduce infiltration and promote disturbance-adapted species (Williams et al. 2005; Sukopp 2004; Pickett et al. 2011). Natural events (windstorms, intense rain, seasonal extremes) further generate colonization microsites (Kowarik and Lippe 2018). In aggregate, these regimes repeatedly favour ruderal–nitrophilous taxa and stabilize novel ecosystems (Hobbs et al. 2006).

2.2.3.3 Fragmentation, Isolation, and Connectivity

Urban matrices fragment habitat into small, isolated patches embedded in infrastructure, constraining dispersal, colonization, and gene flow and increasing local extinction risk. Consequent shifts often favour generalists. Mitigation includes wildlife corridors (vegetated strips, riparian buffers, street trees), green roofs/vertical greening as stepping stones, and ecological network planning to rebuild connectivity (Tzoulas et al. 2007; Aronson et al. 2017; Kowarik and Lippe 2018).

2.2.3.4 Species Interactions, Invasions, and Microclimate

Spatial constraints and altered abiotic regimes intensify competition, while invasive plants—highly plastic and fast-reproducing—often displace natives and reshape ecosystem functions. The urban heat island extends growing seasons and reduces frost, assisting *thermophilous* taxa; built forms create light mosaics that benefit shade-tolerant or light-flexible species and accelerate novel network assembly (Grime 1979; Kowarik 2011; Aronson et al. 2017).

2.2.3.5 Human-Mediated Dispersal and Horticultural Escape

Human movement supplies powerful **dispersal vectors**: diaspores of *Ambrosia artemisiifolia* and *Bidens frondosa* adhere to vehicles, while *Taraxacum officinale* and *Plantago major* often hitchhike on footwear or pet fur (Ansong and Pickering 2013). Road networks act as corridors, and turbulence can deposit diaspores onto verges, where disturbance and high nutrients favour establishment (Ansong and Pickering 2013). Ornamental plantings repeatedly introduce non-native taxa, with frequent escapes by *Ailanthus altissima*, *Robinia pseudoacacia*, *Buddleja davidii*, *Ligustrum vulgare* (Aronson et al. 2017).

2.2.3.6 Urban Soils and Pollution

Urban soils are often compacted, heterogeneous, and contaminated; yet many species persist using traits that tolerate these stresses: *Plantago major*, *Taraxacum officinale*, *Polygonum aviculare* under compaction; *Ailanthus*, *Robinia*, *Populus alba* in metal-rich substrates; *Chenopodium album*, *Rumex crispus* in nutrient-poor/disturbed soils (Pouyat et al. 2010; Sukopp 2004). These filters/traits drive distinct urban floras.

2.2.3.7 Climate and Microclimate in Cities

The **UHI** reduces frost and extends the growing season (e.g., *Taraxacum officinale, Prunus cerasifera),* favouring rapid-growing invasive trees (e.g., *Ailanthus, Robinia) and* taxa with high physiological plasticity under drought, heat, and nutrient imbalance (Neil and Wu 2006). Built surfaces create shaded niches and reflective microhabitats that benefit species such as *Hedera helix* and *Lamium purpureum, accelerating departure from classi*cal rural succession (Kowarik 2011).

2.2.4 From Framework to Practice: Steering Trajectories and Restoring Functions

Cities can turn dynamics into design by **accelerating succession** and tailoring vegetation to local series/*geoseries*. Practical measures include staged establishment (pioneer grasses → shrubs → native trees) to achieve semi-natural woodlands with improved carbon storage, air quality, and thermal regulation (Tzoulas et al. 2007; Aronson et al. 2017). Green infrastructure and nature-based solutions—green roofs, vertical greening, wildflower meadows—raise biodiversity and connectivity where fragmentation is acute (Oberdorfer et al. 2007). Use dynamic-catenal guidance to match species to regional PNV and to select *minor—series* strategies for extreme sites. Finally, implement adaptive management with monitoring of composition, soils, and functions (e.g., infiltration, cooling) and engage stakeholders for durable outcomes (Rivas-Martínez 2011).

2.3 Research on Urban Flora and Vegetation: Concepts, Approaches, and Ecological Implications (E.R., F.B.)

Urban environments combine high habitat heterogeneity, strong disturbance, altered climate, and continuous species introductions, creating a complex mosaic of ecological niches. Urban flora and vegetation represent the plant communities that emerge in these settings (Fig. 2.5). Long-term research shows that cities can host remarkably high plant species richness—often matching or exceeding that of nearby rural areas (Wittig 2004). Urban plant assemblages result from interactions between regional biogeographical processes and intense human influence, making cities

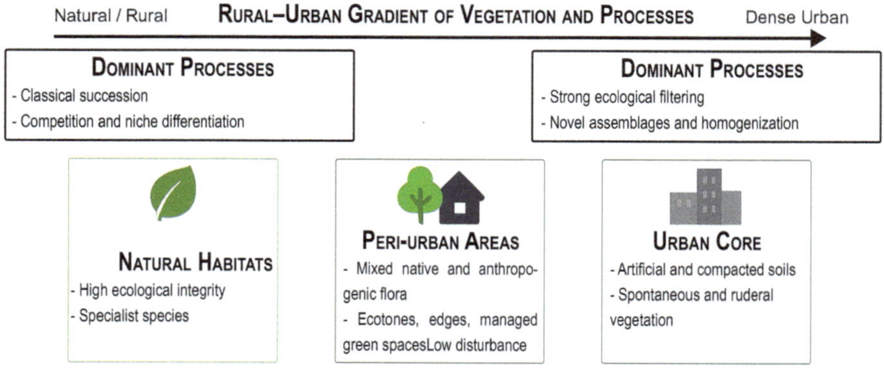

Fig. 2.5 Plant communities and dominant ecological processes shift along a gradient from natural rural habitats to dense urban areas. (Image by F. Bartoli)

valuable laboratories for studying ecological dynamics, species turnover, and plant adaptation (Zerbe et al. 2004; Celesti-Grapow and Blasi 1998).

Research in this field examines how species establish, persist, and interact in cities; how vegetation responds to urbanization and climate change; and how urban assemblages differ from those in non-urban ecosystems. It integrates floristics, vegetation science, biogeography, functional ecology, and conservation biology, offering essential knowledge for urban planning, green infrastructure design, and biodiversity conservation (La Sorte et al. 2014). The study of native and alien species in urban floras is crucial for understanding urban ecology and conservation. Cities are not passive recipients of invasions but active environments where ecological filtering, adaptation, and novel species interactions occur—challenging the view that alien species automatically signify degradation.

Urban floras should be seen as dynamic assemblages shaped by multiple drivers at multiple scales. Understanding how native persistence, alien establishment, and functional convergence interact is essential for interpreting urban ecosystem functioning and guiding effective management and conservation strategies (Zerbe et al. 2004).

2.3.1 Definitions and Terminology

Clear terminology is essential in urban botanical research, as cities host diverse plant assemblages that differ in origin, structure, and ecological function. Urban studies typically distinguish between urban flora and urban vegetation, two related but conceptually distinct units requiring different methods (Wittig 2004; Zerbe et al. 2004).

Urban flora encompasses all plant species occurring within a city—cultivated and spontaneous—reflecting interactions between regional native pools and human-mediated introductions (Wittig 2004; Celesti-Grapow et al. 2006). Native species persist in remnant semi-natural habitats or recolonize disturbed sites, while many adopt synanthropic strategies (Wittig 2004; Pyšek et al. 2004). Non-native species arrive through horticulture, trade, and transport, ranging from *archaeophytes* to recent *neophytes*, some of which become invasive under high propagule pressure and disturbance (Rat et al. 2017; Pyšek et al. 2004). As a result, urban floras are highly dynamic, with long-term studies documenting substantial turnover alongside stable or increasing richness (Salinitro et al. 2019; Fratarcangeli et al. 2022).

Urban vegetation refers to these species organized into plant communities—cultivated, semi-spontaneous, and spontaneous—forming a fragmented mosaic shaped by management, disturbance, and substrate sealing (Zerbe et al. 2004). Spontaneous vegetation is especially important as a biodiversity reservoir, a driver of succession, and a sensitive indicator of soil and microclimatic conditions, often supporting high native diversity even in highly urbanized areas (Salinitro et al. 2018; Fratarcangeli et al. 2022).

2.3.2 *Urban Habitats as Ecological Filters*

Urban habitats differ from natural ecosystems in substrate composition, disturbance regimes, and microclimatic conditions, favouring species tolerant of physical stress, nutrient imbalance, and drought. Their vertical and horizontal stratification creates numerous ecological niches within small areas. Studies in European cities show that distinct plant assemblages consistently correspond to specific urban habitat types—such as semi-natural soils, paved surfaces, walls, roofs, and underground structures—each defined by environmental constraints and associated patterns of species richness, life-forms, and functional traits (Salinitro et al. 2018) (Fig. 2.6).

Paved areas and walls are typically colonized by stress-tolerant plants with shallow roots, high desiccation tolerance, and minimal substrate requirements, whereas semi-natural soils in parks, road verges, and abandoned sites support more diverse communities, including competitive perennials and woody species (Salinitro et al. 2018; Zerbe et al. 2004). These contrasts illustrate how urban habitats function as ecological filters, selectively favouring specific traits and strategies.

The walls of Lucca offer an emblematic example of this filtering process: their extent, exposure gradients, microclimatic variability, and material composition create a wide range of **microhabitats** (Fig. 2.7), each of which favours specific groups of organisms. Differences between cool, shaded sides (rich in moisture) and warm, xeric sides (with intense solar exposure) select distinct communities, demonstrating how the urban environment shapes biological composition (Caneva and Tomei 2017).

On wall surfaces, biological colonization follows an ecological sequence regulated by these environmental filters. Cyanobacteria, algae, and lichens act as pioneers in the driest conditions, forming black patinas on south-facing walls due to

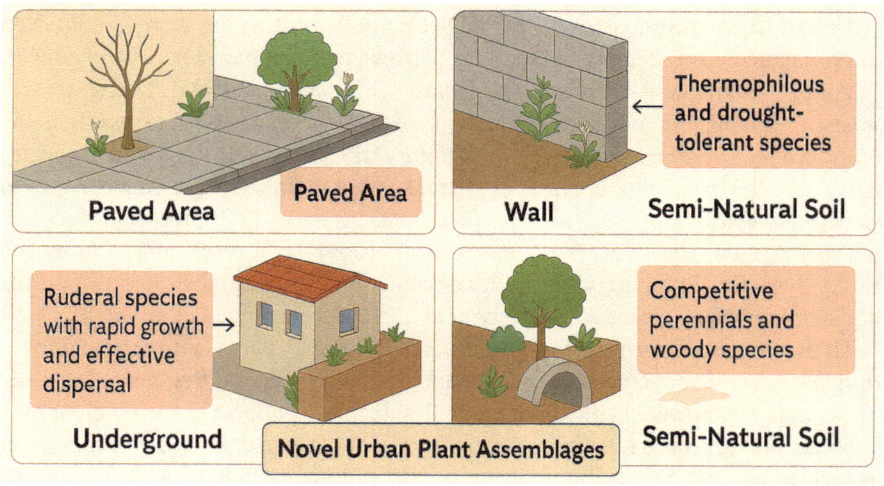

Fig. 2.6 Different urban substrates host distinct plant assemblages shaped by microhabitat conditions. (Image AI-generated with Microsoft 365 Copilot)

Fig. 2.7 Vegetation on the Lucca walls reflects how urban structures create distinct microhabitats, where exposure, moisture, and substrate selectively favour drought-tolerant pioneers on sunny faces and moisture-dependent species in shaded, humid zones. (Photos by G. Caneva)

their high tolerance to heat and desiccation. Conversely, bryophytes prefer cool, moist exposures and are almost absent on the more arid southern sections. Vascular plants, about 150 species, also respond to urban filtering: northern exposures host communities linked to *Parietarietea judaicae*, while more arid sectors favour rocky species such as *Sedum dasyphyllum*, *Clinopodium nepeta*, *Bromus madritensis*, and *Campanula erinus*. Only species with specific adaptive strategies—drought tolerance, rapid life cycles, root systems exploiting minimal fissures—are able to persist. Synanthropic and invasive species (e.g., *Ailanthus altissima*, *Erigeron sumatrensis*) also thrive thanks to their ecological plasticity. Overall, these data show that cities are not merely degraded environments but complex ecosystems where selective pressures shape functional traits and produce biological communities distinct from those of surrounding natural habitats. In this sense, the walls of Lucca function as a living laboratory of urban ecology, where microclimatic gradients, material structure, and human pressures create a spontaneous "vertical garden" shaped by the interaction between nature and the city.

Spontaneous vegetation is a defining feature of urban botanical systems, developing in areas with little or irregular human management. These communities are typically dominated by ruderal species with rapid growth, early reproduction, high seed output, and effective dispersal—traits well suited to disturbance-driven, unstable urban habitats (Pyšek et al. 2004; Kühn and Klotz 2006). Such species readily exploit temporary opportunities created by construction, soil movement, trampling, and vegetation removal.

However, spontaneous vegetation is not limited to short-lived opportunists. Unmanaged urban habitats can also host perennial herbs, shrubs, and even trees, especially where disturbance intensity declines or semi-natural substrates persist (Zerbe et al. 2004; Salinitro et al. 2018). In Mediterranean and southern European cities, these communities may additionally include *thermophilous* and drought-tolerant species adapted to shallow, stony soils and extreme temperature fluctuations, often found on walls, pavements, and roofs (Salinitro et al. 2018).

2.3.3 Novel Urban Plant Assemblages: From Spontaneous Vegetation to Biotic Homogenization

Urban vegetation often forms plant assemblages with no direct counterparts in nearby natural landscapes. Native and non-native species commonly coexist, creating **novel ecosystems** shaped by the distinctive selective pressures of urban environments (Wittig 2004; Kowarik 2011). Comparative studies across cities show a trend toward **biotic homogenization**, as a relatively small group of cosmopolitan ornamental and disturbance-tolerant species spreads widely, making urban floras increasingly similar even across different biogeographical regions. As a result, distant cities may share many species despite environmental differences (La Sorte et al. 2014). However, homogenization is not uniform. Within the city, heterogeneity remains high, shaped by local conditions, land-use history, and management practices. Spontaneous vegetation in marginal habitats often preserves elements of regional floras and contributes to maintaining urban biodiversity (Fratarcangeli et al. 2022).

2.3.4 Functional Traits and Life-Form Spectra

Functional trait approaches are highly effective for understanding the composition and dynamics of urban plant communities. Traits linked to dispersal, growth form, life span, and stress tolerance strongly influence which species thrive in urban environments (Pyšek et al. 2004; Kühn and Klotz 2006).

Urban floras commonly feature high proportions of therophytes and hemicryptophytes, reflecting adaptation to disturbance and seasonal variability. Long-term studies, however, show rising numbers of phanerophytes and geophytes, driven by urban greening, ornamental plantings, and climate warming (Salinitro et al. 2019). These shifts illustrate how urban vegetation responds to both management and broader environmental change.

Functional convergence across cities further indicates that urban environments act as strong ecological filters: species from different biogeographical regions often share similar trait combinations suited to urban selective pressures (Ricotta et al. 2009; La Sorte et al. 2014).

2.3.5 Spontaneous Vegetation as an Indicator and Ecological Resource

Spontaneous urban vegetation offers key insights into ecological processes in cities. Because these communities respond quickly to disturbance, soil conditions, and microclimate, they act as sensitive indicators of environmental quality and land-use change (Wittig 2004; Salinitro et al. 2018).

Ecologically, spontaneous vegetation contributes to urban biodiversity, supports pollinators and other fauna, and initiates successional processes in abandoned or restored sites. Long-term surveys show that it enhances habitat heterogeneity and helps sustain high species richness even in densely built environments (Fratarcangeli et al. 2022).

Rather than signalling neglect, spontaneous vegetation represents a fundamental component of urban ecosystems and a valuable subject for urban botanical research, offering insights into plant adaptation, community assembly, and the ecological effects of urbanization.

2.3.6 Native and Alien Species in Urban Floras: Patterns, Processes, and Biogeographical Context

Native and alien species jointly shape urban floras, with cities acting as major hubs of plant introduction and as environments where both groups coexist under intense disturbance and environmental filtering, offering insights into establishment, persistence, and spread in human-dominated landscapes (Pyšek et al. 2004; Celesti-Grapow et al. 2006). Across Europe, alien species are proportionally more frequent in urban floras than in surrounding non-urban areas, although native species still constitute most of the total richness, reflecting regional history, biogeography, land-use patterns, and socio-economic contexts (Celesti-Grapow and Blasi 1998; Rat et al. 2017).

High propagule pressure from horticulture, trade, and transport, together with frequent disturbance, promotes alien establishment by creating open niches and reducing competition from late-successional natives, yet urban floras remain complex assemblages in which native, *archaeophyte*, and *neophyte* species interact under multiple ecological drivers (Wittig 2004; Celesti-Grapow et al. 2006).

Despite homogenizing forces driven largely by widespread alien and synanthropic species, climate, regional species pools, and historical land-use patterns

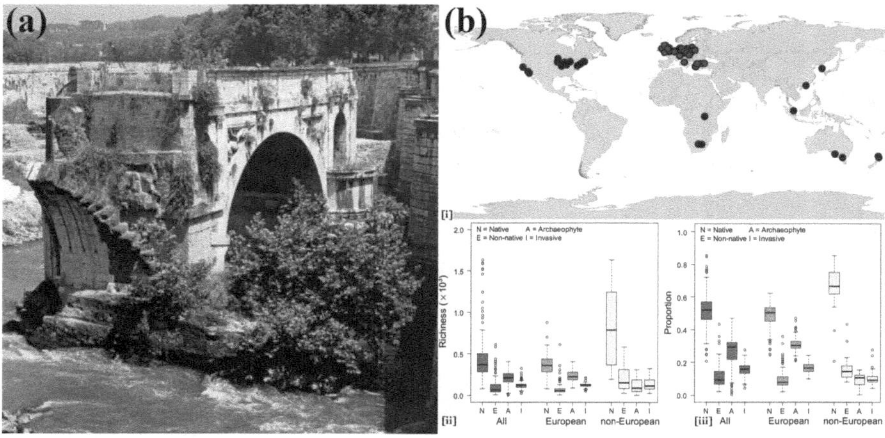

Fig. 2.8 (**a**) Contrasting components of Rome's urban flora: native *Capparis spinosa*, in rocky Mediterranean habitats; non-native *Platanus* × *hybrida* naturalized along the banks of the Tiber River (from Celesti-Grapow and Blasi 1998) (**b**) Geographic distribution of sampled cities (i) and variation in urban floristic composition (ii–iii) proportions of native (N), non-native (E), archaeophyte (A), and invasive (I) species. (From La Sorte et al. 2014)

continue to give each city a distinct floristic identity (Celesti-Grapow and Blasi 1998) (Fig. 2.8a), even as beta-diversity studies reveal increasing similarity among cities worldwide due to a shared subset of species adapted to urban conditions (La Sorte et al. 2014; Fig. 2.8b).

Nevertheless, within-city heterogeneity remains high, with habitat-level differences often exceeding those between cities at comparable latitudes. Studies from Southeast Europe further highlight the importance of regional context: cities in this region tend to have high overall species richness and lower proportions of neophytes, reflecting later and less intensive urbanization and stronger persistence of native vegetation (Rat et al. 2017), underscoring that urbanization does not produce uniform floristic outcomes but interacts with biogeographical legacies to shape urban plant composition.

2.3.7 Temporal Dynamics and Long-Term Change

Urban floras are dynamic systems marked by continuous change. Long-term surveys show that species turnover—driven by both gains and losses—is a defining feature of urban plant communities, yet overall species richness often remains stable or even increases despite shifts in composition (Pyšek et al. 2004).

Italian case studies illustrate these dynamics clearly. In Bologna, more than a century of data reveals increased total species richness, largely due to alien species, alongside shifts in life-form spectra and ecological preferences (Salinitro et al. 2019). In Rome, grid-based surveys over 23 years document substantial floristic change, with spontaneous vegetation helping maintain overall diversity (Fratarcangeli et al. 2022).

Furthermore, previous studies on the Colosseum's plant communities (Caneva 2004), spanning about four centuries, highlighted significant changes in ecological succession driven by microhabitats, microclimatic conditions, the degree of anthropogenic disturbance, and the evolution of substrates. Through the phytosociological approach, it was possible distinguish several vegetational stages (Fig. 2.9), with their characteristic species.

In this case, such stages were defined by pioneer communities, which colonizes rock surfaces and masonry poor in soil, until two different final forest stages, (*Quercetea ilicis* and *Querco-Fagetea*), combined with ruderal species, indicators of disturbance and intense human presence, and escaped cultivated species, It highlighted the historical connections between the Colosseum and surrounding orchards, gardens, and anthropized areas, revealing non-native elements introduced through accidental dispersal (Caneva 2004).

Temporal changes in urban floras are shaped not only by urban development but also by broader drivers such as climate change. Rising temperatures, altered rainfall, and the urban heat island effect favour *thermophilus* and drought-tolerant species, contributing to shifts in composition and functional traits—patterns evident in Mediterranean cities and increasingly observed elsewhere (Salinitro et al. 2018, 2019).

Fig. 2.9 Conceptual vision of natural urban succession (**a**) overtaking ancient Colosseum structures and (**b**) representative stages from pioneer colonizers on masonry to ruderal species, xerophilous Mediterranean communities, and mature nemoral vegetation. (Photos by G. Caneva)

2.3.8 Phylogenetic and Functional Perspectives

Beyond taxonomic patterns, urban floras have increasingly been studied through phylogenetic and functional lenses. Phylogenetic analyses show that alien species in cities are often non-randomly clustered, reflecting shared evolutionary histories and trait syndromes linked to successful urban establishment (Ricotta et al. 2009).

Functional studies reveal strong trait convergence among urban plant communities: traits related to disturbance tolerance, dispersal ability, and stress resistance are consistently overrepresented, regardless of species origin. Consequently, urban floras may differ taxonomically across regions yet remain functionally similar (Kühn and Klotz 2006; La Sorte et al. 2014).

These findings underscore that urban floras cannot be understood solely through species lists. Integrating taxonomic, phylogenetic, and functional perspectives provides a more complete understanding of the processes shaping urban plant communities and the patterns of homogenization and differentiation they produce.

2.3.9 Ecological Role in Cities

Urban vegetation is essential for sustaining ecological processes in cities, benefiting both ecosystem functioning and human well-being (Fig. 2.10). Beyond aesthetics, plant communities influence microclimate, biogeochemical cycles, hydrology, habitat availability, and landscape connectivity. These ecological contributions arise from the combined roles of cultivated, semi-spontaneous, and spontaneous vegetation, each shaping urban ecosystem dynamics in distinct ways (Zerbe et al. 2004; Wittig 2004).

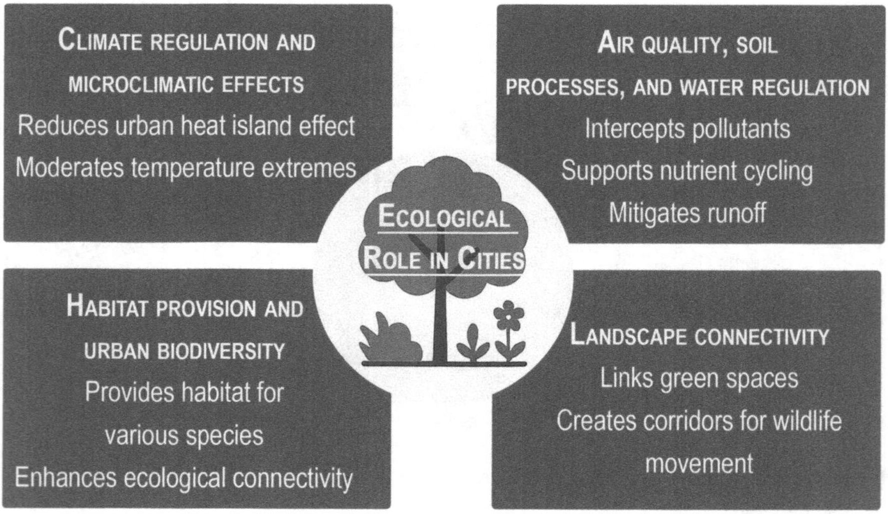

Fig. 2.10 Key ecological roles of urban vegetation in cities. (Image by F. Bartoli)

2.3.9.1 Climate Regulation and Microclimatic Effects

Urban vegetation plays a major role in regulating the climate by reducing the urban heat island effect through shading, evapotranspiration, and altered surface energy balance. Areas with greater vegetation cover consistently show lower surface and air temperatures, especially during heatwaves (Salinitro et al. 2018, 2019).

Trees, shrubs, and herbaceous plants moderate temperature extremes at scales ranging from individual streets to entire neighbourhoods. Spontaneous and semi-spontaneous vegetation is increasingly recognized as important in this process, as unmanaged green spaces often provide higher structural complexity and greater evapotranspiration capacity than intensively maintained lawns (Zerbe et al. 2004).

2.3.9.2 Air Quality, Soil Processes, and Water Regulation

Urban vegetation significantly improves air quality by intercepting particulate matter and absorbing gaseous pollutants, with leaves, bark, and soil acting as sinks for airborne contaminants (Wittig 2004). Although removal efficiency differs among species and habitats, the combined effect of widespread vegetation can meaningfully enhance air quality in cities with extensive green infrastructure. Urban soils associated with vegetation also support key biogeochemical processes, including nutrient cycling, organic matter accumulation, and microbial activity. Spontaneous vegetation stabilizes disturbed substrates and promotes gradual soil development, aiding successional recovery in degraded areas (Zerbe et al. 2004; Salinitro et al. 2018). Vegetated surfaces improve rainwater infiltration, reduce runoff, and help mitigate flood risk during intense rainfall events. With climate change increasing extreme weather, even small patches of spontaneous vegetation can contribute to local hydrological regulation by slowing water flow and enhancing infiltration (Fratarcangeli et al. 2022).

2.3.9.3 Habitat Provision and Urban Biodiversity

Urban vegetation supports a wide range of organisms by providing food, shelter, and breeding sites, thereby sustaining complex trophic interactions. Across cities, higher plant diversity consistently correlates with higher faunal diversity, underscoring the key role of vegetation in maintaining urban biodiversity (Rat et al. 2017). Spontaneous vegetation is especially valuable. Marginal habitats—such as brownfields, railway corridors, and riverbanks—often host structurally diverse plant communities that act as refuges for species rare or declining in surrounding agricultural areas (Zerbe et al. 2004). These sites also serve as ecological stepping stones, improving connectivity within fragmented urban landscapes and supporting the persistence of regional biodiversity.

2.3.10 Cultural, Educational, and Social Dimensions

Urban vegetation not only supports ecological processes but also enriches the cultural and social fabric of cities. Green spaces improve aesthetic quality, provide recreational opportunities, and enhance physical and psychological well-being, with exposure to diverse plant communities linked to reduced stress and stronger human–nature connections (Zerbe et al. 2004).

From an educational standpoint, urban vegetation offers accessible ways to observe plant diversity, ecological dynamics, and environmental change. Spontaneous vegetation enables direct observation of succession, adaptation, and species interactions in everyday settings, underscoring the value of cities as open-air laboratories for ecological education and research (Wittig 2004).

2.3.11 Implications for Conservation and Urban Planning

Urban vegetation plays a key role in biodiversity conservation and should be fully integrated into urban planning. Research shows that cities can support species of conservation concern and significantly contribute to broader biodiversity goals when habitat heterogeneity and ecological continuity are preserved (Zerbe et al. 2004; Rat et al. 2017). Effective conservation in urban areas requires moving beyond a narrow focus on "naturalness" and recognising the value of novel and spontaneous plant communities. Rather than indiscriminately removing non-native species, management should prioritise functional roles, ecological impacts, and long-term dynamics, as highlighted by comparative studies of urban floras (Wittig 2004). Incorporating vegetation into planning processes, protecting unmanaged or lightly managed spaces, and acknowledging the ecological contribution of spontaneous flora are essential for building resilient, functional, and ecologically rich cities. Urban vegetation is therefore not marginal, but a structural component of sustainable urban ecosystems.

2.4 Nature Conservation Strategies and Habitat Restoration Actions (G.A.R., F.B.)

2.4.1 Importance of Nature Conservation in Urban Ecosystems

Urban ecosystems play a critical role in biodiversity conservation, significantly affecting local and global ecological dynamics. As urbanization accelerates globally, with more than 55% of the world's population residing in urban areas, integrating green spaces and conserving natural habitats within cities is essential for maintaining biodiversity and ecosystem services (Ward et al. 2021; Monteiro et al. 2020). These urban green spaces—parks, gardens, and urban forests—serve as vital refugia for various species, favouring the persistence of native flora and fauna that would otherwise be edged out by urban development (Sirakaya et al. 2018). The recognition of urban areas as biodiversity hotspots introduces new paradigms in conservation strategies, highlighting the need for urban ecological frameworks to facilitate species conservation within fragmented landscapes (Salviano et al. 2021). Moreover, urban areas can provide essential ecosystem services such as air purification, temperature regulation, and recreational spaces, reaffirming the importance of conservation within urban planning processes (De Luca et al. 2021). The integration of nature into cities not only strengthens biodiversity conservation but also improves human well-being, establishing a symbiotic (rather than parasitic) relationship between urban residents and their environment (Belaire et al. 2011). This duality calls for proactive conservation strategies that leverage urban biodiversity to create healthier, more resilient cities (Fumagalli et al. 2014).

2.4.2 Challenges Posed by Fragmentation and Biodiversity Loss

Urbanization inevitably leads to habitat fragmentation, disrupting ecological networks and causing significant biodiversity loss. Fragmentation can isolate populations, reduce genetic diversity, and impede species movement, which are crucial for maintaining healthy ecosystems (Rudd et al. 2002; Cai et al. 2021). Additionally, the rapid expansion of urban landscapes often leads to the degradation of natural habitats, further exacerbating the challenges faced by urban biodiversity (Xie et al. 2023). Invasive species thrive in disturbed habitats, competing with native species and altering ecosystem functionalities (Abella and Chiquoine 2018). Furthermore, the socio-economic dynamics of urban environments often prioritize development over ecological integrity, complicating conservation efforts (Weng 2015; Cai et al. 2021). The need for integrated environmental planning that addresses these challenges is critical to reverse biodiversity decline and ensure the sustainability of urban ecosystems. Achieving effective habitat connectivity is imperative not only for facilitating wildlife movement but also for restoring ecological processes that have been compromised by urban encroachment (Saura et al. 2013). Only through a comprehensive understanding of these challenges can urban planning systems incorporate robust conservation strategies that mitigate habitat fragmentation and enhance biodiversity resilience (Salviano et al. 2021).

2.4.3 Principles of Urban Nature Conservation

2.4.3.1 Goals & Objectives: Biodiversity and Ecosystem Services

Urban conservation addresses rapid urbanization, biodiversity loss, and climate change by preserving biodiversity and enhancing ecosystem services that sustain ecological stability and human well-being (Sirakaya et al. 2018). Key services include clean air and water, heat mitigation, recreation, and mental-health benefits, all of which improve urban resilience and liveability (Štrbac et al. 2023) (Fig. 2.11).

2.4.3.2 Policy Frameworks: From Global to Local

International agreements (e.g., CBD, UNESCO-MAB, SDGs—Goal 11) call for mainstreaming biodiversity across sectors and scales (Perrings et al. 2011; Hanna and Comìn 2021). In Europe, the EU Biodiversity Strategy 2030 and the Nature Restoration Regulation (2024/1991) align city planning with restoration targets (De Luca et al. 2021; Štrbac et al. 2023). National statutes and local regulations translate these aims into enforceable standards and projects, while also catalysing public stewardship.

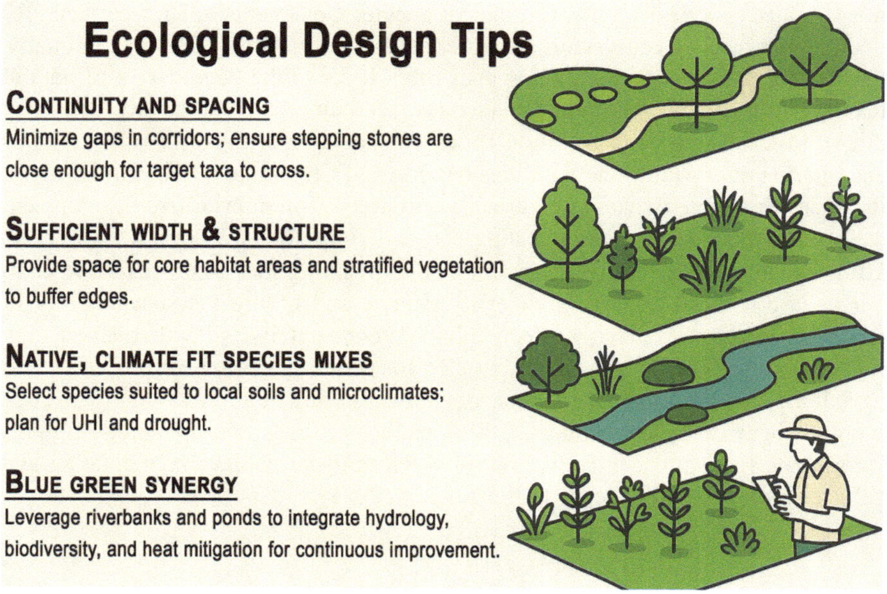

Fig. 2.11 Key design principles for planning functional green infrastructure. (Image by F. Bartoli)

2.4.3.3 Integration in Urban Planning

Green infrastructure (GI)—networks of natural and semi-natural areas—connects isolated habitats and delivers multiple services (Monteiro et al. 2020; Wang et al. 2024). Coupled with **nature-based solutions (NbS) and well-designed green spaces (parks, street trees, green roofs), GI improves biodiversity, air quality, and heat mitigation, advancing climate-resilient** urbanism (Štrbac et al. 2023; Hölscher et al. 2023).

2.4.3.4 Habitat Restoration: Concepts and Approach

Restoration ecology seeks to **recover degraded ecosystems** by reinstating native species and processes, guided by reference ecosystems and implemented through adaptive management (plan–do–monitor–adjust), which is critical under rapidly changing urban conditions (Cortina et al. 2021; Hüse et al. 2016; Xie et al. 2023).

Core Techniques

**Soil remediation* (phytoremediation, bioremediation, soil washing) to reduce contaminants and rebuild edaphic functions—prerequisite for successful revegetation (Belaire et al. 2011; Ward et al. 2021). **Vegetation reintroduction* with native, locally adapted species to restore habitat quality and ecological processes (pollination, soil stabilization) (García-Antón et al. 2021; Rudd et al. 2002). **Use of non-natives only with caution* (e.g., temporary nurse plants that are demonstrably non-invasive), given invasion risks documented in urban parks (Abella and Chiquoine 2018; Hüse et al. 2016). **Monitoring & evaluation* with clear success indicators (composition, richness, functional metrics) and socio-ecological outcomes to support adaptive management and long-term sustainability (Brown et al. 2016; Zhou et al. 2023).

Three Take-Home Practices

**Plan with policy coherence:* Map local goals to CBD/SDGs/EU directives and municipal rules; pair regulation with public engagement (Perrings et al. 2011; Belaire et al. 2011). **Design for connectivity and climate:* Build GI/NbS networks (corridors, steppingstones, green roofs) to boost biodiversity and urban resilience (Monteiro et al. 2020; Hölscher et al. 2023). **Restore adaptively:* Remediate soils, prioritize natives, limit non-natives to controlled, temporary roles, and monitor–adjust for durable outcomes (Cortina et al. 2021; Brown et al. 2016).

2.4.4 Role of Ecological Connectivity in Sustaining Urban Biodiversity

Ecological connectivity is a core principle for sustaining urban biodiversity, as corridors and stepping stones between fragmented habitats facilitate species movement, genetic exchange, and long-term *population viability* (García-Antón et al. 2021). Integrating such ecological networks into urban planning allows cities to embed nature within built environments, enhancing biodiversity while supporting essential ecosystem services. Connectivity involves not only physical links but also the functional quality of habitats, underscoring the value of multifunctional green spaces that benefit both wildlife and urban communities (Fischer et al. 2012; Dong et al. 2023). Research shows that green areas connected through ecological corridors strengthen urban ecological resilience and improve key services such as climate regulation and pollination (Fig. 2.12).

Developing robust connectivity strategies is therefore essential for cities aiming to expand while maintaining their role as ecological nodes in broader landscapes (Salviano et al. 2021). By enhancing connectivity, cities can reconcile human activities with natural processes and contribute to more sustainable urban futures (Kowarik and Lippe 2018).

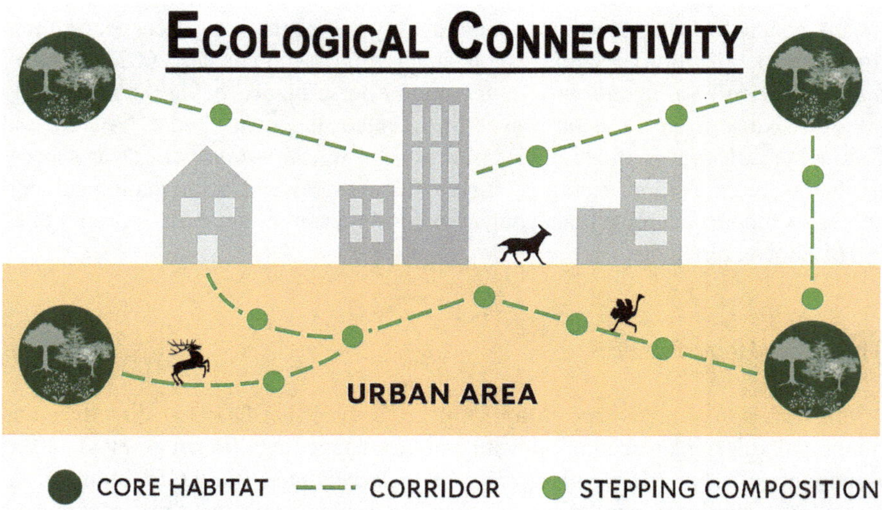

Fig. 2.12 Core habitat patches, ecological corridors, and stepping-stone elements function together to facilitate wildlife movement across fragmented urban environments. (Image by F. Bartoli)

2.4.5 Ecological Corridors

Ecological corridors are linear landscape elements that connect otherwise isolated habitat patches, enabling movement, dispersal, and gene flow among wildlife populations—functions that are especially critical where urbanization fragments the matrix. By providing safe routes across hostile surfaces (roads, walls, sealed soils), corridors reduce the functional effects of habitat loss, allowing species to track resources seasonally, recolonize after local extinctions, and maintain metapopulation stability in rapidly changing cityscapes (Xie et al. 2023).

Design quality determines corridor performance. Three levers are consistently important:

- *Width*—Wider corridors support a larger spectrum of species, reduce edge effects, and buffer microclimate, which is vital in heat-prone, wind-exposed urban canyons (Rudd et al. 2002; Rouget et al. 2006).
- *Continuity & permeability*—Uninterrupted stretches (or carefully designed crossing solutions) reduce mortality and energetic costs of movement; few, shorter gaps are preferable to many, long ones (Rouget et al. 2006).
- *Habitat quality & structure*—Stratified, multi-layered vegetation (trees–shrubs–herb layer) offers foraging, cover, and nesting niches while delivering co-benefits such as shade, wind-breaks, evapotranspiration cooling, and nutrient cycling (Semerdzhieva and Borisova 2021). A native-species palette aligned with local soils and microclimates improves resilience and supports local fauna (Rouget et al. 2006).

Corridors can act as habitat in their own right—not merely conduits—allowing access to food and shelter and, in some cases, establishment of new subpopulations in previously unsuitable or isolated areas. In aggregate, well-designed networks increase ecological resilience and help systems adapt to stressors, including climate-related heat and hydrological extremes (García-Antón et al. 2021).

2.4.5.1 Urban Examples

- *River corridors* in cities simultaneously support aquatic and riparian biodiversity, provide travel routes for fish, birds, and semi-aquatic mammals, and deliver ecosystem services—filtration, erosion control, flood buffering—that improve urban waterway health (Francis et al. 2019).
- *Greenways*—continuous, vegetated routes connecting parks and open spaces—double as recreation spines for people while providing shaded, cooler paths for wildlife to move across neighbourhoods; when integrated with green-infrastructure (GI) components, they also contribute to heat mitigation and stormwater management (Horte and Eisenman 2020).

- *Railway verges* are often undervalued; yet, with appropriate management (reduced mowing frequencies, native mixes, invasive control), these long, linear strips can function as movement corridors and habitat for plants, invertebrates, small vertebrates, and some birds—crucial in districts where every meter of green edge matters (Ouédraogo et al. 2020).

2.4.6 Stepping Stones

Stepping stones are **small habitat patches** that facilitate movement between larger core areas—particularly effective where continuous corridors are unfeasible (dense built fabric, land-use conflicts) (Lapin et al. 2024). They increase functional connectivity by creating "hops" across hostile matrices, thereby helping maintain gene flow and reducing the demographic risks of small, isolated populations. Their value rises further under climate change, as species adjust their ranges and phenologies and require more frequent access to microrefugia (Saura et al. 2013). In practice, steppingstones span pocket parks, roadside greens, green roofs, vegetated courtyards, and urban ponds. Their performance depends on placement (spatial configuration and spacing), size and shape (edge-to-area ratio), and habitat quality (structural diversity, native flora, microclimatic buffering). When strategically sited, they support pollinator circuits, urban bird flyways, and small-mammal traverses, effectively "stitching" the urban fabric together at fine scales (Lapin et al. 2024).

2.4.6.1 Urban Examples

- *Pocket parks*—small, distributed green islands within dense blocks—boost neighbourhood biodiversity and human access to nature; using **local native species** and layered plantings maximizes habitat value (Semerdzhieva and Borisova 2021).
- *Rooftop gardens/green roofs* add three-dimensional connectivity, cool rooftops, and create nectar/**pollen stations** in otherwise barren airspaces; they are particularly useful along pollinator routes and bird flyways (Rudd et al. 2002).
- *Urban ponds* link blue-green infrastructure, enhance stormwater regulation, and provide **breeding and foraging habitat** for amphibians, insects, and birds (Hill et al. 2016; Horte and Eisenman 2020).

For both corridors and stepping stones, monitoring and adaptive management are essential: track use (camera traps, acoustic sensors, pollinator transects), habitat quality (native cover, structural layers), and connectivity performance (e.g., least-cost paths), then adjust widths, plant palettes, mowing regimes, and crossing designs accordingly (Wang et al. 2024).

References

Abella S, Chiquoine L (2018) The good with the bad: when ecological restoration facilitates native and non-native species. Restor Ecol 27(2):343–351. https://doi.org/10.1111/rec.12874

Ansong M, Pickering C (2013) Are weeds hitchhiking a ride on your car? A systematic review of seed dispersal on cars. PLoS One 8(11):e80275. https://doi.org/10.1371/journal.pone.0080275

Aronson MF, Lepczyk CA, Evans K, Goddard M, Lerman S, MacIvor J et al (2017) Biodiversity in the city: key challenges for urban green space management. Front Ecol Environ 15(4):189–196. https://doi.org/10.1002/fee.1480

Attorre F, Stanisci A, Bruno F (1997) The urban woods of Rome (Italy). Plant Biosyst Int J Dealing Aspects Plant Biol 131(2):113–135. https://doi.org/10.1080/11263504.1997.10654173

Belaire J, Dribin A, Johnston D, Lynch D, Minor E (2011) Mapping stewardship networks in urban ecosystems. Conserv Lett 4(6):464–473. https://doi.org/10.1111/j.1755-263x.2011.00200.x

Biondi E (2011) Phytosociology today: methodological and conceptual evolution. Plant Biosyst 145:19–29. https://doi.org/10.1080/11263504.2011.602748

Brown C, Shaker R, Das R (2016) A review of approaches for monitoring and evaluation of urban climate resilience initiatives. Environ Dev Sustain 20(1):23–40. https://doi.org/10.1007/s10668-016-9891-7

Cai Z, Page J, Cvetković V (2021) Urban ecosystem vulnerability assessment of support climate-resilient city development. Urban. Planning 6(3):227–239. https://doi.org/10.17645/up.v6i3.4208

Caneva G (ed) (2004) Amphitheatrum naturae: the Colosseum—history and environment interpreted through its flora. Electa. (Original work in Italian)

Caneva G, Tomei PE (eds) (2017) Un giardino verticale sulle Mura di Lucca: Mi/u-rabilia. Gangemi Editore spa

Capotorti G, Del Vico E, Lattanzi E, Tilia A, Celesti-Grapow L (2013) Exploring biodiversity in a metropolitan area in the Mediterranean region: the urban and suburban flora of Rome (Italy). Plant Biosyst Int J Dealing Aspects Plant Biol 147(1):174–185. https://doi.org/10.1080/11263504.2013.771715

Celesti-Grapow L, Blasi C (1998) A comparison of the urban flora of different phytoclimatic regions in Italy. Glob Ecol Biogeogr Lett 7(5):367–378. https://doi.org/10.1046/j.1466-822x.1998.00304.x

Celesti-Grapow L, Pyšek P, Jarošík V, Blasi C (2006) Determinants of native and alien species richness in the urban flora of Rome. Divers Distrib 12(5):490–501

Cortina J, García-Sánchez I, Grace M, Andrés P, Baker S, Bullock C et al (2021) Barriers to ecological restoration in Europe: expert perspectives. Restor Ecol 29(4). https://doi.org/10.1111/rec.13346

De Luca C, Naumann S, Davis M, Tondelli S (2021) Nature-based solutions and sustainable urban planning in the European environmental policy framework: analysis of the state of the art and recommendations for future development. Sustainability 13(9):5021. https://doi.org/10.3390/su13095021

Del Río S, Alonso-Redondo R, González-Pérez A, Álvarez-Santacoloma A, Breogán Ferreiro Lera G, Penas Á (2022) Dynamic-Catenal phytosociology for evaluating vegetation. In: Cano Carmona E, Cano Ortiz A, Quinto Canas R, Maria Musarella C (eds) Vegetation index and dynamics. IntechOpen. https://doi.org/10.5772/intechopen.97426

Del Tredici P (2010) Spontaneous urban vegetation: reflections of change in a globalized world. Nat Cult 5(3):299–315. https://doi.org/10.3167/nc.2010.050305

Dong J, Guo R, Guo F, Guo X, Zhang Z (2023) Pocket parks-a systematic literature review. Environ Res Lett 18(8):083003. https://doi.org/10.1088/1748-9326/ace7e2

Fischer L, Lippe M, Kowarik I (2012) Urban grassland restoration: which plant traits make desired species successful colonizers? Appl Veg Sci 16(2):272–285. https://doi.org/10.1111/j.1654-109x.2012.01216.x

Francis R, Chadwick M, Turbelin A (2019) An overview of non-native species invasions in urban river corridors. River Res Appl 35(8):1269–1278. https://doi.org/10.1002/rra.3513

Fratarcangeli C, Fanelli G, Testolin R, Buffi F, Travaglini A (2022) Floristic changes of vascular flora in the city of Rome through grid-cell census over 23 years. Urban Ecosyst 25:1851–1864

Fumagalli N, Colombo C, Ferrario P, Senes G, Toccolini A (2014) Suburban waterfront with ecological and recreational function: planning based on network analysis. J Agric Eng 44(4):141. https://doi.org/10.4081/jae.2013.183

García-Antón A, Garza V, Traba J (2021) Connectivity in the Spanish metapopulation of DuPont's lark may be maintained by dispersal over a medium-distance range and stepping stones. PeerJ 9:e11925. https://doi.org/10.7717/peerj.11925

Godefroid S, Koedam N (2007) Urban plant species patterns are highly driven by density and function of built-up areas. Landsc Ecol 22:1227–1239. https://doi.org/10.1007/s10980-007-9102-x

Grime JP (1979) Plant strategies and vegetation processes. Wiley, London, p 222

Hanna E, Comìn F (2021) Urban green infrastructure and sustainable development: a review. Sustainability 13(20):11498. https://doi.org/10.3390/su132011498

Hill M, Biggs J, Thornhill I, Briers R, Gledhill D, White J et al (2016) Urban ponds as an aquatic biodiversity resource in modified landscapes. Glob Chang Biol 23(3):986–999. https://doi.org/10.1111/gcb.13401

Hobbs RJ, Arico S, Aronson J, Baron JS, Bridgewater P, Cramer VA et al (2006) Novel ecosystems: theoretical and management aspects of the new ecological world order. Glob Ecol Biogeogr 15(1):1–7. https://doi.org/10.1111/j.1466-822X.2006.00212.x

Hölscher K, Frantzeskaki N, Collier M, Connop S, Kooijman E, Lodder M et al (2023) Strategies for mainstreaming nature-based solutions in urban governance capacities in ten European cities. NPJ Urban Sustain 3(1). https://doi.org/10.1038/s42949-023-00134-9

Horte O, Eisenman T (2020) Urban greenways: a systematic review and typology. Land 9(2):40. https://doi.org/10.3390/land9020040

Hüse B, Szabó S, Deák B, Tóthmérész B (2016) Mapping an ecological network of green habitat patches and their role in maintaining urban biodiversity in and around Debrecen city (eastern Hungary). Land Use Policy 57:574–581. https://doi.org/10.1016/j.landusepol.2016.06.026

Itani M, Al Zein M, Nasralla N, Talhouk SN (2020) Biodiversity conservation in cities: defining habitat analogues for plant species of conservation interest. PLoS One 15(6):e0220355. https://doi.org/10.1371/journal.pone.0220355

Kalusová V, Čepková N, Lososová Z (2016) Which traits influence the frequency of plant species occurrence in urban habitat types? Urban Ecosyst 20:65–75. https://doi.org/10.1007/s11252-016-0588-3

Kowarik I (2011) Novel urban ecosystems, biodiversity, and conservation. Environ Pollut 159(8–9):1974–1983

Kowarik I, Lippe M (2018) Plant population success across urban ecosystems: a framework to inform biodiversity conservation in cities. J Appl Ecol 55(5):2354–2361. https://doi.org/10.1111/1365-2664.13144

Kühn I, Klotz S (2006) Urbanization and homogenization – comparing the floras of urban and rural areas in Germany. Biol Conserv 127(3):292–300. https://doi.org/10.1016/j.biocon.2005.06.033

La Sorte FA, Aronson MFJ, Williams NSG, Celesti-Grapow L, Cilliers S, Clarkson BD et al (2014) Beta diversity of urban floras among European and non-European cities. Glob Ecol Biogeogr 23(7):769–779. https://doi.org/10.1111/geb.12159

Lapin K, Hoffmann J, Braun M, Oettel J (2024) Identification and prioritization of stepping stones for biodiversity conservation in forest ecosystems. Conserv Sci and Pract 6(7). https://doi.org/10.1111/csp2.13161

Latzel V, Mihulka S, Klimešová J (2008) Plant traits and regeneration of urban plant communities after disturbance: does the bud bank play any role? Appl Veg Sci 11(3):387–394. https://doi.org/10.3170/2008-7-18487

MacArthur RH, Wilson EO (1967) The theory of Island biogeography. Princeton University Press, Princeton

Maehara K, Iwachido Y, Katsuhara H, Tomitaka M, Seto K, Ushio M, Sasaki T (2024) Plant and soil microbial community assembly processes across urban vacant lots. J Veg Sci 35(3):e13262. https://doi.org/10.1111/jvs.13262

Monteiro R, Ferreira J, Antunes P (2020) Green infrastructure planning principles: an integrated literature review. Land 9(12):525. https://doi.org/10.3390/land9120525

Neil K, Wu J (2006) Effects of urbanization on plant flowering phenology: a review. Urban Ecosyst 9(3):243–257. https://doi.org/10.1007/s11252-006-9354-2

Oberdorfer E, Lundholm J, Bass B, Coffman RR, Doshi H, Dunnett N, Gaffin S, Kohler M, Liu KK, Rowe B (2007) Green roofs as urban ecosystems: ecological structures, functions, and services. Bioscience 57(10):823–833. https://doi.org/10.1641/B571005

Ouédraogo D, Villemey A, Vanpeene S, Coulon A, Azambourg V, Hulard M et al (2020) Can linear transportation infrastructure verges constitute a habitat and/or a corridor for vertebrates in temperate ecosystems? A systematic review. Environ Evid 9(1). https://doi.org/10.1186/s13750-020-00196-7

Pausas JG, Keeley JE (2014) Evolutionary ecology of resprouting and seeding in fire-prone ecosystems. New Phytol 204(1):55–65. https://doi.org/10.1111/nph.12921

Perrings C, Naeem S, Ahrestani F, Bunker D, Burkill P, Canziani G et al (2011) Ecosystem services, targets, and indicators for the conservation and sustainable use of biodiversity. Front Ecol Environ 9(9):512–520. https://doi.org/10.1890/100212

Pickett ST, Cadenasso ML, Grove JM, Boone CG, Groffman PM, Irwin E et al (2011) Urban ecological systems: scientific foundations and a decade of progress. J Environ Manag 92(3):331–362. https://doi.org/10.1016/j.jenvman.2010.08.022

Pouyat RV, Szlavecz K, Yesilonis ID, Groffman PM, Schwarz K (2010) Chemical, physical, and biological characteristics of urban soils. Urban Ecosyst Ecol 55:119–152. https://doi.org/10.2134/agronmonogr55.c7

Pyšek P (1991) Succession in urban habitats: An analysis of phytosociological data. Preslia 63:125–138

Pyšek P, Chocholoušková Z, Pyšek A, Jarošík V, Chytrý M, Tichý L et al (2004) Trends in species diversity and composition of urban vegetation over three decades. J Veg Sci 15(6):781–788

Quinto-Canas R (2014) Flora y Vegetación de la Serra do Caldeirão. Phd thesis, Facultad Ciencias Experimentales, Universidad de Jaén, Jaén, Spain

Rat MM, Gavrilović MT, Radak BĐ, Bokić BS, Jovanović SD, Božin BN et al (2017) Urban flora in Southeast Europe and its correlation with urbanization. Urban Ecosyst 20(4):811–822. https://doi.org/10.1007/s11252-017-0645-6

Ricotta C, Celesti-Grapow L, Blasi C (2009) Phylogenetic analysis of urban alien floras. J Ecol 97(6):1243–1251

Rivas-Martínez S (2011) Mapa de series, geoseries y geopermaseries de vegetación de España. Asociacion Española de Fitosociologia (AEFA)

Rouget M, Cowling R, Lombard A, Knight A, Kerley G (2006) Designing large-scale conservation corridors for pattern and process. Conserv Biol 20(2):549–561. https://doi.org/10.1111/j.1523-1739.2006.00297.x

Rudd H, Vala J, Schaefer V (2002) Importance of backyard habitat in a comprehensive biodiversity conservation strategy: a connectivity analysis of urban green spaces. Restor Ecol 10(2):368–375. https://doi.org/10.1046/j.1526-100x.2002.02041.x

Salinitro M, Alessandrini A, Zappi A, Melucci D, Tassoni A (2018) Floristic diversity in different urban ecological niches of a southern European city. Sci Rep 8:15110

Salinitro M, Alessandrini A, Zappi A, Tassoni A (2019) Impact of climate change and urban development on the flora of a southern European city: analysis of biodiversity change over a 120-year period. Sci Rep 9:9464

Salviano I, Gardon F, Santos R (2021) Ecological corridors and landscape planning: a model to select priority areas for connectivity maintenance. Landsc Ecol 36(11):3311–3328. https://doi.org/10.1007/s10980-021-01305-8

Saura S, Bodin Ö, Fortin M (2013) Editor's choice: stepping stones are crucial for species' long-distance dispersal and range expansion through habitat networks. J Appl Ecol 51(1):171–182. https://doi.org/10.1111/1365-2664.12179

Semerdzhieva L, Borisova B (2021) Urban ecosystems assessment: an integrated approach to maintenance of habitats and their biodiversity. J Bulgarian Geogr Soc 45:99–106. https://doi.org/10.3897/jbgs.e78975

Sirakaya A, Cliquet A, Harris J (2018) Ecosystem services in cities: towards the international legal protection of ecosystem services in urban environments. Ecosyst Serv 29:205–212. https://doi.org/10.1016/j.ecoser.2017.01.001

Štrbac S, Kašanin-Grubin M, Pezo L, Stojić N, Lončar B, Ćurčić L et al (2023) Green infrastructure designed through nature-based solutions for sustainable urban development. Int J Environ Res Public Health 20(2):1102. https://doi.org/10.3390/ijerph20021102

Sukopp H (2004) Human-caused impact on preserved vegetation. Landsc Urban Plan 68(4):347–355. https://doi.org/10.1016/S0169-2046(03)00152-X

Tzoulas K, Korpela K, Venn S, Yli-Pelkonen V, Kaźmierczak A, Niemela J, James P (2007) Promoting ecosystem and human health in urban areas using green infrastructure: a literature review. Landsc Urban Plan 81(3):167–178. https://doi.org/10.1016/j.landurbplan.2007.02.001

Van der Maarel E (2005) Vegetation ecology – an overview. In: van der Maarel E (ed) Vegetation ecoloy. Blackwell, p 395

Vega KA, Schläpfer-Miller J, Kueffer C (2021) Discovering the wild side of urban plants through public engagement. Plants People Planet 3:389–401. https://doi.org/10.1002/ppp3.10191

Wang D, Xu P, An B, Guo Q (2024) Urban green infrastructure: bridging biodiversity conservation and sustainable urban development through adaptive management approach. Front Ecol Evol 12. https://doi.org/10.3389/fevo.2024.1440477

Ward E, Doroski D, Felson A, Hallett R, Oldfield E, Kuebbing S et al (2021) Positive long-term impacts of restoration on soils in an experimental urban forest. Ecol Appl 31(5). https://doi.org/10.1002/eap.2336

Weng Y (2015) Contrasting visions of science in ecological restoration: expert-lay dynamics between professional practitioners and volunteers. Geoforum 65:134–145. https://doi.org/10.1016/j.geoforum.2015.07.023

Williams NSG, Morgan JW, McDonnell MJ, McCarthy MA (2005) Plant traits and local extinctions in natural grasslands along an urban–rural gradient. J Ecol 93:1203–1213

Williams NS, Hahs AK, Vesk PA (2015) Urbanisation, plant traits and the composition of urban floras. Perspect Plant Ecol Evol Syst 17(1):78–86. https://doi.org/10.1016/j.ppees.2014.10.002

Wittig R (2004) The origin and development of the urban flora of Central Europe. Urban Ecosyst 7:323–339. https://doi.org/10.1007/s11252-005-6833-9

Xie Y, Wang Q, Xie M, Shibata S (2023) Construction feasibility evaluation for potential ecological corridors under different widths: a case study of Chengdu in China. Landsc Ecol Eng 19(3):381–399. https://doi.org/10.1007/s11355-023-00547-9

Xu W, Dai W, Ding Y, Song S, Liu Q, Yang W (2024) Drivers of spontaneous plant communities in urban parks: a case from Nanjing, China. Sustainability 16(9):3841

Zerbe S, Maurer U, Peschel T, Schmitz S (2004) Diversity of flora and vegetation in European cities as a potential for nature conservation in urban-industrial areas. Landsc Urban Plan 65(3):105–116

Zhou Y, Yao J, Chen M, Tang M (2023) Optimizing an urban green space ecological network by coupling structural and functional connectivity: a case for biodiversity conservation planning. Sustainability 15(22):15818. https://doi.org/10.3390/su152215818

Chapter 3
Alien Species in Cities: Issues and Threats

Carmelo Maria Musarella, Ricardo José Quinto Canas, Giulia Caneva, and Valerio Lazzeri

3.1 Urbanization and New Ecological Spaces for Alien Species (C.M.M.)

Urban expansion and the continuous transformation of landscapes are giving rise to new urban ecosystems characterized by increasing habitat fragmentation, altered microclimates, and persistent anthropogenic disturbances such as air pollution, traffic, and construction activities. Although these ecosystems are artificial environments, they are evolving in ways that generate ecological dynamics significantly different from those of natural ecosystems, creating conditions that facilitate the adaptation of both local and introduced species. The concept of "novel urban ecosystems" describes these emerging systems, which host biodiversity adapted to urban conditions while presenting new challenges for conservation. These systems require innovative approaches to manage the interaction between urbanization and nature (Ahern 2016).

Rapid urbanization is creating these complex ecosystems that demand integrated urban planning policies capable of reconciling development with ecological conservation (Kowarik 2011). Within this context, alien species find new ecological spaces.

C. M. Musarella (✉)
Department of AGRARIA, Mediterranea University of Reggio Calabria, Reggio Calabria, Italy
e-mail: carmelo.musarella@unirc.it

R. J. Q. Canas
Faculty of Sciences and Technology, University Algarve, CCMAR, Faro, Portugal

G. Caneva
Department of Sciences, University Roma Tre, Rome, Italy

V. Lazzeri
National Research Council, Institute of Research on Terrestrial Ecosystems (IRET), Pisa, Italy

G. Caneva et al. (eds.), *Urban Botany*, Future City 22,
https://doi.org/10.1007/978-3-032-28986-5_3

Alien species (AS) refers to non-native species (including subspecies or lower taxa), i.e., species out of their biogeographical natural distribution and growing into new areas by human actions, intentional or unintentional, that might survive and subsequently reproduce (CBD 2002; Essl et al. 2018; IPBES 2023; Richardson et al. 2010).

Invasive Alien Species (IAS) are alien species whose introduction or spread has been found to threaten or adversely impact upon biodiversity and related local ecosystem services (European Commission 2014).

In urban areas, generalist **Alien Plant Species (APS)**, capable of tolerating a wide range of ecological conditions and resistant to environmental stress, find particularly favorable conditions for settling and spreading. Their ability to thrive promotes **biotic homogenization**, reducing ecological diversity and increasing the similarity between urban and natural ecosystems. As McKinney (2006) notes, the increasing spread of alien species significantly contributes to these homogenization processes, representing one of the most pressing ecological challenges for modern urban biodiversity.

Rapid urbanization, in fact, profoundly alters local ecological dynamics, creating biological landscapes that become increasingly similar across cities and reducing native species diversity. This requires a re-evaluation of urban management policies to integrate biodiversity conservation into urban planning.

To develop effective conservation and management strategies, it is essential to have a deep understanding of the ecological dynamics that govern the distribution and spread of species within ecosystems, particularly in urbanized areas, where the interaction between natural factors and anthropogenic influences is especially complex and intense. Understanding how these factors interact and how they modify ecological processes is crucial for identifying the most appropriate intervention measures and ensuring the protection and sustainability of biodiversity. This requires adapting conservation policies to local specificities and the ecological challenges arising from urbanization (Aronson et al. 2014).

Cities provide highly favorable ecological conditions for the establishment and spread of alien species due to the combination of several factors. First, the presence of distinctive microclimates—often modified by buildings, roads, and other artificial structures—creates environments that may be less hostile for some species than natural habitats. In addition, the heterogeneous structure of urban surfaces, including gardens, rooftops, courtyards, and other green spaces, provides a wide range of ecological niches where alien species can readily adapt and proliferate. This is further enhanced by the presence of dispersal corridors, such as parks, canals, and interconnected green spaces, which facilitate species movement across the urban matrix and enable rapid, continuous expansion. These areas also represent key sites for the introduction and establishment of exotic ornamental species, which are frequently planted for decorative or functional purposes. Finally, competitive pressure in cities is generally lower than in natural ecosystems, as local biodiversity is often reduced; consequently, alien species, more tolerant of environmental disturbances, can establish and adapt more easily (Williams et al. 2015). Studies focusing on ornamental introductions and urban "propagule pressure" highlight the central role of urban green spaces in the spread of APS, accelerating their establishment and, in many cases, promoting their adaptation and stabilization in novel environments (Lozano et al. 2023; Pyšek and Hulme 2010; Reichard and White 2001).

Some species, such as the tree of heaven (*Ailanthus altissima* (Mill.) Swingle, Simaroubaceae) or the Fountain Grass (*Cenchrus setaceus* (Forssk.) Morrone, *Poaceae*), show remarkable ecological plasticity and effectively colonize road verges, railway networks, archaeological sites, abandoned industrial areas, and other disturbed habitats, often becoming dominant (Gaertner et al. 2017; Regione Calabria 2023).

Ailanthus altissima (Fig. 3.1) was introduced to Europe by Jesuits, who considered it a potential substitute for the silkworm, and later became one of the most invasive species, threatening native ecosystems (Hu 1979). Its spread has become so problematic that the European Union has implemented specific measures to support containment and eradication efforts. This species is listed among the alien species of Union concern under Regulation (EU) No. 1143/2014 on IAS (EU source). Today, it has been introduced to all continents except Antarctica (ISPRA source) and is absent only from the coldest regions, such as some Scandinavian countries.

The pathways for introducing APSs are diverse and closely linked to global flows of goods and people. The main vectors include: (a) the international trade of ornamental plants, which often transports non-native species beyond their natural ranges; (b) the movement of materials such as substrates, soils, and forest products, which may inadvertently carry seeds, rhizomes, or other propagules; and (c) more broadly, global trade and human-mediated transport systems, which increase opportunities for APS introduction and spread.

Even activities that may appear harmless, such as the movement of contaminated soils or the distribution of substrates for construction and urban agriculture, can facilitate the accidental spread of IASs. When not adequately monitored and regulated, these pathways represent one of the primary drivers of alien species

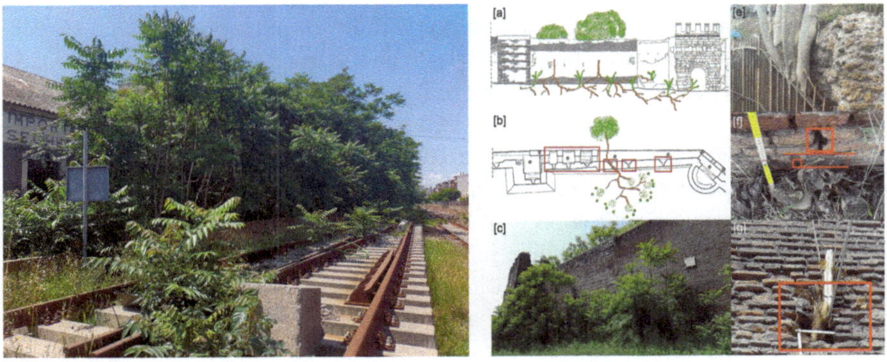

Fig. 3.1 Invasion of *Ailanthus altissima* (Mill.) Swingle (Simaroubaceae): (**a**) in the surroundings of abandoned industrial buildings and along disused railway tracks in Siderno, Reggio Calabria (Italy). (Photo by C.M.M. Musarella); (**b**) along the Aurelian Walls of Rome, where it creates severe cracks to the archaeological buildings. (Modified from Trotta et al. 2020)

expansion worldwide, highlighting the importance of preventive strategies and the careful management of trade flows to reduce the risk of introduction and spread (Hulme 2009).

Predicting the invasive behaviour of APSs remains highly complex, as their success depends on a network of interactions among multiple ecological factors. These include species functional traits—such as ecological plasticity, germination capacity, propagule production, and growth rate—which determine their ability to adapt to novel environments. At the same time, habitat characteristics, including frequent disturbances, the availability of open spaces, substrate heterogeneity, and competition with native species, influence the likelihood of establishment and proliferation. Finally, local climatic conditions, such as temperature, precipitation regimes, and seasonality, further modulate the performance and competitive ability of introduced species. This complex combination of interacting factors makes it difficult to reliably predict which species will become invasive and under which ecological conditions, highlighting the need for integrated theoretical approaches and predictive models that simultaneously consider both species properties and the characteristics of the recipient ecosystems (Catford et al. 2009).

In urban environments, IASs exert multiple and often significant impacts on local ecosystems. Among the most widely documented effects are direct competition with native species, which can lead to reductions in native biodiversity, alterations in vegetation structure and composition, and disruptions of biogeochemical cycles, with potential consequences for soil fertility and quality. These changes also affect key ecosystem services essential for urban well-being, such as shading, microclimate regulation, water retention, and soil stabilization, thereby reducing the capacity of green spaces to provide ecological and social benefits (Capuano and Caruso 2023). Such impacts, widely supported by global comparative studies and meta-analyses, highlight the need for targeted management and control strategies to preserve the functionality of urban ecosystems (Vilà et al. 2011).

In addition to ecological effects, IASs in urban environments can generate significant socio-economic impacts. These include increased costs for public green space management and maintenance, interference with urban infrastructure such as sidewalks, pipelines, and power lines, and risks to public health, for instance, in the case of allergenic, toxic, or vector-supporting species. These effects fall within the categories of socio-economic impacts standardized by the SEICAT framework, which provides a tool to assess and compare the consequences of alien species on human activities and social well-being (Bacher et al. 2018). Well-documented examples of ecological and socio-health impacts of IAPSs include the tree black locust (*Robinia pseudoacacia* L., Fabaceae), which in European semi-natural habitats can significantly alter plant community composition and diversity, often becoming dominant and reducing native species presence, and *Ambrosia artemisiifolia* L. (Asteraceae), whose spread in many urban areas is associated with serious public health problems due to the production of highly allergenic pollen, causing conditions such as rhinitis and asthma in local populations (Essl et al. 2015; Vítková et al. 2020).

Monitoring APSs requires up-to-date tools and integrated approaches that combine periodic checklists, field floristic surveys, citizen science contributions, and predictive modelling based on climatic, environmental, and land-use data. In particular, the use of Species Distribution Models (SDMs) allows the identification of areas at risk of invasion, supporting management decisions and targeted conservation strategies by integrating information on species ecological traits and local environmental factors (Gallien et al. 2012; Guisan et al. 2013). In some European and extra-European countries, these tools are already widely employed in urban planning and conservation policies (Orozco-Meza and Frazão-Cancela 2023; Shen et al. 2025; Thakur et al. 2023).

Regulation (EU) No. 1143/2014, adopted by the European Union (EU) in 2014, provides a legal framework to prevent the introduction and spread of IASs that have significant impacts on biodiversity, the environment, the economy, and human health. The regulation identifies species considered threats at the EU level and establishes mandatory measures for all Member States. These include monitoring to detect invasive species, management actions to limit their spread, and communication measures to ensure timely and transparent sharing of information among Member States and competent authorities. The overall objective is to protect European ecosystems and agricultural resources from external biological threats, promoting coordinated and harmonized management across the EU (European Commission 2014). The effectiveness of these policies, however, depends heavily on the availability of up-to-date and reliable data, as well as on structured, ongoing collaboration among public authorities, the scientific community, and environmental management professionals (Pergl et al. 2017). Furthermore, adopting an integrated approach that links invasive species management with urban and territorial planning is crucial, as urban areas often serve as key points of introduction and spread, and represent strategic locations for implementing biodiversity protection measures (Gaertner et al. 2016).

Case Study

***Cenchrus setaceus* in Calabria (Italy): Prevention, Eradication, and Assessment of Potential Habitat Expansion** (Fig. 3.2)

Cenchrus setaceus (synonym: *Pennisetum setaceum* (Forssk.) Chiov.), is an invasive herbaceous plant native to the Middle East, the Arabian Peninsula, and North Africa, belonging to the family Poaceae (Plants of the World Online 2025). In Italy, this species has been recorded mainly in southern regions (Laface et al. 2020; Lozano et al. 2024). Recently, Galasso et al. (2024) updated the checklist of alien vascular plants in the country, confirming the presence and increasing spread of *C. setaceus* across Italian territory. The species was first reported in Calabria by Castellano and Marino (2007). Over the last few decades, its expansion has been substantial, representing a serious threat to Mediterranean ecosystems. Its ability to compete with native species and alter local biodiversity has led to its classification as a significant ecological threat (Spampinato et al. 2019).

Musarella et al. (2020) provided updated information on the alien flora of Calabria, highlighting the urgent need for management measures to limit the spread of *C. setaceus*. Spampinato et al. (2022) subsequently confirmed the species' invasiveness in the region, and recent studies in Mediterranean xerothermic pastures

Fig. 3.2 Two connected scenes show efforts to control an invasive grass species. (**a**) Delimitation of the experimental plot at the Macellari Stream in Pellaro (Reggio Calabria, Italy). (**b**) Information board placed at the centre of each plot, containing project details and identifying data, to raise awareness among the local population about the ongoing experiment. (Photos by C.M.M. Musarella)

have shown that *C. setaceus* negatively impacts local biodiversity (Musarella et al. 2025). To mitigate further spread, integrating expert knowledge with species distribution models has been identified as a key approach for designing targeted control strategies and effective management interventions (Lozano et al. 2023, 2024). Predicting the potential distribution of the species also provides critical insights into its dynamics and invasion timeline.

The present case study aims to address these challenges through a multi-faceted approach. The primary objective is to reduce the risk of new invasions through continuous monitoring and targeted interventions. Concurrently, the project seeks to eradicate existing *C. setaceus* populations in invaded areas by testing control techniques in experimental plots. MaxEnt modelling was employed to identify areas at highest risk of invasion in Calabria. Additionally, the project develops guidelines for long-term management and disseminates findings to the scientific community, management authorities, and the public, thereby raising awareness and supporting informed decision-making.

The first phase of the project focused on a preliminary mapping of areas invaded by *C. setaceus*. This was carried out using scientific literature, aerial photographs, and GPS data to accurately locate where the species is present. At the same time, areas for experimental plots were identified in collaboration with local authorities and managers of invaded sites. During the prevention phase, the project aimed to raise awareness among local communities, farmers, and other stakeholders. This included working with professional associations and municipal offices. Meetings, seminars, and training courses have been organized to teach people how to recognize the Fountain Grass and prevent its spread. In addition, practical guidelines for early control of the species are produced and shared, using websites and social media to reach a wide audience. In selected areas, manual removal of the invasive plants was carried out during bioblitz events, engaging citizens directly in the work. The removed plants were then destroyed by burning, which is an effective way to prevent the species from spreading further.

The eradication phase involved the creation of experimental plots in which different control techniques were tested, including mechanical, chemical, and environmentally friendly approaches. The effectiveness of invasive vegetation removal will be continuously monitored by observing the regeneration of native plants. The experimental plots, which may vary in size, allowed the techniques to be applied in a controlled way and provide meaningful data. The approach also included the use of selective herbicides in critical areas. Each intervention was carefully monitored to assess its environmental and economic impacts.

At the same time, the project collected environmental, ecological, and species distribution data, using the MaxEnt© software to create predictive models of the potential distribution of *C. setaceus*. This approach helped identify the most vulnerable areas and plan targeted interventions in Calabria. The data collected were also be used to produce maps highlighting high-risk zones, providing valuable information for management and control strategies.

So far, the project has begun with an experiment in a plot located within the Macellari Stream in Pellaro, in the Metropolitan City of Reggio Calabria. First, phytosociological surveys were conducted in 50 m^2 subplots following the Braun-Blanquet methodology (1964) to identify the natural vegetation present and assess how it may have been influenced by the presence of the species. Then, within a delimited and marked area of approximately 100 m^2, 80 clumps of *C. setaceus* were removed (Fig. 3.2).

Next, all mature and immature inflorescences were manually removed, collected in a bag, and safely burned. Subsequently, all clumps found within the plot were eradicated both mechanically and manually (Fig. 3.3a). Ten days later, the final phase involved burning the removed clumps (Fig. 3.3b). Continuous monitoring of the experimental plot will follow to detect the possible emergence of new clumps. Additional plots will be established in different habitats and under varying environmental conditions to better understand how to effectively remove this invasive alien species in diverse settings.

Fig. 3.3 (**a**) Mechanical and manual removal of Fountain Grass (*Cenchrus setaceus* (Forssk.) Morrone, Poaceae) clumps within the experimental plot, (**b**) Burning clumps in a pit excavated within it. (Photos by C.M.M. Musarella)

3.2 Natural Spreading and Biological Invasions (R.J.Q.C.)

A scientific assessment of the state of knowledge regarding IASs and the risks of biological invasions, in which cities often play important roles as centers for the spread of alien species, will help counteract the biological invasion process at local, regional, and national scales. It will also contribute to understanding the role of cities as sites for the introduction of IASs and their subsequent spread into adjacent areas and natural ecosystems.

The impacts of invasive species introduced through human activities are widely recognized as a major global threat to nature and to nature's contributions to people. They represent one of the main drivers to biodiversity loss, with one million plant and animal species now at risk of extinction (European Commission 2014; IPBES 2023).

Introducing measures to reduce the impacts of IASs on biodiversity and ecosystem services will contribute to the effective achievement of Sustainable Development Goal 15 of the 2030 Agenda. These actions also make a significant—though not exclusive—contribution to Target 15.8, which calls for: *"introducing measures to prevent the introduction and significantly reduce the impact of invasive alien species on land and water ecosystems and control or eradicate the priority species,"* and they are relevant as well to sustainable cities (Goal 11).

Addressing this need, the Kunming–Montreal Global Biodiversity Framework, adopted by the Conference of the Parties to the Convention on Biological Diversity, includes in its Target 6 the goal of eliminating, minimizing, reducing, or mitigating the impacts of IASs on biodiversity and ecosystem services, and of reducing by at least 50% the rates of introduction and establishment of other known or potential IASs by 2030. For the purposes of this assessment, actions to achieve this ambitious target should be integrated into governance approaches through specific strategic measures that promote transformative change to meet the goals of IAS prevention, control, and biological invasion management.

Currently, many conservation policies and international and national legislative instruments already address IAS-related issues; however, up-to-date scientific assessments remain essential to provide an adequate knowledge base for addressing the challenges posed by IAPSs in urban contexts.

The movement of native species out of their natural range into new areas has resulted in the rise of biological invasions, also known as the invasion process (Mack 2003). According to IPBES (2023), the process of biological invasion involves a series of stages—transport, introduction, establishment, and spread or dispersal—and depends on the ability of a species to overcome successive barriers (geographic, captive or cultivated, survival, reproduction, dispersal, and environmental) that impede movement from one stage to the next. Human activities, such as agricultural practices and ornamental planting, can enable species to bypass the biotic and abiotic barriers that define their natural limits.

Some APSs' introductions have become invasive, as anthropogenic factors or drivers associated with urban or semi-urban areas may facilitate the introduction and dissemination of IASs through pathways such as trade, tourism, transport networks, linear infrastructure, horticulture, and green spaces. Urban areas are well documented as hotspots of APSs establishment and spread because of high colonization and propagule pressures (Hulme 2009; Kühn et al. 2017; Vantarová et al. 2023).

The annual rate of first records worldwide has increased since 1800, and is currently at its highest level, with 37% of documented alien species introductions over the last two centuries occurring since 1970. Introductions for ornamental purposes or via international transport have become increasingly important pathways and, unfortunately, continue to rise, with approximately 200 new alien species now recorded annually worldwide (IPBES 2023; Pyšek et al. 2020; Seebens et al. 2017). In Europe alone, approximately 20 new annual records of IASs are reported (Seebens et al. 2025).

Historically, many plant invasions began with the intentional introduction of ornamental species into cities and urban areas. These plants subsequently escaped from gardens for centuries and spread into natural or protected areas, continuing to be a major driver of alien plant invasions today (Čeplová et al. 2017; Pyšek et al. 2020; van Kleunen et al. 2018). According to Chytrý et al. (2012), projected trends in plant invasions in Europe under different future land-use scenarios indicate that urban areas have the second-highest level of expected invasion.

Both indirect and direct drivers play a crucial role in facilitating biological invasions. Direct drivers, notably fragmented ecosystems, land-use changes as well as climate change, become increasingly influential in the later stages of the invasion process, promoting the establishment and spread of IASs. With the global increasing of urbanization and land-use changes, we are witnessing the destruction or replacement of native natural vegetation, resulting in habitat degradation that promotes colonization and plant invasion (Didham et al. 2005; Li and Yu 2023; Liu et al. 2021; MacDougall and Turkington 2005). For example, in southern Portugal (Algarve region), IAPSs represent a major threat to biodiversity not only in heavily transformed landscape areas but also in natural habitats of high conservation value (Canas et al. 2024) (Fig. 3.4). The successful establishment and spread of invasive species, especially in artificially disrupted habitats, may be linked to their highly competitive capacity compared to native species, due to their strong reproductive ability and adaptability to these environments (Bradley et al. 2010; Valladares et al. 2015). In addition to direct competition with potential natural plant communities, plant invasion can also affect soil properties, nutrient availability, water use, and the productivity of agricultural landscapes (Pyšek and Richardson 2010; Vantarová et al. 2023). In this context, urban green areas are essential for increasing biodiversity in cities, however urban ecosystems are also hotspots for biological invasions (Gaertner et al. 2017).

Fig. 3.4 (a) Clump of *Cenchrus setaceus* (Forssk.) Morrone (Poaceae) spontaneously growing between the sidewalk and margins along the streets of the Urbanização Vale do Garrão, Vale do Lobo, Almancil (Algarve, Portugal). (Photo by C.M. Musarella). (b) *Cenchrus setaceus* in urban green spaces in the Algarve region (Portugal). (Photo by R.J. Quinto Canas)

3.3 In "Praise of Vagabonds" and Risks (G.C., V.L.)

The title of this contribution is inspired by a similarly titled essay by the highly renowned French landscape architect Gilles Clément, a contemporary writer and botanist who has designed numerous gardens and participated in countless art exhibitions (from the Centre Pompidou Museum in Paris to the Melle Contemporary Art Biennial, *A Vast Circle*, Arc et Senans, Fig. 3.5). In his books, which serve as manifestos of green design, he has chosen evocative titles (Clément 2001, 2015): *Planetary Garden*, *Garden in Motion*. In these works, he expressed important concepts—such as the notion of the *Third Landscape*—and his ideas have become key reference material for teaching contemporary garden design in architecture faculties.

Starting from the observation that the natural landscape is never fixed or immobile but is instead in constant flux, he urges gardeners to respect species that settle independently and praises these "vagabonds," as he has done in his previous works.

Recounting their history, origins, and his own encounters with them, he explains how humans, herbicides, concrete, tilling, and industrial farming have enabled these stray plants to settle and grow, praising their role as species that help nature reclaim abandoned and disused areas.

In the name of defending planetary diversity—which perhaps fascinates us because it serves as an "**invitation to tolerance**"—he expresses his philosophy of the landscape and of the relationship between humans and plants, inviting us to appreciate and value these citizens of the world.

However, while one can agree with several aspects of his perspective, it is also necessary to distance oneself from some of the concepts he emphasizes. As

Fig. 3.5 The Project Saline Royale (from the UNESCO salt factory), inspired by the Ideal City of Chaux imagined by Claude-Nicolas Ledoux, in the eighteenth century, and now by Gilles Clément. It covers 13 hectares by 30 gardens, which accompany and do not domesticate nature. (https://www.salineroyale.com/en/gardens/a-vast-circle/)

previously highlighted, urban areas represent habitats with several distinctive features that set them apart from natural environments, often making them more susceptible to invasions by alien species. The **threats posed by APSs to nature, human health, and societal systems cannot be overlooked**.

Nevertheless, APSs can also provide benefits. It is important to note that many advantages and disadvantages associated with APSs are also shared by native plant species. On the one hand, APSs offer benefits that are generally unrelated to their non-native status. On the other hand, the disadvantages they cause are often attributable to the species' inherent characteristics, rather than solely to their origin.

As shown in Chap. 4, cities have been adorned with plants since ancient times, reflecting the deep connection between humans and nature, as well as our inherent affinity for plants and other living beings, a concept known as **biophilia** (Kellert and Wilson 1993). Based on this biophilia, living in close contact with nature (both native and alien species) is well known to have a positive impact on human well-being (Grinde and Patil 2009). Nevertheless, it is important to consider that, **with respect to native biodiversity, the negative impacts of APSs often far outweigh their benefits** (Makhabu et al. 2025).

It is reasonable to assert that a highly vegetated area, where many APSs thrive, is preferable in many respects to bare or sparsely vegetated ground. The literature includes numerous studies on citizens' perceptions of urban meadows. For example, urban meadows dominated by perennial plants—plants that retain their green color for extended periods and are managed to be structurally diverse—tend to show relatively high perceived species richness and are preferred by citizens (Southon et al. 2017). Naturalized APSs can form part of such meadows and contribute to the overall benefits they provide.

Among the effects of urbanization is the depletion of natural vegetation, resulting in relatively large areas with little or no plant cover. Depending on the activities carried out on the land, significant negative impacts may occur on soil horizons and structure, as well as on microbial communities and nutrient cycles. Consequently, soil properties may decline, accompanied by phenomena such as soil compaction, severe moisture reduction, and low organic matter content (Sheoran et al. 2010).

Soils with little or no vegetation are also more susceptible to erosion, which has many negative impacts on soil stability, biodiversity, and productivity. This is particularly concerning inhabited areas, such as cities, where flooding and landslides can result in fatalities. Soil covered by dense vegetation is often less susceptible to hydrogeological instability, since plants intercept raindrops, transpire soil water, facilitate water infiltration, enhance organic matter deposition, and increase surface roughness. This is especially true for perennial plants, regardless of their origin, which produce a highly developed root network that persists even during the dry season (Durán Zuazo and Rodríguez Pleguezuelo 2008) or that colonize stream banks, making these areas more stable. Soil resistance to erosion can also depend on plant biodiversity, and therefore, it is important to verify that APSs contribute positively to this factor without reducing overall plant species richness (Berendse et al. 2015).

It is unavoidable that APSs become part of the wild flora of any given area, especially within urban boundaries. Thus, it is important to highlight all the real and potential benefits deriving from their presence in cities. Some APSs have been found to produce litter of better quality, characterized by higher nutrient content (C, N, and P) and lower levels of structural defence compounds. According to Patil et al. (2020), the litter decomposition velocity of some alien species was higher than that of native ones. On the one hand, this can stimulate microorganisms and detritivores to produce new soil and facilitate nutrient cycling; on the other hand, it can enhance soil fertility, potentially favouring further invasion by APSs. A well-known invasive tree, *Robinia pseudoacacia*, apart from its ability to colonize vast areas, outcompete native species, and increase soil nitrogen content, is also useful for slope stabilization and soil erosion control.

Some flower-bearing APSs can support pollinators through the production of nectar and pollen and, therefore, can enhance honey production in urbanized areas (Hausmann et al. 2016). The effects of APSs on plant-pollinator relationships are particularly complex. Nevertheless, some authors have pointed out that such impacts may also be positive: depending on the pollinator class, APSs can enhance the abundance, species richness, and diversity in the case of generalist pollinators such as hoverflies or increase abundance alone in more specialized groups such as wild bees (Kovács-Hostyánszki et al. 2022). Wild woody APSs can also support avian presence by providing suitable nesting sites; however, *Acacia saligna* (Labill.) H.L.Wendl. (Fabaceae) has been found to be detrimental to bird species richness in the South African urban area of Cape Town (Dures and Cumming 2010).

Some APSs and certain IAPSs are aesthetically pleasing and have ornamental potential. Examples include, among others, extremely invasive species such as butterfly bush (*Buddleja davidii* Franch), water primrose (*Ludwigia grandiflora*

(Michx.) Greuter & Burdet), and Himalayan balsam (*Impatiens glandulifera* Royle), which were evaluated positively from an aesthetic perspective by Lindemann-Matthies (2016). An emblematic species, African wood-sorrel (*Oxalis pes-caprae* L.) (Fig. 3.6), native to South Africa, combines a pleasing morphological appearance with the capacity to rapidly invade vast surfaces through vegetative growth, thereby outcompeting native species (Di Meo et al. 2025). Therefore, it is not surprising that highly vegetated areas—even when densely colonized by APSs—may be considered more aesthetically pleasing than bare or sparsely vegetated ground, and that some APSs possess ornamental value.

Many aquatic plants are known to purify water (Guittonny-Philippe et al. 2015). In some cases, APSs growing in riverbeds, although posing threats to native biodiversity, can nevertheless improve water quality by forming vegetated structures.

In general, as further shown in Chap. 5, green areas help improve air quality, mitigate certain weather extremes by acting as windbreaks, reduce urban noise generated by human activities, and provide better living conditions for people. Finally, many wild plants, including some non-native species, are considered edible.

However, we cannot ignore that severe threats to native biodiversity are posed by IAPSs through various mechanisms, due to their high capacity to invade large areas.

Although urban environments are frequently species-poor or have undergone homogenization compared to more natural habitats, cities can be more biodiverse than expected and sometimes harbour endemic or rare plant species (Planchuelo

Fig. 3.6 *Oxalis pes-caprae* L. (Photo by F. Bretzel & V. Lazzeri)

et al. 2019). In some cases, cities serve as unique refuges for plant species with highly disjunct distributions separated by vast distances, such as *Athamanta macedonica* (L.) Spreng., which grows in the Balkans but also in the Italian city of Campobasso. Additionally, cities may host exclusive narrow endemics, exemplified by *Hieracium australe* Fr. subsp. *australe* in Milan. This phenomenon may be facilitated by the presence of natural areas within urban boundaries, whose importance for biodiversity has been thoroughly emphasized (Aronson et al. 2014). Urban residents often perceive IAPSs negatively, and their management is considered important, although some interventions may be viewed as not strictly necessary (Potgieter et al. 2019).

It is sometimes difficult to establish causality between invasion and declines in native biodiversity, and no extinctions in the wild or total extinctions of native plants due to APSs have been recorded to date (Downey and Richardson 2016). A meta-analysis showed that, in general, IAPSs cause a decline in native species richness and abundance when comparing invaded plots with either non-invaded plots or those where alien species removal was performed (Andreu and Vilà 2011). Given the characteristics of cities, this effect can be especially pronounced in urban environments, where space for wild plants is limited and the presence of APSs is particularly abundant.

Regarding plant–pollinator networks, flowering **APSs**, in addition to outcompeting native plant species, **can also have detrimental effects on pollinators by favouring certain groups**. This is highlighted by Kovács-Hostyánszki et al. (2022), who found that late-flowering IAPSs supported generalist, late-flying pollinators. Similarly, certain APSs, due to specific morphological features of their flowers, can be detrimental to some pollinator groups that cannot feed because of the short length of their tongues.

APSs can also threaten native biodiversity through hybridization. A textbook example is the newly described species *Sporobolus anglicus* (C.E. Hubb.). P.M. Peterson & Saarela (formerly *Spartina anglica* C.E. Hubb.), which appeared in England due to hybridization between the native *S. maritimus* (Curtis) P.M. Peterson & Saarela [formerly *Spartina maritima* (Curtis) Fernald] and the introduced *S. alterniflorus* (Loisel.) P.M. Peterson & Saarela (formerly *Spartina alterniflora* Loisel.) (Gray et al. 1991). Cities, where the abundance of alien species is often relatively high, could facilitate hybridization and introgression events, since gene flow is generally thought to occur from the more abundant to the rarer taxon (Largiadèr 2007).

Another threat posed by **APSs is the facilitation of disease transmission to humans and native species**. In this regard, a possible association has been identified between IAPSs and malaria, with the association appearing stronger for aquatic IAPSs. This issue is particularly concerning in densely populated urban areas. Evidence also suggests that managing aquatic IAPSs, such as water hyacinth (*Pontederia crassipes* Mart.) could contribute positively to malaria control, as reviewed by Stone et al. (2018). Other concerning diseases, such as trypanosomiasis and Lyme disease, are linked to the presence of IAPSs like *Lantana* spp. and *Berberis thunbergii* DC., respectively. These species can be found in anthropized environments as well as in natural and semi-natural areas near cities or within urban boundaries (Mack and Smith 2011; Musarella et al. 2024). Some IAPSs pose threats to human and animal health through their physical characteristics, such as spines, thorns, or other stinging structures (Musarella 2020).

Among species that can become naturalized, especially in anthropized sites, members of the *Cactaceae* family—particularly the subfamily *Opuntioideae*—are well-known for causing injuries that can sometimes be quite severe (Miller et al. 2000). Also, many naturalized and invasive grasses in the Poaceae family, such as *Bromus s.l.* and the spiny *Cenchrus*, can potentially be physically harmful to animals, mainly due to the awns of their spikelets or the spiny, fused outer bristles surrounding their spikelets, respectively. Certain biochemical properties of some IAPSs also pose threats to human and animal health, especially in densely populated areas such as cities. For example, accidental contact between human skin and the sap of these plants, followed by exposure to sunlight, can lead to photosensitization. Species containing furanocoumarin derivatives, such as *Heracleum* spp. (Pfurtscheller and Trop 2014), along with other genera producing different compounds, can cause severe phytophotodermatitis (Grosu et al. 2024).

Plants, as sessile organisms, have developed a wide range of defence compounds primarily to protect themselves from herbivorous animals. Some of these compounds are toxic to animals and humans when ingested. Consumption of such molecules can occur either accidentally or intentionally. Accidental ingestion of poisonous alien plants may happen when individuals are unaware that a particular species contains toxic substances or mistake it for an edible plant. For example, borage (*Borago officinalis* L.), native to the central and western Mediterranean but now nearly cosmopolitan, is common in urban and suburban areas and is often regarded as an edible species in phytoalimurgy (see Chap. 4). However, like most taxa in the family Boraginaceae, it contains pyrrolizidine alkaloids, a group of compounds capable of causing severe liver damage (Wiedenfeld and Edgar 2011). Similarly, comfrey (*Symphytum officinale* L. s.l.,) and coltsfoot (*Tussilago farfara* L.), both found in or near cities, have long been considered therapeutic plants but are now recognized as highly poisonous (Avila et al. 2020). Other pyrrolizidine alkaloid-producing plants belong to the tribe Senecioneae Cass. (Asteraceae), which includes ornamental IAPs such as *Senecio angulatus* L.f., *S. mikanioides* Otto ex Walp., and *Delairea odorata* Lem. These species can spread into natural and semi-natural habitats near urban settlements, and the alkaloids they produce can contaminate honey.

Poisoning can also occur when alien plants are ingested by mistake, as in the case of autumn crocus (*Colchicum autumnale* L., Colchicaceae), which inexperienced gatherers may confuse with saffron crocus (*Crocus sativus* L., Iridaceae) or garlic (*Allium sativum* L., Amaryllidaceae). Other poisonous plants are intentionally ingested for their known psychotropic properties. A well-known example of an alien species frequently found in and around cities is *Datura stramonium* L. (Solanaceae) and related taxa (Kerchner and Farkas 2020). Although these species may have therapeutic applications, their consumption is also responsible for severe poisoning cases that can be fatal.

As a further example, the castor bean plant (*Ricinus communis* L., Euphorbiaceae), cultivated for ornamental purposes and oil production, tends to become naturalized, usually in anthropized areas. This species accumulates the toxic lectin ricin, an agglutinin and the alkaloid ricinine, and is responsible for both accidental and deliberate poisonings, the latter usually related to suicide attempts (Worbs et al. 2011).

In cities, APSs can pose an additional threat to human health due to the allergenic potential of some species, such as the well-known case of pollen allergy caused by the North American common ragweed (*Ambrosia artemisiifolia*), an annual plant that thrives in human-disturbed habitats, including those commonly found in urban environments (Fig. 3.7). Moreover, milder temperatures and locally increased CO_2 concentrations typical of cities can enhance the growth of pollen-producing plants and increase pollen production, while wetter conditions can extend the pollen release period. In addition, a synergy between several atmospheric pollutants and pollen can occur, especially in cities, worsening the clinical manifestations of respiratory symptoms (Schiavoni et al. 2017). Other APSs can produce biogenic volatile organic compounds (BVOCs) that, in turn, can lead to increased concentrations of ozone and secondary organic aerosols both of which are harmful to humans and animals.

IASs in cities can also act as reservoirs of plant pathogens (Najberek et al. 2022). APSs can also pose a threat to infrastructure and to the conservation of archaeological remains. Woody species, through their robust root systems, can contribute to the destruction of ancient architectural structures (Celesti-Grapow and Blasi 2004). Among these species, *Ailanthus altissima* poses the greatest threat to

Fig. 3.7 Allergenic species: **A–D** *Ambrosia artemisiifolia* L. (Asteraceae) and **E–H**: *Parietaria judaica* L. (Photo by F. Bretzel & V. Lazzeri)

buildings. Nevertheless, it should also be acknowledged that species native to their ranges can cause similar damage. This is the case, for example, of common ivy (*Hedera helix* L. subsp. *helix*) (Bartoli et al. 2017).

If, on the one hand, as already mentioned, APSs can stabilise riverbanks, on the other hand, alien plants growing in riverbeds can also pose a threat by reducing the space available for water flow and increasing the risk of flooding under certain conditions (Musarella et al. 2020; Siclari et al. 2026). Certain riparian IAPs, such as the visually attractive *Impatiens glandulifera*, can also contribute to soil erosion in rivers (Greenwood and Kuhn 2015).

The presence of alien plants in cities and peri-urban areas is unavoidable, and they should not be demonized, but may sometimes be accepted. However, defending native biodiversity remains a primary goal for better protecting nature and avoiding the negative effects of anthropization, and it has become a cornerstone of the European Strategy for Nature Conservation through PNRR actions.

References

Ahern J (2016) Novel urban ecosystems: concepts, definitions and a strategy to support urban sustainability and resilience. Lands Archit Front 4(1)

Andreu J, Vilà M (2011) Native plant community response to alien plant invasion and removal. Manag Biol Invasions 2(1):81–94. https://doi.org/10.3391/mbi.2011.2.1.08

Aronson MFJ, La Sorte FA, Nilon CH, Katti M, Goddard MA et al (2014) A global analysis of the impacts of urbanization on bird and plant diversity reveals key anthropogenic drivers. Proc Royal Soc B Biol Sci 281(1780):20133330. https://doi.org/10.1098/rspb.2013.3330

Avila C, Breakspear I, Hawrelak J, Salmond S, Evans S (2020) A systematic review and quality assessment of case reports of adverse events for borage (Borago officinalis), coltsfoot (Tussilago farfara) and comfrey (Symphytum officinale). Fitoterapia 142:104519. https://doi.org/10.1016/j.fitote.2020.104519

Bacher S, Blackburn TM, Essl F, Genovesi P, Heikkilä J et al (2018) Socio-economic impact classification of alien taxa (SEICAT). Methods Ecol Evol 9(1):159–168. https://doi.org/10.1111/2041-210X.12844

Bartoli F, Romiti F, Caneva G (2017) Aggressiveness of *Hedera helix* L. growing on monuments: evaluation in Roman archaeological sites and guidelines for a general methodological approach. Plant Biosyst 151(5):866–877. https://doi.org/10.1080/11263504.2016.1218969

Berendse F, Van Ruijven J, Jongejans E, Keesstra S (2015) Loss of plant species diversity reduces soil erosion resistance. Ecosystems 18(5):881–888. https://doi.org/10.1007/s10021-015-9869-6

Bradley BA, Blumenthal DM, Wilcove DS, Ziska LH (2010) Predicting plant invasions in an era of global change. Trends Ecol Evol 25(5):310–318. https://doi.org/10.1016/j.tree.2009.12.003

Braun-Blanquet J (1964) Pflanzensoziologie. Springer, Vienna. https://doi.org/10.1007/978-3-7091-8110-2

Canas RQ, Neiva J, Xarepe D, Cano-Ortiz A, Raposo M, Gomes CP (2024) Alien and invasive Flora in Algarve (Southern Portugal). In: Calabrò F, Madureira L, Morabito FC, Piñeira Mantiñán MJ (eds) Networks, markets & people, vol 1183. Springer, pp 395–411. https://doi.org/10.1007/978-3-031-74501-0_41

Capuano A, Caruso G (2023) New records for the alien vascular flora of Calabria (S-Italy). Res J Ecol Environ Sci *3*(2):1–35. https://doi.org/10.31586/rjees.2023.723

Castellano G, Marino P (2007) Segnalazione di Pennisetum setaceum (Poaceae) in Calabria. Collana Sicil For 34:295–295. https://iris.unipa.it/handle/10447/40216

Catford JA, Jansson R, Nilsson C (2009) Reducing redundancy in invasion ecology by integrating hypotheses into a single theoretical framework. Divers Distrib 15(1):22–40. https://doi.org/10.1111/j.1472-4642.2008.00521.x
CBD (2002) Alien species that threaten ecosystems, habitats or species (UNEP/CBD/COP/6/20). Convention on Biological Diversity. https://www.cbd.int/doc/decisions/cop-06-dec-23-en.pdf
Celesti-Grapow L, Blasi C (2004) The role of alien and native weeds in the deterioration of archaeological remains in Italy. Weed Technol 18:1508–1513
Čeplová N, Lososová Z, Kalusová V (2017) Urban ornamental trees: a source of current invaders; a case study from a European City. Urban Ecosyst 20(5):1135–1140. https://doi.org/10.1007/s11252-017-0665-2
Chytrý M, Wild J, Pyšek P, Jarošík V, Dendoncker N et al (2012) Projecting trends in plant invasions in Europe under different scenarios of future land-use change. Glob Ecol Biogeogr 21(1):75–87. https://doi.org/10.1111/j.1466-8238.2010.00573.x
Clément G (2001) Le jardin en mouvement: De la vallée au jardin planétaire (4e éd. augm). Sens & Tonka
Clément G (2015) The planetary garden: and other writings. Univ. Pennsylvania
Di Meo MC, Napoletano P, De Marco A, Bianchi AR et al (2025) The diffuse and invasive weed Oxalis pes-caprae L.: from botany to chemical, functional and ecological interactions with soil and organisms. Heliyon 11(13):e43603. https://doi.org/10.1016/j.heliyon.2025.e43603
Didham RK, Tylianakis JM, Hutchison MA, Ewers RM, Gemmell NJ (2005) Are invasive species the drivers of ecological change? Trends Ecol Evol 20(9):470–474. https://doi.org/10.1016/j.tree.2005.07.006
Downey PO, Richardson DM (2016) Alien plant invasions and native plant extinctions: a six-threshold framework. AoB PLANTS 8:plw047. https://doi.org/10.1093/aobpla/plw047
Durán Zuazo VH, Rodríguez Pleguezuelo CR (2008) Soil-erosion and runoff prevention by plant covers. A review. Agron Sustain Dev 28(1):65–86. https://doi.org/10.1051/agro:2007062
Dures SG, Cumming GS (2010) The confounding influence of homogenising invasive species in a globally endangered and largely urban biome: does habitat quality dominate avian biodiversity? Biol Conserv 143(3):768–777. https://doi.org/10.1016/j.biocon.2009.12.019
Essl F, Bacher S, Genovesi P, Hulme PE, Jeschke JM et al (2018) Which taxa are alien? Criteria, applications, and uncertainties. Bioscience 68(7):496–509. https://doi.org/10.1093/biosci/biy057
Essl F, Biró K, Brandes D, Broennimann O, Bullock JM et al (2015) Biological Flora of the British Isles: *Ambrosia artemisiifolia*. J Ecol 103(4):1069–1098. https://doi.org/10.1111/1365-2745.12424
European Commission (2014) Regulation (EU) No 1143/2014 of the European Parliament and of the Council of 22 October 2014 on the prevention and management of the introduction and spread of invasive alien species. http://data.europa.eu/eli/reg/2014/1143/oj
Gaertner M, Larson BMH, Irlich UM, Holmes PM, Stafford L et al (2016) Managing invasive species in cities: a framework from Cape Town, South Africa. Landsc Urban Plan 151:1–9. https://doi.org/10.1016/j.landurbplan.2016.03.010
Gaertner M, Wilson JRU, Cadotte MW, MacIvor JS, Zenni RD, Richardson DM (2017) Non-native species in urban environments: patterns, processes, impacts and challenges. Biol Invasions 19(12):3461–3469. https://doi.org/10.1007/s10530-017-1598-7
Galasso G, Conti F, Peruzzi L, Alessandrini A, Ardenghi NMG et al (2024) A second update to the checklist of the vascular flora alien to Italy. Plant Biosyst 158(2):297–340. https://doi.org/10.1080/11263504.2024.2320129
Gallien L, Douzet R, Pratte S, Zimmermann NE, Thuiller W (2012) Invasive species distribution models—how violating the equilibrium assumption can create new insights. Glob Ecol Biogeogr 21(11):1126–1136. https://doi.org/10.1111/j.1466-8238.2012.00768.x
Gray AJ, Marshall DF, Raybould AF (1991) A century of evolution in Spartina anglica. Adv Ecol Res 21:1–62. https://doi.org/10.1016/S0065-2504(08)60096-3

Greenwood P, Kuhn N (2015) The annual invasive plant Impatiens glandulifera as a trigger for high-magnitude soil erosion in temperate river systems. Geophys Res Abstr 17
Grinde B, Patil GG (2009) Biophilia: does visual contact with nature impact on health and well-being? Int J Environ Res Public Health 6(9):2332–2343. https://doi.org/10.3390/ijerph6092332
Grosu C, Jîjie A-R, Manea H, Moacă E-A, Iftode A et al (2024) New insights concerning phytophotodermatitis induced by phototoxic plants. Life 14(8):1019. https://doi.org/10.3390/life14081019
Guisan A, Tingley R, Baumgartner JB, Naujokaitis-Lewis I, Sutcliffe PR et al (2013) Predicting species distributions for conservation decisions. Ecol Lett 16(12):1424–1435. https://doi.org/10.1111/ele.12189
Guittonny-Philippe A, Petit M-E, Masotti V, Monnier Y, Malleret L et al (2015) Selection of wild macrophytes for use in constructed wetlands for phytoremediation of contaminant mixtures. J Environ Manag 147:108–123. https://doi.org/10.1016/j.jenvman.2014.09.009
Hausmann SL, Petermann JS, Rolff J (2016) Wild bees as pollinators of city trees. Insect Conserv Divers 9(2):97–107. https://doi.org/10.1111/icad.12145
Hu SY (1979) Ailanthus. Arnoldia 39(2):29–50
Hulme PE (2009) Trade, transport and trouble: managing invasive species pathways in an era of globalization. J Appl Ecol 46(1):10–18. https://doi.org/10.1111/j.1365-2664.2008.01600.x
IPBES (2023) Thematic assessment report on invasive alien species and their control. IPBES Secretariat. https://doi.org/10.5281/ZENODO.7430692
Kellert SR, Wilson EO (1993) The Biophilia hypothesis. Island Press/Shearwater Books
Kerchner A, Farkas Á (2020) Worldwide poisoning potential of Brugmansia and Datura. Forensic Toxicol 38(1):30–41. https://doi.org/10.1007/s11419-019-00500-2
Kovács-Hostyánszki A, Szigeti V, Miholcsa Z, Sándor D, Soltész Z, Török E, Fenesi A (2022) Threats and benefits of invasive alien plant species on pollinators. Basic Appl Ecol 64:89–102. https://doi.org/10.1016/j.baae.2022.07.003
Kowarik I (2011) Novel urban ecosystems, biodiversity, and conservation. Environ Pollut 159(8–9):1974–1983. https://doi.org/10.1016/j.envpol.2011.02.022
Kühn I, Wolf J, Schneider A (2017) Is there an urban effect in alien plant invasions? Biol Invasions 19(12):3505–3513. https://doi.org/10.1007/s10530-017-1591-1
Laface VLA, Musarella CM, Cano Ortiz A, Quinto Canas R, Cannavò S et al (2020) Three new alien taxa for Europe and a chorological update on the alien vascular Flora of Calabria (Southern Italy). Plants 9(9):1181. https://doi.org/10.3390/plants9091181
Largiadèr CR (2007) Hybridization and introgression between native and alien species. In: Nentwig W (ed) Biological invasions, vol 193. Springer, pp 275–292. https://doi.org/10.1007/978-3-540-36920-2_16
Li Y, Yu F-H (2023) Managing the risk of biological invasions. iScience 26(11):108221. https://doi.org/10.1016/j.isci.2023.108221
Lindemann-Matthies P (2016) Beasts or beauties? Laypersons' perception of invasive alien plant species in Switzerland and attitudes towards their management. NeoBiota 29:15–33. https://doi.org/10.3897/neobiota.29.5786
Liu C, Diagne C, Angulo E, Banerjee A-K, Chen Y, Cuthbert RN et al (2021) Economic costs of biological invasions in Asia. NeoBiota 67:53–78. https://doi.org/10.3897/neobiota.67.58147
Lozano V, Di Febbraro M, Brundu G, Carranza ML, Alessandrini A et al (2023) Plant invasion risk inside and outside protected areas: propagule pressure, abiotic and biotic factors definitively matter. Sci Total Environ 877:162993. https://doi.org/10.1016/j.scitotenv.2023.162993
Lozano V, Marzialetti F, Acosta ATR, Arduini I, Bacchetta et al (2024) Prioritizing management actions for invasive non-native plants through expert-based knowledge and species distribution models. Ecol Indic 166:112279. https://doi.org/10.1016/j.ecolind.2024.112279
MacDougall AS, Turkington R (2005) Are invasive species the drivers or passengers of change in degraded ecosystems? Ecology 86(1):42–55. https://doi.org/10.1890/04-0669

Mack RN (2003) Phylogenetic constraint, absent life forms, and preadapted alien plants: a prescription for biological invasions. Int J Plant Sci 164(S3):S185–S196. https://doi.org/10.1086/368399

Mack R, Smith M (2011) Invasive plants as catalysts for the spread of human parasites. NeoBiota 9:13–29. https://doi.org/10.3897/neobiota.9.1156

Makhabu SW, Tshireletso K, Setlalekgomo MR, Seletlo Z, Legwatagwata et al (2025) Impact of invasive alien species on biodiversity and ecosystem services. In: Hambira WL, Abdeta AA, Moalafhi DB, Muposhi VK, Mosepele K (eds) Environmental change and biodiversity conservation in sub-Saharan Africa, vol 75. Springer, pp 299–319. https://doi.org/10.1007/978-3-031-73136-5_16

McKinney ML (2006) Urbanization as a major cause of biotic homogenization. Biol Conserv 127(3):247–260. https://doi.org/10.1016/j.biocon.2005.09.005

Miller EB, Gilad A, Schattner A (2000) Cactus thorn arthritis: case report and review of the literature. Clin Rheumatol 19(6):490–491. https://doi.org/10.1007/s100670070014

Musarella CM (2020) *Solanum torvum* Sw. (Solanaceae): a new alien species for Europe. Genet Resour Crop Evol 67(2):515–522. https://doi.org/10.1007/s10722-019-00822-5

Musarella CM, Laface VLA, Angiolini C, Bacchetta G, Bajona E, Banfi et al (2024) New alien plant taxa for Italy and Europe: an update. Plants 13(5):620. https://doi.org/10.3390/plants13050620

Musarella CM, Sciandrello S, Domina G (2025) Competition between alien and native species in xerothermic steno-Mediterranean grasslands: Cenchrus setaceus and Hyparrhenia hirta in Sicily and southern Italy. Vegetos 38(3):1055–1062. https://doi.org/10.1007/s42535-024-00871-x

Musarella CM, Stinca A, Cano-Ortíz A, Laface VLA, Petrilli R et al (2020) New data on the alien vascular flora of Calabria (Southern Italy). Annali Di Botanica 10:2020. https://doi.org/10.13133/2239-3129/14838

Najberek K, Olszańska A, Tokarska-Guzik B, Mazurska K, Dajdok Z, Solarz W (2022) Invasive alien species as reservoirs for pathogens. Ecol Indic 139:108879. https://doi.org/10.1016/j.ecolind.2022.108879

Orozco-Meza M, Frazão-Cancela J (2023) SDMS, herramienta para la conservación de la biodiversidad en ciudades: Caso de estudio: Ponte verde de Queluz (Sintra, Portugal). Seminario Internacional de Investigación en Urbanismo. https://doi.org/10.5821/siiu.12772

Patil M, Kumar A, Kumar P, Cheema NK, Kaur R et al (2020) Comparative litter decomposability traits of selected native and exotic woody species from an urban environment of north-western Siwalik region, India. Sci Rep 10(1):7888. https://doi.org/10.1038/s41598-020-64576-2

Pergl J, Pyšek P, Bacher S, Essl F, Genovesi P et al (2017) Troubling travellers: are ecologically harmful alien species associated with particular introduction pathways? NeoBiota 32:1–20. https://doi.org/10.3897/neobiota.32.10199

Pfurtscheller K, Trop M (2014) Phototoxic plant burns: report of a case and review of topical wound treatment in children. Pediatr Dermatol 31(6). https://doi.org/10.1111/pde.12396

Planchuelo G, Von Der Lippe M, Kowarik I (2019) Untangling the role of urban ecosystems as habitats for endangered plant species. Landsc Urban Plan 189:320–334. https://doi.org/10.1016/j.landurbplan.2019.05.007

Plants of the World Online (2025) Cenchrus setaceus https://powo.science.kew.org

Potgieter LJ, Gaertner M, O'Farrell PJ, Richardson DM (2019) Perceptions of impact: invasive alien plants in the urban environment. J Environ Manag 229:76–87. https://doi.org/10.1016/j.jenvman.2018.05.080

Pyšek P, Hulme PE (2010) Biological invasions in Europe 50 years after Elton: time to sound the ALARM. In: Richardson DM (ed) Fifty years of invasion ecology. Wiley, pp 71–88. https://doi.org/10.1002/9781444329988.ch7

Pyšek P, Hulme PE, Simberloff D, Bacher S, Blackburn TM, Carlton JT et al (2020) Scientists' warning on invasive alien species. Biol Rev 95(6):1511–1534. https://doi.org/10.1111/brv.12627

Pyšek P, Richardson DM (2010) Invasive species, environmental change and management, and health. Annu Rev Environ Resour 35(1):25–55. https://doi.org/10.1146/annurev-environ-033009-095548

Regione Calabria (2023) Carta Natura—Geografia degli habitat. Arti Grafiche Cardamone

Reichard SH, White P (2001) Horticulture as a pathway of invasive plant introductions in the United States. Bioscience 51(2):103. https://doi.org/10.1641/0006-3568(2001)051[0103:HAAPOI]2.0.CO;2

Richardson PJ, Lundholm JT, Larson DW (2010) Natural analogues of degraded ecosystems enhance conservation and reconstruction in extreme environments. Ecol Appl 20(3):728–740. https://doi.org/10.1890/08-1092.1

Schiavoni G, D'Amato G, Afferni C (2017) The dangerous liaison between pollens and pollution in respiratory allergy. Ann Allergy Asthma Immunol 118(3):269–275. https://doi.org/10.1016/j.anai.2016.12.019

Seebens H, Blackburn TM, Dyer EE, Genovesi P, Hulme et al (2017) No saturation in the accumulation of alien species worldwide. Nat Commun 8(1):14435. https://doi.org/10.1038/ncomms14435

Seebens H, Meyerson LA, Richardson DM, Lenzner B et al (2025) Biological invasions: a global assessment of geographic distributions, long-term trends, and data gaps. Biol Rev 100(6):2542–2583. https://doi.org/10.1111/brv.70058

Shen C, Chen X, Zhou C, Xu L, Qian M, Zhao H, Li K (2025) Predicting suitable spatial distribution areas for urban trees under climate change scenarios using species distribution models: a case study of Michelia chapensis. Land 14(3):638. https://doi.org/10.3390/land14030638

Sheoran V, Sheroan A, Poonia P (2010) Soil reclamation of abandoned mine land by revegetation. Int J f Soil Sediment Water 3:1–20

Siclari A, Raimondo M, Musarella CM, Foti G, Catalfamo D, Postorino G, Barbaro G, Minniti F, Novembre F, Bombino G (2026) Control of invasive species in urban channelized watercourses of the metropolitan area of Reggio Calabria (Italy) for their management and hydraulic protection. Proc. of the Mediterranean Life Sciences Union Annual Meet. Na, Italy, 9–11 Dec. 2025

Southon GE, Jorgensen A, Dunnett N, Hoyle H, Evans KL (2017) Biodiverse perennial meadows have aesthetic value and increase residents' perceptions of site quality in urban green-space. Landsc Urban Plan 158:105–118. https://doi.org/10.1016/j.landurbplan.2016.08.003

Spampinato G, Cannavò S, Cano-Ortiz A, Caruso G, Laface V et al (2019) Invasività di Cenchrus setaceus (Forssk.) Morrone in Italia. Notiziario della Società Botanica Italiana 3:1–10

Spampinato G, Laface VLA, Posillipo G, Cano Ortiz A, Quinto Canas R et al (2022) Alien flora in Calabria (Southern Italy): an updated checklist. Biol Invasions 24(8):2323–2334. https://doi.org/10.1007/s10530-022-02800-y

Stone CM, Witt ABR, Walsh GC, Foster WA, Murphy ST (2018) Would the control of invasive alien plants reduce malaria transmission? A review. Parasit Vectors 11(1):76. https://doi.org/10.1186/s13071-018-2644-8

Thakur S, Rai ID, Singh B, Dutt HC, Musarella CM (2023) Predicting the suitable habitats of *Elwendia persica* in the Indian Himalayan Region (IHR). Plant Biosyst 157(4):769–778. https://doi.org/10.1080/11263504.2023.2204090

Trotta G, Savo V, Cicinelli E, Carboni M, Caneva G (2020) Colonization and damages of Ailanthus altissima (Mill.) Swingle on archaeological structures: evidence from the Aurelian Walls in Rome (Italy). Int Biodeterior Biodegradation 153:105054. https://doi.org/10.1016/j.ibiod.2020.105054

Valladares F, Bastias CC, Godoy O, Granda E, Escudero A (2015) Species coexistence in a changing world. Front Plant Sci 6. https://doi.org/10.3389/fpls.2015.00866

van Kleunen M, Essl F, Pergl J, Brundu G, Carboni et al (2018) The changing role of ornamental horticulture in alien plant invasions. Biol Rev 93(3):1421–1437. https://doi.org/10.1111/brv.12402

Vantarová KH, Eliáš P, Jiménez-Ruiz J, Tokarska-Guzik B, Cires E (2023) Biological invasions in the twenty-first century: a global risk. Biologia 78(5):1211–1218. https://doi.org/10.1007/s11756-023-01394-7

Vilà M, Espinar JL, Hejda M, Hulme PE, Jarošík V et al (2011) Ecological impacts of invasive alien plants: a meta-analysis of their effects on species, communities and ecosystems: ecological impacts of invasive alien plants. Ecol Lett 14(7):702–708. https://doi.org/10.1111/j.1461-0248.2011.01628.x

Vítková M, Sádlo J, Roleček J, Petřík P, Sitzia T et al (2020) Robinia pseudoacacia-dominated vegetation types of Southern Europe: species composition, history, distribution and management. Sci Total Environ 707:134857. https://doi.org/10.1016/j.scitotenv.2019.134857

Wiedenfeld H, Edgar J (2011) Toxicity of pyrrolizidine alkaloids to humans and ruminants. Phytochem Rev 10(1):137–151. https://doi.org/10.1007/s11101-010-9174-0

Williams NSG, Hahs AK, Vesk PA (2015) Urbanisation, plant traits and the composition of urban floras. Perspect Plant Ecol Evol Syst 17(1):78–86. https://doi.org/10.1016/j.ppees.2014.10.002

Worbs S, Köhler K, Pauly D, Avondet M-A, Schaer M, Dorner MB, Dorner BG (2011) Ricinus communis intoxications in human and veterinary medicine—a summary of real cases. Toxins 3(10):1332–1372. https://doi.org/10.3390/toxins3101332

Part II
Plants and People in Urban Botany

Chapter 4
Traditional Ecological Knowledge Versus Plant Blindness: Plants and Humans in Urban Areas

Giulia Caneva, Ana Cano-Ortiz, Zohreh Hosseini, Carmelo Maria Musarella, Fabio Orlandi, Miriam Patti, Luigia Ruga, Nicolò Soldovieri, and Giovanni Spampinato

Abstract This chapter analyses the change of view and relevance of nature in urbanised context and the present need of changing approach.

The popular knowledge of plants is progressively decreasing in the modern society and the urban ethnobotany", proposes a paradigm shift in the neglecting view of plants that can have the beneficial effect of reconnecting citizens with biological rhythms and traditions increasingly threatened by modernization. Ethnobotanical Greening Solutions (EGS) give Specific Benefits, that can be performed trough Urban Foodscapes, Healing Gardens and Living Pharmacies, Traditional Craftsmanship Zones, Ritual, Magical, and Symbolic Spaces can be settled in cities in order to stimulate such ecological and cultural need. A Methodological Approach for Plant Taxa Selection, such as the Maintenance and Monitoring Strategies, is also given. The case study of the Graecanic Area of Reggio Calabria is also illustrated.

A second paragraph analyses the cultural factors and reasons that gave rise to the evolution in the view of the plants and to the Plant blindness in the modern society.

G. Caneva (✉) · Z. Hosseini · N. Soldovieri
Department of Sciences, University Roma Tre, Rome, Italy
e-mail: giulia.caneva@uniroma3.it

A. Cano-Ortiz
Departamento de Didáctica de las Ciencias Experimentales, Sociales y Matemáticas, Facultad de Educación - Centro de Formación Profesorado - Universidad Complutense de Madrid, Madrid, Spain

C. M. Musarella
Department of AGRARIA, Mediterranean University of Reggio Calabria, Reggio Calabria, Italy

M. Patti · G. Spampinato
Department of AGRARIA, Università degli Studi Mediterranea di Reggio Calabria, Reggio Calabria, Italy

F. Orlandi · L. Ruga
Department of Civil and Environmental Engineering, University of Perugia, Perugia, Italy

G. Caneva et al. (eds.), *Urban Botany*, Future City 22,
https://doi.org/10.1007/978-3-032-28986-5_4

The change from the pagan religions to the monotheistic one, where God, and humans, as its most perfect creature, can explain the cultural startup of Anthropocene, where man assumed the central role in the universe. To contrast the common idea that humans can be separated from the nature and that all the other living things exist for humans, a growing number of initiatives are growing that can stimulate the plant 'attention among the general public.

Citizen science and Urban Educational Spaces are based on *involvement* (of people), *awareness* (of nature conservation issues), *collaboration* (in developing and sharing effective solutions), and *change* (in human behaviour towards the environment). Scientific projects and Urban Educational Spaces encourage cooperative and experiential learning, promoting the development of attitudes and values oriented toward environmental conservation and respect.

Keywords Ethnobotany · Ethnobotanical Greening Solutions · Plant blindness · plant awareness disparity · Citizen science · Urban Educational Spaces

4.1 Introduction (G.C., N.S.)

Early civilizations—and many indigenous peoples still today—maintained a close, interdependent relationship with nature. This connection led to careful observation of natural phenomena, not only for practical purposes, as further discussed in the ethnobotanical section, but also for the creation of myths linked to magic, spirituality, and religious beliefs (Brosse 1991; Frazer 1890).

This view greatly changed during the century, and the Traditional Ethnobotanical Knowledge, i.e., the Popular Knowledge of plants, is today at serious risk of erosion within modern society (Balick and Cox 2020) (Fig. 4.1).

The presence of plants in urban areas is undoubtedly beneficial for environmental health. Nevertheless, contemporary urban societies often hold a strongly anthropocentric perception of the natural world. Many people do not fully recognize

Fig. 4.1 Popular traditions, which link people with Nature. (**a**) Procession of vegetalized men covered with mosses on Corpus Christi Celebration in Bajar (Spain); (**b**) The fest of "Rumit- The walking forest" in Satriano di Lucania (Italy). (Photo by Fausto Podavini)

plants as living organisms, nor understand their fundamental role in sustaining life on Earth (Wandersee and Schussler 1999).

A shift in the neglectful view of plants has been observed in recent years, and urban citizens appear to be developing a new, more attentive approach, one that must be strengthened through improved education on the values of nature. Indeed, the concept of **biophilia** (Kellert and Wilson 1993) has increasingly entered public consciousness, highlighting the positive impact of nature on human well-being (Fig. 4.2).

The ongoing ecological crises have repositioned urban environments as crucial spaces for rethinking how societies perceive, value, and interact with the living world (Fig. 4.3). Although often described as artificial and ecologically impoverished, cities are dynamic socio-ecological systems, where spontaneous vegetation, ornamental species, and designed green infrastructures coexist with dense human activity (Kowarik 2011). Developing ecological awareness in these contexts requires

Fig. 4.2 The "need" of green, as expressed in architectural projects and in the imaginary of indoor or outdoor spaces. (Coupled elaborations by G. Caneva by free images)

Fig. 4.3 Different views of modern cities planning and their relationship with green spaces: (**a**) project of an "Ideal city" for workers' dwellings in 1833 (Milan); (**b**) Norman Foster's Santa Giulia Masterplan, Milan (2003)

more than technical knowledge: it calls for a cultural shift in how residents interpret the presence of plants and ecological processes in everyday life.

Citizen science plays a central role in this transformation. By bridging scientific knowledge and lived experience, it offers a concrete pathway to counteract plant blindness and foster ecological awareness within urban societies.

4.2 Ethnobotany in Urban Areas (C.M.M., M.P., G.S.)

4.2.1 Definition, Historical Context, and Classification

Ethnobotany, a discipline that investigates the complex interactions between human societies and the plant kingdom, goes beyond the mere cataloguing of uses and customs. It represents a holistic study of resource-management systems, symbolic beliefs, and vernacular classifications of nature (Balick and Cox 2020; Prance 2007). Although the term was formally coined by Harshberger (1896), plant-use practices have been intrinsic to human experience since the origins of civilization, forming the basis for survival, medicine, and ritual practices (Gilani and Rahman 2005).

In urban contexts, the "urban ethnobotany", proposes a paradigm shift: plants are no longer considered mere aesthetic elements but rather "biocultural infrastructures" capable of reconnecting citizens with biological rhythms and traditions increasingly threatened by modernization (Patti et al. 2025b; Plieninger et al. 2023).

Within the Mediterranean Basin, and particularly in southern Italy, the cultural stratification resulting from millennia of exchanges among Greeks, Romans, Byzantines, Arabs, and Normans has generated an exceptional botanical and cultural biodiversity (Caneva et al. 2013; Gentile et al. 2022; Pieroni and Quave 2005). Integrating this heritage into urban "Greening Solutions or infrastructures (GI)" responds to cultural and practical values.

4.2.1.1 Urban Foodscapes and Urban Alimurgy (Edible GI Spaces)

This category concerns the design of spaces dedicated to the conservation and active harvesting (foraging) of edible spontaneous species (Wild Food Plants). It extends beyond urban agriculture by recovering synanthropic and ruderal species - often dismissed as "weeds" - that historically constituted the backbone of the so-called "poor Mediterranean diet." (The Local Food-Nutraceuticals Consortium 2005). These plants, such as *Sonchus oleraceus* L. or *Cichorium intybus* L., sometimes provide higher micronutrient contents compared to commercial cultivars and enhance food security in times of crisis, as they are self-sustaining resources that require minimal agronomic inputs (Asfaw et al. 2023; Grivetti and Ogle 2000).

4.2.1.2 Healing Gardens and Living Pharmacies (Ethnomedicinal Spaces)

Medicinal plants have played a crucial role in the history of pharmacology and remain a primary source of new bioactive compounds (Heinrich and Gibbons 2001; Süntar 2020). The design of urban ethnomedicinal gardens serves not only aesthetic but also educational and conservation purposes: preserving in situ the germplasm of species such as *Malva sylvestris* L. or *Sambucus nigra* L., traditionally used to treat respiratory or dermatological ailments (Patti et al. 2025b; Vitalini et al. 2013). A nature-based approach to health is promoted in these spaces, which also serve as "living museums" of folk medicine. People in highly stressful environments, such as urban areas, can therefore access natural and affordable well-being (Jamal 2023).

4.2.1.3 Traditional Craftsmanship Zones (Areas for Material Culture)

The use of plants for producing tools, textile fibres, and musical instruments represents a pre-industrial technological heritage of considerable value (Maruca et al. 2019). Urban greening solutions that integrate species, such as *Phragmites australis* L. (or also *Arundo donax* L.) (used for basketry and musical instruments) or *Spartium junceum* L. (used for textile fibres) help preserve traditional craftsmanship skills. These areas may host open-air workshops where expertise passes across generations, counteracting the loss of cultural memory (Bellusci 2017; Patti et al. 2025a) (Fig. 4.4).

4.2.1.4 Ritual, Magical, and Symbolic Spaces (Intangible Heritage Spaces)

Plants often possess a symbolic dimension that transcends their material use, being associated with apotropaic practices (e.g., protection against the evil eye) or religious rituals (e.g., *Corpus Domini* flower carpets) (González-Tejero et al. 2008; Musarella et al. 2019; Patti et al. 2025b). Integrating species such as *Olea europaea* L. or

Fig. 4.4 The Ethnobotanical Garden at Rottaja (Pisa, Italy) and example of a parcel. (From Caneva et al. 2013)

Fig. 4.5 Sacred and profane celebrations linked to ethnobotanical traditions (**a**) The flower festival on the occasion of the Corpus Domini celebration of Genzano (Rome, Italy). (Photo by P. Guarrera); (**b**) The Karnival in Liguria (Italy) with flowered wagons. (Photo by E. Mariotti)

Hypericum perforatum L. into public spaces fosters social cohesion by reviving collective rituals that reinforce community identity (Pieroni and Cattero 2019). One example is the making of "*Pupazze di Bova*", which is still practiced today in southern Calabria (Italy) using *Arundo donax* L. stems, and *Olea europaea* L. leaves. During annual religious festivals, female doll-like figures are created: the framework is constructed from *A. donax* stems, while the "*stiddhe*" (stars), obtained by intertwining pairs of olive leaves, are fixed to the structure (Patti et al. 2025a) (Fig. 4.5).

4.2.2 Methodological Approach for Plant Taxa Selection

The selection of taxa for urban ethnobotanical projects must move beyond a purely ornamental perspective and adopt quantitative scientific criteria to evaluate both the cultural value and ecological suitability of plant species. The proposed methodological framework is based on rigorous bibliographic research and field analyses, validated through standardized ethnobotanical indices, and articulated into the following operational phases:

4.2.2.1 Quantitative Ethnobotanical Analysis

The following indices are among the most important for identifying "key species":

Cultural Importance Index (CI): the sum of the proportion of informants citing each use of a species. High CI values indicate deep cultural embeddedness (Pardo-de-Santayana et al. 2007; Sujarwo et al. 2014).

Cultural Value (CV): a composite index integrating the number of uses, citation frequency, and species versatility (Reyes-García et al. 2006). Species with high CV values represent priority candidates.

Relative Frequency of Citation (RFC): measures species popularity. Values close to 1 indicate widespread knowledge, ensuring immediate social recognition (Tardío and Pardo-de-Santayana 2008).

4.2.2.2 Ecological and Chorological Assessment (Ecological Suitability)

Urban green sustainability depends on the **adaptation of species to local pedoclimatic conditions**. Priority should be given to **native taxa**, which in regions such as Calabria (southern Italy) constitute more than 80% of the ethnobotanical flora (Patti et al. 2025b). Chorotype analysis should favour **Steno-Mediterranean** and **Euri-Mediterranean** species, which are typically drought-tolerant and well-suited to Mediterranean urban environments. From a life-form perspective, **hemicryptophytes** and **chamaephytes** are particularly suitable for resilient, low-maintenance ground cover.

4.2.2.3 Harvest Sustainability (Plant Part Value—PPV)

In urban settings, harvesting must be managed with care to avoid resource depletion. The **PPV index** helps identify species from which renewable parts—such as leaves, flowers, or fruits—are harvested, rather than destructive parts such as roots or entire plants. Recent data indicate that **leaves** are the most frequently harvested plant parts (PPV $\approx$ 0.30), followed by fruits, confirming the feasibility of non-invasive management in productive green spaces (Chaachouay et al. 2019; Patti et al. 2025b).

4.2.3 Specific Benefits of Ethnobotanical Greening Solutions (EGS)

Ethnobotanical greening solutions provide a broader range of ecosystem and social benefits (Ecosystem Services) than conventional urban green spaces, contributing to the development of healthier and more resilient cities.

4.2.3.1 Biodiversity and Genetic Resource Conservation

Cities can function as "Noah's Ark" for ancient local varieties (landraces) and crop wild relatives (CWR). These plants possess genetic traits conferring resistance to pests and climatic stress, which have often been lost in industrial cultivars, thus representing a strategic resource for future food security (Abenavoli et al. 2021; Perrino 2022).

4.2.3.2 Nutraceutical Valorization and Public Health

Wild edible plants, particularly wild leafy vegetables, often exhibit high nutritional profiles, with elevated levels of fibre, polyphenols, and antioxidants (Romojaro et al. 2013). Integrating these species into urban green areas diversifies diets and provides free access to local "superfoods," helping counteract modern malnutrition-related diseases such as diabetes and obesity (Bonaccio et al. 2018; Guarrera and Savo 2016).

4.2.3.3 Cultural Resilience and Territorial Identity

In an era of homogenizing globalization, the recovery and enhancement of ethnobotanical traditions strengthen community identity and the sense of belonging. For minority or marginalized groups—such as the Grecanic or Occitan communities in Italy—plants serve as carriers of language, history, and collective memory (Bellia and Pieroni 2015; Mattalia et al. 2020). Urban green spaces thus become instruments of historical narration and cultural resilience.

4.2.3.4 Pharmacological Potential and Innovation

Documenting traditional plant uses often represents the first step in bioprospecting, the search for new bioactive molecules for pharmaceutical development (Chaachouay and Zidane 2024; Domingo-Fernández et al. 2023). Species cultivated in urban ethnobotanical gardens may also support research in phototherapeutics and medicinal plant authentication (Liu et al. 2023).

4.2.4 *Maintenance and Monitoring Strategies*

The management of EGS requires a dynamic and participatory approach, distinct from the standardized maintenance practices typical of ornamental urban green spaces.

4.2.4.1 Phenological Maintenance and Regulated Foraging

Mowing and pruning interventions must follow natural biological cycles to enhance the production of useful plant parts (e.g., avoiding cutting during the flowering period of melliferous or officinal species). To prevent overexploitation, regulated foraging protocols should be implemented while simultaneously educating citizens on sustainable harvesting practices (Kalle and Sõukand 2016).

4.2.4.2 Participatory Monitoring (Citizen Science)

Engaging local communities - particularly elderly knowledge holders and schools - is essential. Citizen science programs can be implemented to monitor plant phenology, pollinator presence, and actual resource use, transforming maintenance into an educational and social activity (Plieninger et al. 2023) (see Par. 4.4).

4.2.4.3 Environmental Safety

Given the potential food use of many ethnobotanical species, systematic monitoring of pollutants (e.g., heavy metals, PAHs) in soils and plant tissues is mandatory. Areas intended for consumption should be located away from direct pollution sources, such as high-traffic roads or industrial zones, and should undergo periodic analyses to ensure public health protection (Sánchez-Mata et al. 2016).

4.2.5 *Challenges and Limitations*

The implementation of urban ethnobotany faces several significant obstacles that require targeted mitigation strategies.

4.2.5.1 Knowledge Erosion

The most serious threat is the rapid loss of Traditional Ethnobotanical Knowledge (TEK). Several studies show that such knowledge is predominantly held by individuals over 60, while younger generations are increasingly detached from nature.

Without active intergenerational knowledge transfer, plants progressively lose their cultural meaning and associated traditional uses.

4.2.5.2 Aesthetic Perception and Social Conflicts

Many ethnobotanical species (e.g., *Urtica dioica* L., *Cichorium intybus* L.) are spontaneous or ruderal plants with a "disorderly" appearance. This often conflicts with conventional aesthetic expectations in urban green spaces, which tend to favour manicured lawns and exotic ornamental species. Environmental education is therefore essential to promote appreciation of the functional beauty and ecological importance of native biodiversity (Menale et al. 2016).

4.2.5.3 Socio-Economic Changes

The abandonment of agro-pastoral practices and the depopulation of rural and inner areas have disrupted the oral transmission of ethnobotanical knowledge (Alhajj Ali et al. 2024). Reintroducing these practices in urban contexts requires not only adequate physical spaces but also the reconstruction of a social fabric that values self-production, sustainability, and traditional ecological knowledge.

4.2.6 Case Study: The Graecanic Area of Reggio Calabria (GARC)

The Graecanic Area of southern Calabria (Southern Italy) represents an ideal open-air laboratory for observing the persistence and evolution of ethnobotanical knowledge within a linguistic and cultural minority context (Fig. 4.6).

This geographically isolated territory has maintained a symbiotic relationship with the local flora, rooted in the ancient heritage of Magna Graecia (Pieroni and Cattero 2019; Spampinato et al. 2017).

A recent in-depth study conducted in the area (Patti et al. 2025a) documented 157 taxa belonging to 50 families, revealing a remarkable richness of uses. This case study provides replicable models for urban ethnobotany:

4.2.6.1 Multifunctional Use and Value

Clinopodium nepeta (L.) Kuntze subsp. *nepeta* ("*nipiteddha*") emerged as the most culturally significant taxon, used simultaneously as a culinary spice, medicinal remedy (digestive, vulnerary), and domestic deodorant. Its hardiness and versatility make it a flagship species for low-maintenance aromatic urban beds. Other

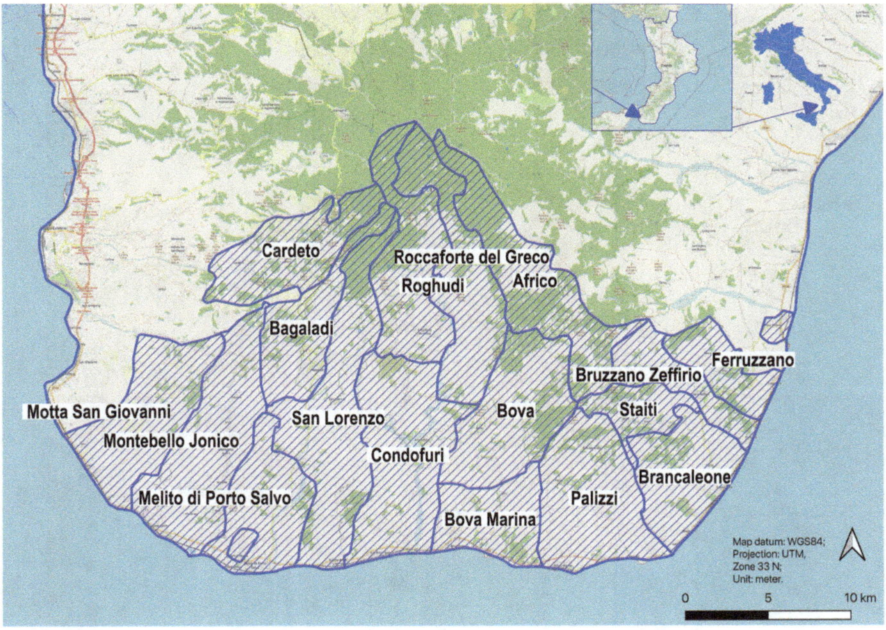

Fig. 4.6 Municipalities belonging to the Graecanic area of the Metropolitan City of Reggio Calabria (Italy). (In blue, the borders- elaboration in Q. Gis by M. Patti)

Lamiaceae with similar value are the shrubby plants *Salvia rosmarinus* Spenn. (syn. *Rosmarinus officinalis* L.) and *Salvia officinalis* L.

4.2.6.2 Valorisation of the Food Landscape

The widespread consumption of common wild plants (e.g., *Cichorium intybus* L., *Anethum piperitum* Ucria, *Arbutus unedo* L.) (Fig. 4.7a)—raw as snacks or prepared in soups—demonstrates how green spaces can contribute to local diets. However, the presence of edible endemics, such as *Crepis aspromontana* Brullo, Scelsi & Spamp., suggests *ex situ* conservation strategies through sustainable gastronomic use (Patti et al. 2026) (Fig. 4.7b).

4.2.6.3 Craftsmanship and Playfulness as Educational Tools

The traditional use of *Opuntia ficus-indica* (L.) Mill. to create children's toys ("carrocciulu") and *Spartium junceum* L. (Fig. 4.7c) for weaving exemplifies how plants can stimulate manual creativity, intergenerational transmission of skills, and outdoor play. These traditions offer nature-based educational opportunities that can be readily adapted to urban contexts (Musarella et al. 2019) (Fig. 4.7).

Fig. 4.7 Six close-up views of different plants. (**a**) Immature fruits of *Arbutus unedo* L. (**b**) Capitula of *Crepis aspromontana* Brullo, Scelsi & Spamp. (**c**) Flowering branches of *Spartium junceum* L. (**d**) Flowering branches of *Malva sylvestris* L. (**e**) Immature fruits of *Ficus carica* L. (**f**) Flowering of *Prunus amygdalus* Batsch. (Photos by C.M. Musarella)

4.2.6.4 Folk Medicine and Resilience

The extensive use of medicinal species for treating common ailments (digestive, dermatological) highlights the healthcare resilience of the local community. Integrating these species (e.g., *Ficus carica* L., *Laurus nobilis* L., *Malva sylvestris* L., *Matricaria chamomilla* L., *Prunus amygdalus* Batsch) into public green spaces enhances the local ethnopharmacological heritage, also supported by recent phytochemical studies (Beddiar et al. 2021; Patti et al. 2025a,b) (Fig. 4.7d–f).

In conclusion, the study carried out in the Graecanic Area and the systematic review of regional knowledge (Patti et al. 2025a,b) confirm that ethnobotany is not a discipline of the past but a key resource for the future of sustainable cities, capable of integrating biological conservation, cultural identity, and human well-being with the occurrence of specific adapt taxa cultivable in the modern cities.

4.3 Plant Blindness in the Modern Society (G.C., Z.H.)

4.3.1 An Evolution of the View of the Plants: Cultural Factors and Reasons

In ancient cultures all around the world, natural elements were perceived as creations and processes governed by divine will, and all natural phenomena were observed to uncover hidden divine messages (Day 2013; Kandeler and Ullrich 2008).

Plants with practical uses or with unique biological features acquired specific associations with abstract ideas and beliefs, which contributed to their symbolic meaning. Their representation was integrated into natural motifs, especially in funerary or religious buildings, as a tool of communication for people unable to read written language (Caneva 2010; Caneva et al. 2014) (Fig. 4.8).

Starting from the medieval period, with the shift from pagan religions to monotheism—where God and humans, as His most perfect creatures, assumed the central role in the Universe—the allegorical religious interpretation dominated artistic

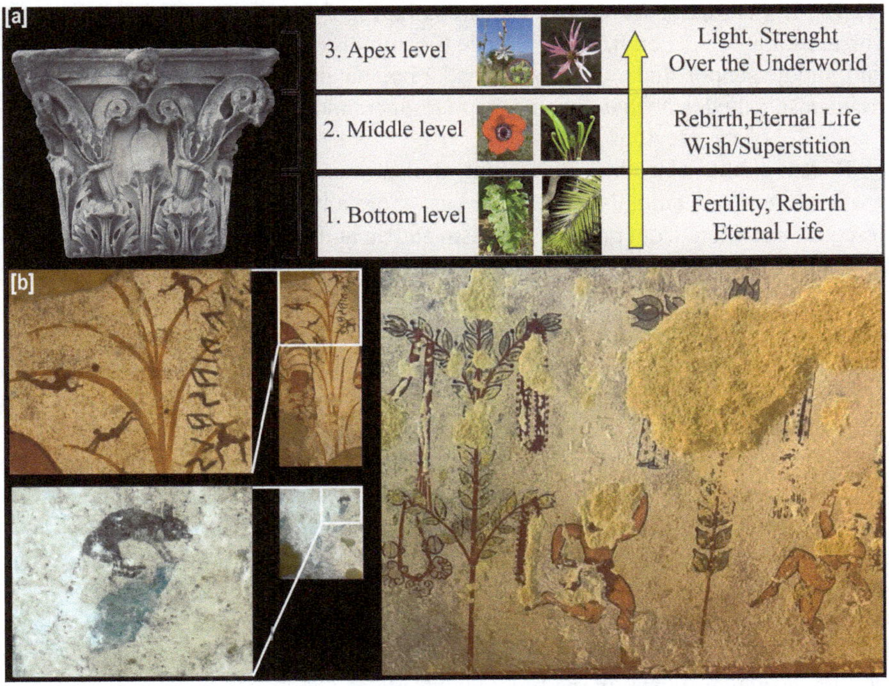

Fig. 4.8 Plants elements derived from artistic representation, where they were used as a tool of communication, (**a**) Corynthian capitals and their symbolic meaning; (**b**) Paintings from Etruscan tombs of Tarquinia, where humans appear very small with respect to plant elements, due to their divine values. (Elaboration by G. Caneva)

expression. Sometimes, distorted natural forms, such as those in ancient painted herbaria, were used to create cryptic elements whose meaning was known only to select people.

In modern and contemporary societies, this sense of the sacred has been weakened, as has the symbolic language expressed through images of nature. A progressive disconnection from nature has also characterized modern society, and this phenomenon is particularly evident in plants, which are often perceived merely as a green background. Therefore, it is not surprising that the term "Plant Blindness" was introduced (Wandersee and Schussler 1999). It describes the correlated inabilities of modern society "to see or notice the plants in one's environment; to recognize the importance of plants in the biosphere and in human affairs; to appreciate the aesthetic and unique biological features of the life forms that belong to the Plant Kingdom; and the misguided anthropocentric ranking of plants as inferior to animals and thus unworthy of consideration". More synthetically, it refers to the tendency of people to overlook plants in their everyday lives (Allen 2003).

Parsley (2020) enriched the discussion on plant blindness by emphasising the different levels of attention given to the plant world compared to the zoological one. He proposed replacing the term "plant blindness" with "plant awareness disparity" (PAD). In fact, he observed that when comparing the common attention to plants versus animals, a generalized preference for animals exists. Among U.S. elementary and middle school students, only 7% had a botanical preference, and such zoocentrism, also called "zoochauvinism", was stronger among males than among females (about 2/3 of the previous group).

Recent studies also show that young students associate the concept of life with motion; therefore, animals are regarded as "more alive" than plants, and these differences may stem from residual animism and an anthropocentric view. As a consequence, moving non-living objects may even be regarded as alive (Yorek et al. 2009) (Fig. 4.9).

Indeed, when looking at life through a phylogenetic and evolutionary perspective, plants and animals belong to sister groups (Bouteau et al. 2021). Some Authors (Thomas et al. 2022) link the low valuation of plants to Aristotle's view, in which animals possess a soul ("*anima*" in Latin and "*anemos*", i.e., wind in Greek) superior to that of plants. However, we believe that the ancient people's view was more complex, and that, when examining Aristotle's philosophy carefully, despite the differences he emphasized, such as the reduced sensitivity of plants and their lower ability to interact with their environment, his perspective was not entirely dismissive of them.

The hierarchical view of nature probably supersedes the ancient vision and can be traced back to the medieval and modern ages. In fact, ancient Greek and Roman culture, as well as others in the Euro Mediterranean context, had the idea of a continuous metamorphosis in the natural world, and this concept was reflected in artistic representation (Caneva 2010), such as in the Latin literature (e.g., *Metamorphosis* by Ovid, and *De Rerum Natura* by Lucretius).

More generally, why do humans now overlook plants? According to the observations reported in Wandersee and Schussler (1999), there are

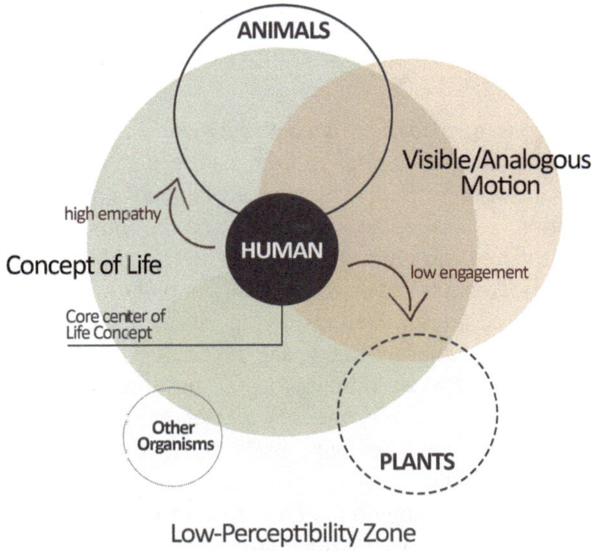

Fig. 4.9 Animistic-anthropocentric construction model of the life concept. (Modified from Yorek et al. (2009))

physiological and psychological reasons for this. There is a very high selection of information in the visual process, and the brain extracts only a minimal portion of the bits that reach conscious attention (from ten million to about 16 bits per second). This selection process seems linked to the search for objects that show movement, conspicuous colours and patterns, unfamiliarity, or potential threats. Consequently, plants do not receive visual attention because they are static (their movements are usually very slow), do not appear to pose a risk to humans, and often blend into the background. This occurs especially when plants are uniformly green, without coloured flowers, such as before spring blooming, fruiting in summer or later, or during the fall season when leaves change colour and assume a new and charming appeal—explaining the recent phenomenon of "the foliage" as a touristic attraction.

When comparing the knowledge and the cultural and "religious or sacral" roles of plants in modern society with those of the ancients or of indigenous people who still live closely connected to nature, we can observe a dramatic reduction in both aspects. As noted in the previous paragraph, in the present loss of Traditional Ethnobotanical Knowledge (TEK), plant awareness remains significantly higher in rural areas than in urban areas. Today, people typically know less about plants than about animals, and in modern society, several indicators show that both people and policymakers often perceive plants simply as the background to other living organisms (Burke et al. 2022). The knowledge of plants, even if not the interest in them, seems to be constantly disappearing in younger generations, following a parallel trend with the decline in the appreciation of their unique cultural identity (Blue et al. 2023).

4.3.2 *The Need of Changing Approach and Actions to Be Carried Out*

The consequence of such a cultural view in modern society is the common idea that humans can be separated from nature and that all other living things exist for humans. Often, students consider plants as a boring topic, and to counter this perception, several authors have stressed the importance of teachers highlighting the most varied and interesting aspects of plant life (Thomas et al. 2022).

Research has shown that training in plant science makes a great difference in the future recognition, feelings, appreciation, and cultivation of plants. In fact, we recognize only what we know, and psychologists "would say that plants have low signal value for most humans today," and that, starting from the first weeks of life, "humans develop a propensity to respond first to things in their environment that have faces, and animals best meet this search criterion" (Wandersee and Schussler 1999).

Students usually do not have an ecological vision, which consists in viewing all living entities as interacting parts, and therefore they cannot comprehend the value of the "whole" (Yorek et al. 2009). People also do not sufficiently consider that more than 20% of plant species are threatened with extinction; more than all described animal taxa, and consequently, efforts to maintain plant biodiversity are given less relevance (Westwood et al. 2020).

Still in biology courses, the role of plants is often ignored by both teachers and students (Pany et al. 2022). Despite the fundamental ecological role of plants for our life, providing oxygen, food, medicines, fibres, timber, as well as protection, climate mitigation, and accounting for 80% of the total biomass on Earth, plants remain overlooked even in natural documentaries. A recent estimation in the UK showed that plants were, on average, mentioned more than seven times less frequently than animals (Burke et al. 2022). As stressed in the literature, "botanists are in danger of becoming 'the dinosaurs of biology in the 21st century" (Woodland 2007). In the last decades, several authors have highlighted the global reduction of botany students, faculty, plant-focused courses, and botany departments (Crisci et al. 2020).

The need for a changing approach is urgent due to the evidence that plants are an essential element in sustainability education and practice (Amprazis and Papadopoulou 2020). Yet, in the paradox of contemporary culture, we use the word "green" in several aspects of sustainability without fully considering the fundamental role of plants (Thomas et al. 2022). However, even if the lack of interest in plants and in botany seems be a persistent issue, recent studies show that in more recent years, probably influenced by the COVID-19 period, a new awareness has emerged regarding the therapeutic value of interacting with plants and their positive effect on human well-being (Burke et al. 2022). During this time, the rise of the phenomenon of "Plant parenting", which describes the emotional bond between people and plants, spread through social media. Similarly, the increase in allotment gardens and therapeutic gardens (see par. 7.4 and 7.5) represents further evidence of the growing interest and appreciation of the plant world in the cities.

Today, plant scientists and botanists have launched a growing number of initiatives to stimulate public interest in plants (Burke et al. 2022). As examples of such different approaches, we can mention symbolic initiatives of days dedicated to plants, such as the "International Fascination of Plants Day" on 18 May (https://plantday18may.org/, launched by the European Plant Science Organisation, EPSO) and the "International Day of Forests" on 21 March, proclaimed by the United Nations General Assembly in 2012 (https://www.fao.org/international-day-of-forests/en). Similarly, we can also mention several projects, such as "PlantLoveStories" (https://www.plantlovestories.com, launched by the University of Pittsburgh main Oakland campus), or "More Than Weeds" (https://morethanweeds.co.uk/), created in France in 2001 by the French Natural History Museum together with the French botanical organisation Tela Botanica, as well as several citizen-science projects (https://www.ecsa.ngo/working-groups/green-spaces/) (Fig. 4.10).

A way to counteract the reduced attention to nature in urban environments is through practical experiences, such as botanical gardens and direct plant cultivation, which help foster renewed plant awareness. Indeed, the increasing popularity of gardening and houseplant cultivation is certainly linked to the growing awareness of the positive impact these activities may have on mental health and well-being, and interacting with plants is therefore increasingly considered a practice of mindfulness.

A change in the neglecting view of plants can be observed when looking at the results of a Google search analysis carried out in the United Kingdom between 2010 and 2020, which showed a new trend in plant-related topics (using the keywords: Houseplant; plant-based diet; garden; botany (field of study); wildflower), with a general increase in popularity (Burke et al. 2022).

In conclusion, a way to counteract this phenomenon seems to be to focus on a renewed improvement in contact with nature, which should start in childhood to create stronger emotional connections. Even though age, gender, and parental influence play a significant differential role, teachers and media can also have a relevant impact in strengthening such attitudes, and they should contribute to this direction.

Furthermore, plant-focused communities and groups on new social media platforms—which can greatly facilitate engagement with new audiences—should stimulate people through practical, recurring experiences, such as cultivating houseplants and urban gardens, and overall by giving greater attention to natural biodiversity and sustainability.

Fig. 4.10 The famous poster to combat plant blindness made by the artists Jonathan Clyde Frey and Ana Kell, working at the Curtis G. Lloyd Library and Museum in Cincinnati, Ohio, USA (2018)

4.4 Citizen Science and Urban Educational Spaces (A.C., L.R., F.O., N.S.)

4.4.1 Ecological Awareness and Citizen Science

Ecological awareness in the city can be defined as "the ability to perceive, understand, and value ecological processes, species interactions, and environmental dynamics that shape urban landscapes." This competence is particularly relevant in modern societies affected by the already mentioned "plant blindness" (Wandersee and Schussler 1999). In cities, it also manifests in the marginalization of everyday vegetation: weeds along sidewalks, self-seeded trees in vacant lots, or mosses on building surfaces are rarely perceived as meaningful components of urban ecosystems.

Overcoming this perceptual bias is a prerequisite for recognizing the ecological value of urban nature and for understanding that even marginal plant life contributes to biodiversity, microclimate regulation, and urban resilience. By fostering direct encounters, a new ecological awareness in the city can emerge through the interaction of scientific knowledge, cultural perception, and participatory practice, counteracting the "extinction of experience" with nature described by Soga and Gaston (2016).

The relationship between humans and plants can be strengthened through citizen science (CS), which has emerged as a powerful strategy for engaging citizens directly in observing, documenting, and interpreting urban biodiversity. Broadly defined as the participation of non-professionals in scientific research and knowledge production (Fraisl et al. 2022), citizen science in urban settings functions not only as a method of data collection but also as a pedagogical and cultural practice (Kruger and Shannon 2000).

This active collaboration is founded on four paradigms, which represent the pillars of CS: *involvement* (of people), *awareness* (of nature conservation issues), *collaboration* (in developing and sharing effective solutions), and *change* (in human behaviour towards the environment, perceived as a vital asset to be safeguarded in the long term).

By encouraging residents to observe, identify, and record living species in their neighborhoods, CS transforms ecological awareness from an abstract concept into an embodied, everyday practice. This participatory approach resonates with Irwin's (1995) notion of "scientific citizenship," in which members of the public are not passive recipients of expert knowledge but active contributors to scientific agendas and environmental decision-making. In this sense, citizen science represents a democratization of ecological knowledge, narrowing the gap between scientific institutions and urban society (Bonney et al. 2016).

Public participation may be limited to a single experience or may become a regular activity in a person's life, contributing to science in different ways—from collecting data for scientific use to analysing and interpreting it. By engaging citizens in scientific processes, CS has the potential to increase awareness and direct understanding of complex issues such as climate change, biodiversity, and environmental sustainability.

Although the term CS first appeared in the 1990s, its popularity has grown considerably in recent years. In particular, the development of modern CS in the 2000s was driven by technological advances, widespread access to the internet, and the rise of social media (Marcenò et al. 2021), and the proliferation of mobile devices such as smartphones and tablets, which made data collection and sharing easier. Recognition of CS is growing in science, politics, education, and society at large. It is now emerging as both a field of research and of practice, increasing the need for insights, standards, terminology, and general guidelines (Vohland et al. 2021). The European CS Association (ECSA) has defined the field based on its ten principles of good practice (Haklay et al. 2021).

Beyond their scientific value, CS practices profoundly reshape participants' perceptions of urban nature. Digital platforms such as iNaturalist, eBird, or PlantNet enable residents to document plant and animal species, phenological events, and ecological interactions across highly heterogeneous urban landscapes. The resulting datasets provide fine-grained information on species distribution and ecological change that is difficult to obtain through conventional monitoring programs (Aronson et al. 2014). Each georeferenced photograph of a sidewalk plant or seasonal flowering in a courtyard helps make urban vegetation visible and legible, transforming plants from an unnoticed background element into recognizable organisms with names, life cycles, and ecological functions. Empirical studies indicate that participation in biodiversity monitoring enhances environmental literacy and strengthens emotional connections to local ecosystems (Cooper et al. 2007; Shanahan et al. 2020).

Facilitated by mobile technologies and cloud-based platforms, citizen science thus counters the passivity associated with plant blindness by cultivating attentiveness, curiosity, and care toward urban plant life (Fraisl et al. 2022).

4.4.2 Citizen Science and Social Implications

Citizen-generated ecological knowledge also has important implications for urban planning and design. Conventional ecological assessments often overlook small parks, street-tree assemblages, interstitial green spaces, and vacant lots, despite their capacity to host diverse plant communities and provide ecosystem services. Community-based observations help reveal these "hidden" urban ecologies, offering insights into species presence, plant health, and seasonal dynamics. For architects, planners, and engineers, such information can complement traditional site analysis by integrating lived ecological knowledge and temporal dynamics into design processes.

This perspective aligns with the principles of ecological urbanism and nature-based solutions, which emphasize designing with ecological processes rather than treating vegetation as a purely decorative or technical element (Beatley and Newman 2013; Elmqvist et al. 2015). In this sense, ecological awareness becomes a design competence: the ability to read cities as living systems shaped by the co-evolution of human and plant agency.

Within broader sustainability frameworks, citizen science also contributes to environmental justice by making ecological knowledge accessible to communities that have historically been excluded from planning and decision-making. Urban green spaces are unevenly distributed, and marginalized neighbourhoods often face reduced access to vegetation, greater exposure to pollution, and increased vulnerability to climate-related risks (Anguelovski et al. 2020). Participatory monitoring enables residents to document these inequalities and produce evidence supporting a more equitable distribution of green infrastructure.

This approach aligns with the concept of "just sustainability," in which ecological health and social equity are understood as inseparable dimensions of urban development (Agyeman 2013). At the same time, limitations must be acknowledged, including unequal access to digital tools and variability in data quality. Addressing these challenges requires inclusive project design, institutional support, and sustained collaboration among researchers, municipalities, and community organizations. CS aligns with this perspective by legitimizing local ecological knowledge and by making visible forms of urban nature that are often dismissed as informal, undesirable, or unmanaged, yet which frequently host resilient plant communities and provide essential ecosystem services. However, participation tends to favour digitally connected and highly educated groups, and data quality can vary. Overcoming these limitations necessitates inclusive design, long-term institutional commitment, and partnerships between researchers, local authorities, and community groups.

Policy makers and NGOs are increasingly relying on citizen volunteers to expand their capacity to monitor and protect natural resources, assess endangered species, and safeguard protected areas (Silvertown 2009). For example, CS projects that study the spread of invasive plant species by local residents can enhance both scientific knowledge and behavioural change within communities (Crall et al. 2015). Such projects often yield both scientific and educational outcomes, including advances in biological research, biodiversity monitoring, and science education (Vohland et al. 2021). Participants in CS gain not only increased knowledge of natural history and biology but also improved scientific literacy, critical thinking skills, and greater awareness of the importance of urban greening (Dickinson et al. 2010).

Recent studies on climate change and the functions of botanical gardens have also highlighted how environmental education and CS can influence participants' knowledge, attitudes, and beliefs. For instance, pollination activities in botanical gardens have helped transform passive observation of nature into active scientific inquiry, even among children (Eberbach and Crowley 2017). CS also enables the collection and analysis of large datasets and supports research on topics that might otherwise lack sufficient funding. The valuable information obtained is often used in the formulation and implementation of environmental policies—an area where cities are playing an increasingly active role, relying on solid data to develop effective strategies.

In conclusion, CS strengthens the human–plant relationship by raising awareness of the climate-mitigation and adaptation potential of urban vegetation, highlighting the value of native species in cities, and encouraging behavioural change. Ultimately, these experiences contribute to the safety, resilience, and well-being of urban communities.

4.4.3 Scientific Projects

CS has also been applied to urban research. A recent European project (LIFE CLIVUT—LIFE18 GIC/IT/001217) demonstrated how involving "active citizenship" (public administrations, associations, schools, businesses, and citizens) can provide valuable input for decisions regarding the selection and management of urban green areas. Citizens contributed by monitoring green spaces, improving management practices (including low-impact strategies for CO_2 storage, particulate matter absorption, and pollution reduction), and raising awareness of the importance of green areas for urban life. These activities led to the development of an Urban Green and Climate Strategy to improve living conditions by introducing suitable tree species. Shared participation enabled citizens to learn about the climatic and ecosystemic roles of different trees and their contribution to mitigating the urban heat island effect.

Another significant initiative was the creation of phenological monitoring areas (PMAs), comprising different plant species (trees and shrubs) with well-documented thermal requirements across their life stages (phenological phases). The goal of phenology initiatives is to raise awareness—particularly among younger urban generations—of the role of green spaces and to increase understanding of plant adaptation and their contribution to climate-change mitigation in cities. PMAs serve as permanent educational laboratories for studying biological phenomena, providing a hands-on way to learn about plants, biology, and climate change (Fig. 4.11).

Phenology (from *phaino*, "to appear") is the study of the seasonal cycles of plant and animal life stages. Plant development is regulated by both internal factors (biorhythms determined by hereditary traits) and external factors (climate), which induce different phases. The applications of phenological research are numerous, with climate change among the most recent and significant fields (Orlandi et al. 2020). Phenology requires long-term, continuous data series, a field where citizens and students can play a crucial role. Many countries now run observation programmes that rely on trained volunteers to collect and present reliable data to support large-scale scientific research.

Fig. 4.11 Citizen science activity with young students in a PMA. (Photos by the Authors)

4.4.4 Urban Educational Spaces

A lack of interest in learning inevitably leads to poor academic performance. This highlights the need to invigorate educational processes by redirecting them toward sustainability and incorporating educational innovation into classrooms, alongside proactive teaching initiatives. There is an urgent need to adopt new strategies for active learning that establish direct interaction with the natural environment, thereby fostering the development of multiple domains of educational knowledge (Schussler and Olzak 2008).

School gardens, botanical gardens, and peri-urban areas emerge as valuable educational resources that encourage cooperative and experiential learning while promoting attitudes and values oriented toward environmental conservation and respect. According to Palacios et al. (2016), the school garden promotes active teaching by bringing students closer to the natural environment. Through this approach, the teaching–learning process is strengthened, offering a medium for more dynamic and participatory forms of education. Both school gardens and botanical gardens, together with peri-urban areas, constitute practical learning laboratories. Botanical gardens were originally established to safeguard and preserve species of botanical interest—particularly endemic and rare species—while also fostering biodiversity research. In the face of the dramatic environmental changes currently occurring because of human impact on natural ecosystems and global climate change, strengthening such learning has become increasingly urgent (De Rivas et al. 2024).

Botanical gardens, school gardens, and peri-urban areas not only fulfil an educational mission but also serve as essential infrastructures for research on ecosystem services, species conservation, and the eradication of invasive alien species, which pose serious threats to native flora (Wani et al. 2023) (Figs. 4.12 and 4.13).

The creation of botanical gardens is particularly relevant since, in many cases, these collections are composed of non-native species. This condition has

Fig. 4.12 Educational activities in the civic archaeological and ethnological park of Modena (Italy) (courtesy by G. Bosi- Unimore)

Fig. 4.13 Examples of botanical gardens with their educative role: (**a**) the first settlement of a diffuse Botanical Garden of Roma Tre University. (**b**) Cactus Garden of Lanzarote (Spain). Family Cactaceae. *Echinocactus grusonii* (a species native to San Luis Potosí, Mexico)

historically contributed to the origins of biological invasions from educational and botanical gardens, and peripheral urban areas, as environmental resources (Monroe and Krasny 2015), bring students closer to their surrounding environment, allowing them not only to learn through experimentation but also to engage with numerous cross-cutting themes—such as sustainability, nutrition and health, values, and citizenship—within the broader framework of ecosystem services (Chang et al. 2024).

This pedagogical resource fosters the development of competencies such as the ability to promote eco-industries within the context of ecosystem services (Chang et al. 2024), thereby contributing to business sustainability. Such sustainability is achieved through an action–research methodology, which, as highlighted by Da Silva Siltori et al. (2024), is particularly relevant in the training of teachers and researchers. This approach enables them to acquire the competencies necessary to prepare sustainability reports and to incorporate sustainability education into schools. To this end, it is essential to automate, compile, and disseminate research on sustainability, as well as to develop competencies related to mitigating climate change in agriculture through the design and application of specific methodologies for such mitigation.

The use of natural and semi-natural environments as educational resources seeks to invigorate teaching practices by establishing a teaching–learning process that integrates curricular content within contexts that foster experiential learning. This approach stimulates students' interest in learning while creating spaces for interaction, reflection, and experimentation. The contents published in the *Boletín Oficial del Estado* (BOE) were reviewed to assess whether current school teaching practices align with contemporary social realities. Tacca (2021) emphasizes that the teaching of Natural Sciences requires the implementation of creative strategies that stimulate and motivate the development of critical–reflective thinking, while simultaneously considering the student's cognitive development, thus ensuring an appropriate pedagogical intervention.

Within the framework of school gardens, botanical gardens, and natural environments, the teaching–learning process was implemented through the design of didactic strategies emphasizing experimentation and hands-on activities. These strategies provided direct contact with the natural environment and encouraged systematic observation.

By using these spaces as pedagogical strategies, students drew on the natural environment to construct their own learning, with inquiry and experimentation as the primary vehicles for knowledge acquisition. According to Samper and Ramírez (2014), experiential learning directly involves students in constructing their own knowledge, enabling them to acquire new content within a familiar context. Within this teaching–learning process, it is essential to identify students' prior knowledge to achieve the competencies needed to address current educational needs.

The results obtained through this experimental process, using school gardens, botanical gardens, and diverse environments as educational platforms, reveal clear differences in students' knowledge and awareness of the use of these resources as learning laboratories. Through these experiences, students learned to distinguish between local native species and introduced species. An analysis of the curricula published by the Government of Spain shows that there is no reference to the teaching of exotic species that pose a risk to wetland habitats (EEC 1992). Pinto et al. (2017) argue that experiential learning consists of a set of practices and techniques that enable students to learn through direct experience. This claim is supported by the results obtained from the proposed teaching approach.

Considering the existing knowledge gap and the disconnect between students, society, and the environment, Fytopoulou et al. (2024) propose curricular reforms to strengthen student participation in environmental issues. We concur with this perspective, given that in Spain, environmental topics—particularly those related to agriculture and forestry—are only superficially addressed within the current educational system.

In conclusion, the use of these semi-natural spaces has proven to be an effective tool for teacher training. They foster sustainability education and promote eco-responsible awareness among future educators. This approach enables the transversal integration of sustainability education across various areas of the teacher education curriculum by establishing a teaching–learning process that leverages these spaces.

By providing future teachers with practical tools, they are better equipped to understand and address the environmental challenges facing society. Moreover, through their future professional practice, they can promote the inclusion of sustainability concepts in school curricula. Following the development of this line of work, the main conclusion is the reaffirmation of the educational value of school gardens and botanical gardens as fundamental pedagogical resources.

References

Abenavoli L, Milanovic M, Procopio AC, Spampinato G, Maruca G et al (2021) Ancient wheats: beneficial effects on insulin resistance. Minerva Med 112(5). https://doi.org/10.23736/S0026-4806.20.06873-1

Agyeman J (2013) Introducing just sustainabilities: policy, planning, and practice. Zed Books

Alhajj Ali S, Tallou A, Vivaldi GA, Camposeo S, Ferrara G, Sanesi G (2024) Revitalization potential of marginal areas for sustainable rural development in the Puglia Region, Southern Italy: part I: a review. Agronomy 14(3):431. https://doi.org/10.3390/agronomy14030431

Allen W (2003) Plant blindness. Bioscience 53(10):926. https://doi.org/10.1641/0006-3568(2003)053[0926:pb]2.0.co;2

Amprazis A, Papadopoulou P (2020) Plant blindness: a faddish research interest or a substantive impediment to achieve sustainable development goals? Environ Educ Res 26(8):1065–1087. https://doi.org/10.1080/13504622.2020.1768225

Anguelovski I, Brand AL, Connolly JJT, Corbera E, Kotsila P et al (2020) Expanding the boundaries of justice in urban greening scholarship: toward an emancipatory, antisubordination, intersectional, and relational approach. Ann Am Assoc Geogr 110(6):1743–1769. https://doi.org/10.1080/24694452.2020.1740579

Aronson MFJ, La Sorte FA, Nilon CH, Katti M, Goddard MA et al (2014) A global analysis of the impacts of urbanization on bird and plant diversity reveals key anthropogenic drivers. Proc Royal Soc B Biol Sci 281(1780):20133330. https://doi.org/10.1098/rspb.2013.3330

Asfaw A, Lulekal E, Bekele T, Debella A, Tessema S et al (2023) Ethnobotanical study of wild edible plants and implications for food security. Trees For People 14:100453. https://doi.org/10.1016/j.tfp.2023.100453

Balick MJ, Cox PA (2020) Plants, people, and culture: the science of ethnobotany, 2nd edn. CRC Press, Taylor & Francis Group

Beatley T, Newman P (2013) Biophilic cities are sustainable, resilient cities. Sustainability 5(8):3328–3345. https://doi.org/10.3390/su5083328

Beddiar H, Boudiba S, Benahmed M, Tamfu AN, Ceylan Ö et al (2021) Chemical composition, anti-quorum sensing, enzyme inhibitory, and antioxidant properties of phenolic extracts of Clinopodium nepeta L. Kuntze. Plants 10(9):1955. https://doi.org/10.3390/plants10091955

Bellia G, Pieroni A (2015) Isolated, but transnational: the glocal nature of Waldensian ethnobotany, Western Alps, NW Italy. J Ethnobiol Ethnomed 11(1):37. https://doi.org/10.1186/s13002-015-0027-1

Bellusci F (2017) Le piante officinali del Parco Nazionale del Pollino [Master Thesis]. University of Pisa

Blue S, Hargiss CLM, Norland J, Dekeyser ES, Comeau P (2023) Plant blindness represents the loss of generational knowledge and cultural identity. Natl Sci Educ 52(1):e20106. https://doi.org/10.1002/nse2.20106

Bonaccio M, Di Castelnuovo A, Costanzo S, Gialluisi A, Persichillo M et al (2018) Mediterranean diet and mortality in the elderly: a prospective cohort study and a meta-analysis. Br J Nutr 120(8):841–854. https://doi.org/10.1017/S0007114518002179

Bonney R, Phillips TB, Ballard HL, Enck JW (2016) Can citizen science enhance public understanding of science? Public Underst Sci 25(1):2–16. https://doi.org/10.1177/0963662515607406

Bouteau F, Grésillon E, Chartier D, Arbelet-Bonnin D, Kawano T et al (2021) Our sisters the plants? Notes from phylogenetics and botany on plant kinship blindness. Plant Signal Behav 16(12):2004769

Brosse J (1991) Mitologia degli alberi, 0th edn, Rizzoli

Burke R, Sherwood OL, Clune S, Carroll R et al (2022) Botanical boom: a new opportunity to promote the public appreciation of botany. Plants People Planet 4(4):326–334. https://doi.org/10.1002/ppp3.10257

Caneva G (2010) Il codice botanico di Augusto. The Augustus botanical code. Gangemi Editore. https://iris.uniroma3.it/handle/11590/185896

Caneva G, Pieroni A, Guarrera P (eds) (2013) Etnobotanica: Conservazione di un patrimonio culturale come risorsa per uno sviluppo sostenibile. Edipuglia

Caneva G, Savo V, Kumbaric A (2014) Big messages in small details: nature in Roman archaeology. Econ Bot 68(1):109–115. https://doi.org/10.1007/s12231-013-9256-1

Chaachouay N, Benkhnigue O, Fadli M, El Ibaoui H, Zidane L (2019) Ethnobotanical and ethnopharmacological studies of medicinal and aromatic plants used in the treatment of metabolic diseases in the Moroccan Rif. Heliyon 5(10):e02191. https://doi.org/10.1016/j.heliyon.2019.e02191

Chaachouay N, Zidane L (2024) Plant-derived natural products: a source for drug discovery and development. Drugs Drug Candidates 3(1):184–207. https://doi.org/10.3390/ddc3010011

Chang B, Chen B, Chen W, Xu S, He X, Yao J, Huang Y (2024) Analysis of trade-off and synergy of ecosystem services and driving forces in urban agglomerations in Northern China. Ecol Indic 165:112210. https://doi.org/10.1016/j.ecolind.2024.112210

Cooper CB, Dickinson J, Phillips T, Bonney R (2007) Citizen science as a tool for conservation in residential ecosystems. Ecol Soc 12(2):11

Crall AW, Jarnevich CS, Young NE, Panke BJ, Renz M, Stohlgren TJ (2015) Citizen science contributes to our knowledge of invasive plant species distributions. Biol Invasions 17(8):2415–2427. https://doi.org/10.1007/s10530-015-0885-4

Crisci JV, Katinas L, Apodaca MJ, Hoch PC (2020) The end of botany. Trends Plant Sci 25(12):1173–1176. https://doi.org/10.1016/j.tplants.2020.09.012

Da Silva Siltori PF, Lourenzani WL, Satolo EG, Ferreira Caldana AC et al (2024) Training future managers to address the challenges of sustainable development: an innovative, interdisciplinary, and multiregional experience on corporate sustainability education. WORLD 5(2):155–172. https://doi.org/10.3390/world5020009

Day J (2013) Botany meets archaeology: people and plants in the past. J Exp Bot 64(18):5805–5816. https://doi.org/10.1093/jxb/ert068

De Rivas R, Vilches A, Mayoral O (2024) Secondary school students' perceptions and concerns on sustainability and climate change. Climate 12(2):17. https://doi.org/10.3390/cli12020017

Dickinson JL, Zuckerberg B, Bonter DN (2010) Citizen science as an ecological research tool: challenges and benefits. Annu Rev Ecol Evol Syst 41(1):149–172. https://doi.org/10.1146/annurev-ecolsys-102209-144636

Domingo-Fernández D, Gadiya Y, Mubeen S, Bollerman TJ, Healy MD et al (2023) Modern drug discovery using ethnobotany: a large-scale cross-cultural analysis of traditional medicine reveals common therapeutic uses. iScience 26(9):107729. https://doi.org/10.1016/j.isci.2023.107729

Eberbach C, Crowley K (2017) From seeing to observing: how parents and children learn to see science in a botanical garden. J Learn Sci 26(4):608–642

EEC (1992) Council directive 92/43/CEE of 21 May 1992 on the conservation of natural habitats and of wild fauna and flora. Off J L 206:7–50

Elmqvist T, Setälä H, Handel S, Van Der Ploeg S, Aronson J, Blignaut J et al (2015) Benefits of restoring ecosystem services in urban areas. Curr Opin Environ Sustain 14:101–108. https://doi.org/10.1016/j.cosust.2015.05.001

Fraisl D, Hager G, Bedessem B, Gold M, Hsing P-Y et al (2022) Citizen science in environmental and ecological sciences. Nat Rev Methods Primers 2(1):64. https://doi.org/10.1038/s43586-022-00144-4

Frazer JG (1890) The Golden bough: a study in comparative religion, vol 1. Macmillan and Co.

Fytopoulou E, Karasmanaki E, Tsantopoulos G (2024) Comparing the effect of scientific background on university students' environmental views. WORLD 5(1):58–78. https://doi.org/10.3390/world5010004

Gentile C, Spampinato G, Patti M, Astrid Laface VL, Musarella CM (2022) Contribution to the ethnobotanical knowledge of Serre Calabre (Southern Italy). Res J Ecol Environ Sci *2*(3):35–55. https://doi.org/10.31586/rjees.2022.389

Gilani AH, Rahman A (2005) Trends in ethnopharmacology. J Ethnopharmacol 100(1–2):43–49. https://doi.org/10.1016/j.jep.2005.06.001

González-Tejero MR, Casares-Porcel M, Sánchez-Rojas CP, Ramiro-Gutiérrez et al (2008) Medicinal plants in the Mediterranean area: synthesis of the results of the project Rubia. J Ethnopharmacol 116(2):341–357. https://doi.org/10.1016/j.jep.2007.11.045

Grivetti LE, Ogle BM (2000) Value of traditional foods in meeting macro- and micronutrient needs: the wild plant connection. Nutr Res Rev 13(1):31–46. https://doi.org/10.1079/095442200108728990

Guarrera PM, Savo V (2016) Wild food plants used in traditional vegetable mixtures in Italy. J Ethnopharmacol 185:202–234. https://doi.org/10.1016/j.jep.2016.02.050

Haklay M, Fraisl D, Greshake Tzovaras B, Hecker S, Gold M, Hager G et al (2021) Contours of citizen science: a vignette study. R Soc Open Sci 8(8):202108. https://doi.org/10.1098/rsos.202108

Harshberger JW (1896) The purposes of ethno-botany. Bot Gaz 21(3):146–154. https://doi.org/10.1086/327316

Heinrich M, Gibbons S (2001) Ethnopharmacology in drug discovery: an analysis of its role and potential contribution. J Pharm Pharmacol 53(4):425–432. https://doi.org/10.1211/0022357011775712

Irwin A (1995) Citizen science: a study of people, expertise, and sustainable development. Routledge

Jamal A (2023) Embracing nature's therapeutic potential: herbal medicine. Int J Multidiscip Sci Arts 2(3):117–126. https://doi.org/10.47709/ijmdsa.v2i1.2620

Kalle R, Sõukand R (2016) Current and remembered past uses of wild food plants in Saaremaa, Estonia: changes in the context of unlearning debt. Econ Bot 70(3):235–253. https://doi.org/10.1007/s12231-016-9355-x

Kandeler R, Ullrich WR (2008) Symbolism of plants: examples of European-Mediterranean culture presented with biology and history of art: general introduction. J Exp Bot 60(1):5–6. https://doi.org/10.1093/jxb/ern359

Kellert SR, Wilson EO (1993) The Biophilia hypothesis. Island Press/Shearwater Books

Kowarik I (2011) Novel urban ecosystems, biodiversity, and conservation. Environ Pollut 159(8–9):1974–1983. https://doi.org/10.1016/j.envpol.2011.02.022
Kruger LE, Shannon MA (2000) Getting to know ourselves and our places through participation in civic social assessment. Soc Nat Resour 13(5):461–478. https://doi.org/10.1080/089419200403866
Liu C, Zuo Z, Xu F, Wang Y (2023) Authentication of herbal medicines based on modern analytical technology combined with chemometrics approach: a review. Crit Rev Anal Chem 53(7):1393–1418. https://doi.org/10.1080/10408347.2021.2023460
Marcenò C, Padullés Cubino J, Chytrý M, Genduso E, Salemi D et al (2021) Facebook groups as citizen science tools for plant species monitoring. J Appl Ecol 58(10):2018–2028
Maruca G, Spampinato G, Turiano D, Laghetti G, Musarella CM (2019) Ethnobotanical notes about medicinal and useful plants of the Reventino Massif tradition (Calabria region, Southern Italy). Genet Resour Crop Evol 66(5):1027–1040. https://doi.org/10.1007/s10722-019-00768-8
Mattalia G, Corvo P, Pieroni A (2020) The virtues of being peripheral, recreational, and transnational: local wild food and medicinal plant knowledge in selected remote municipalities of Calabria, Southern Italy. Ethnobot Res Appl 19. https://doi.org/10.32859/era.19.38.1-20
Menale B, De Castro O, Cascone C, Muoio R (2016) Ethnobotanical investigation on medicinal plants in the Vesuvio National Park (Campania, Southern Italy). J Ethnopharmacol 192:320–349. https://doi.org/10.1016/j.jep.2016.07.049
Monroe MC, Krasny ME (2015) Across the Spectrum: resources for environmental educators. North American Ass. for Environmental Education
Musarella CM, Paglianiti I, Cano-Ortiz A, Spampinato G (2019) Indagine etnobotanica nel territorio del Poro e delle Preserre Calabresi (Vibo Valentia, S-Italia). Atti Soc Tosc Sci Nat Res Pisa Mem Ser B 126:13–28. https://doi.org/10.2424/ASTSN.M.2018.17
Orlandi F, Rojo J, Picornell A, Oteros J, Pérez-Badia R, Fornaciari M (2020) Impact of climate change on olive crop production in Italy. Atmos 11(6):595. https://doi.org/10.3390/atmos11060595
Palacios J, Amud N, Mendoza D (2016) Implementación de huertas escolares como estrategia de enseñanza-aprendizaje de la biología. Thesis magistral, Univ Pont Bol, Medellín, pp 1–110. https://repository.upb.edu.co/bitstream/handle/20.500.11912/2950/T.G.%20JULIO%20%C3%89DINSON%20PALACIOS%20Y%20OTROS.pdf
Pany P, Meier FD, Dünser B, Yanagida T, Kiehn M, Möller A (2022) Measuring students' plant awareness: a prerequisite for effective botany education. J Biol Educ 58(5):1103–1116. https://doi.org/10.1080/00219266.2022.2159491
Pardo-de-Santayana M, Tardío J, Blanco E, Carvalho AM, Lastra JJ et al (2007) Traditional knowledge of wild edible plants used in the northwest of the Iberian Peninsula (Spain and Portugal): a comparative study. J Ethnobiol Ethnomed 3(1):27. https://doi.org/10.1186/1746-4269-3-27
Parsley KM (2020) Plant awareness disparity: a case for renaming plant blindness. Plants People Planet 2(6):598–601. https://doi.org/10.1002/ppp3.10153
Patti M, Giuffrè AM, Musarella CM, Spampinato G (2026) A comparative morphometric and phytochemical analysis of the Calabrian endemic treasure Crepis aspromontana Brullo, Scelsi & Spamp. With alimentary potential. Genet Resour Crop Evol 73(1):35. https://doi.org/10.1007/s10722-025-02632-4
Patti M, Musarella CM, Laface VLA, Spampinato G (2025a) Ethnobotanical survey in the Graecanic Area of Reggio Calabria (Southern Italy): A treasure chest of biodiversity and traditions at risk of extinction. Ethnobot Res Appl 30. https://doi.org/10.32859/era.30.13.1-29
Patti M, Musarella CM, Spampinato G (2025b) Ethnobotanical knowledge in Calabria (southern Italy): a summary review. Heliyon 11(2):e42050. https://doi.org/10.1016/j.heliyon.2025.e42050
Perrino EV (2022) Ancient and modern grains: effects on human health: a first short review. Res J Ecol Environ Sci 2(1):21–25. https://doi.org/10.31586/rjees.2022.225
Pieroni A, Cattero V (2019) Wild vegetables do not lie: comparative gastronomic ethnobotany and ethnolinguistics on the Greek traces of the Mediterranean Diet of southeastern Italy. Acta Bot Bras 33(2):198–211. https://doi.org/10.1590/0102-33062018abb0323

Pieroni A, Quave CL (2005) Traditional pharmacopoeias and medicines among Albanians and Italians in southern Italy: a comparison. J Ethnopharmacol 101(1–3):258–270. https://doi.org/10.1016/j.jep.2005.04.028

Pinto G, Prolongo M, Alonzo J, Díaz I, Carla O, Díaz F (2017) Fomento del aprendizaje experiencial de la química: estudio del caso de un proyecto de innovación educativa. ALDEQ 32:95–100. https://www.researchgate.net/publication/324208622_Fomento_del_Aprendizaje_Experiencial_de_la_Quimica_Estudio_del_Caso_de_un_Proyecto_de_Innovacion_Educativa

Plieninger T, Abunnasr Y, D'Ambrosio U, Guo T, Kizos T, Kmoch L, Topp E, Varela E (2023) Biocultural conservation systems in the Mediterranean region: the role of values, rules, and knowledge. Sustain Sci 18(2):823–838. https://doi.org/10.1007/s11625-022-01155-6

Prance G (2007) Ethnobotany, the science of survival: a declaration from Kaua'i. Econ Bot 61(1):1–2. https://doi.org/10.1663/0013-0001(2007)61[1:ETSOSA]2.0.CO;2

Reyes-García V, Huanca T, Vadez V, Leonard W, Wilkie D (2006) Cultural, practical, and economic value of wild plants: a quantitative study in the Bolivian Amazon. Econ Bot 60(1):62–74

Romojaro A, Botella MÁ, Obón C, Pretel MT (2013) Nutritional and antioxidant properties of wild edible plants and their use as potential ingredients in the modern diet. Int J Food Sci Nutr 64(8):944–952. https://doi.org/10.3109/09637486.2013.821695

Sánchez-Mata MDC, Matallana-González MC, Morales P (2016) The contribution of wild plants to dietary intakes of micronutrients (I): vitamins. In: Sánchez-Mata MDC, Tardío J (eds) Mediterranean wild edible plants. Springer, pp 111–139. https://doi.org/10.1007/978-1-4939-3329-7_6

Samper A, Ramírez A (2014) Diseño de una propuesta pedagógica de educación para la seguridad vial estructurada bajo el modelo de aprendizaje experiencial, Bogotá. http://hdl.handle.net/10656/2918

Schussler EE, Olzak LA (2008) It's not easy being green: student recall of plant and animal images. J Biol Educ 42(3):112–119. https://doi.org/10.1080/00219266.2008.9656-123

Shanahan EA, Raile ED, Naughton HT, Wallner MP, Houghton KA (2020) Public opinion about management strategies for a low-profile species across multiple jurisdictions: whitebark pine in the northern Rockies. People and Nature 2(3):784–796. https://doi.org/10.1002/pan3.10094

Silvertown J (2009) A new dawn for citizen science. Trends Ecol Evol 24(9):467–471. https://doi.org/10.1016/j.tree.2009.03.017

Soga M, Gaston KJ (2016) Extinction of experience: the loss of human–nature interactions. Front Ecol Environ 14(2):94–101. https://doi.org/10.1002/fee.1225

Spampinato G, Crisarà R, Cannavò S, Musarella C (2017) I fitotoponimi della Calabria meridionale: Uno strumento per l'analisi del paesaggio e delle sue trasformazioni. Atti Soc Tosc Sci Nat Mem Ser B 124:61–72

Sujarwo W, Arinasa IBK, Salomone F, Caneva G et al (2014) Cultural erosion of Balinese indigenous knowledge of food and nutraceutical plants. Econ Bot 68(4):426–437. https://doi.org/10.1007/s12231-014-9288-1

Süntar I (2020) Importance of ethnopharmacological studies in drug discovery: role of medicinal plants. Phytochem Rev 19(5):1199–1209. https://doi.org/10.1007/s11101-019-09629-9

Tacca MCVR (2021) Theory of subjectivity and learning: possibilities and perspectives. In: Goulart DM, Martínez AM, Adams M (eds) Theory of subjectivity from a cultural-historical standpoint, vol 9. Springer, pp 189–202. https://doi.org/10.1007/978-981-16-1417-0_12

Tardío J, Pardo-de-Santayana M (2008) Cultural importance indices: a comparative analysis based on the useful wild plants of Southern Cantabria (Northern Spain). Econ Bot 62(1):24–39. https://doi.org/10.1007/s12231-007-9004-5

The Local Food-Nutraceuticals Consortium (2005) Understanding local Mediterranean diets: a multidisciplinary pharmacological and ethnobotanical approach. Pharm Res 52(4):353–366. https://doi.org/10.1016/j.phrs.2005.06.005

Thomas H, Ougham H, Sanders D (2022) Plant blindness and sustainability. Int J Sustain High Educ 23(1):41–57. https://doi.org/10.1108/IJSHE-09-2020-0335

Vitalini S, Iriti M, Puricelli C, Ciuchi D, Segale A, Fico G (2013) Traditional knowledge on medicinal and food plants used in Val San Giacomo (Sondrio, Italy)—an alpine ethnobotanical study. J Ethnopharmacol 145(2):517–529. https://doi.org/10.1016/j.jep.2012.11.024

Vohland K, Land-Zandstra A, Ceccaroni L, Lemmens R, Perelló J, Ponti M, Samson R, Wagenknecht K (eds) (2021) The science of citizen science. Spring. https://doi.org/10.1007/978-3-030-58278-4

Wandersee JH, Schussler EE (1999) Preventing plant blindness. Am Biol Teach 61(2):82–86. https://doi.org/10.2307/4450624

Wani ZA, Akhter F, Ridwan Q, Rawat YS, Ahmad Z et al (2023) A bibliometric analysis of studies on plant endemism during the period of 1991–2022. J Zool Bot Gard 4:692–710. https://doi.org/10.3390/jzbg4040049

Westwood M, Cavender N, Meyer A, Smith P (2020) Botanic garden solutions to the plant extinction crisis. Plants People Planet 3(1):22–32. https://doi.org/10.1002/ppp3.10134

Woodland DW (2007) Are botanists becoming the dinosaurs of biology in the 21st century? S Afr J Bot 73(3):343–346. https://doi.org/10.1016/j.sajb.2007.03.005

Yorek N, Şahin M, Aydın H (2009) Are animals 'more alive' than plants? Animistic-anthropocentric construction of life concept. EURASIA J Math Sci Technol Educ 5(4). https://doi.org/10.12973/ejmste/75287

Chapter 5
Plants in the View of Ecosystem Services and Their Role in Nature-based Solutions

Fabio Orlandi, Francesca Bretzel, Livia Bonciarelli, Nicoletta Cannone, Simona Ceschin, Luca D'Amato, Desirée Muscas, Marco Fornaciari, Valerio Lazzeri, Francesco Malfasi, Valentina Savo, and Andrea Scartazza

5.1 Nature-based Solutions and Ecosystem Services and Disservices (V.S.)

Adopting an ecosystem services-based approach in urban design provides several benefits to citizens, and Nature-based Solutions (NbSs) often help realize these benefits (Elmqvist et al. 2015; Lafortezza et al. 2018). These benefits encompass social and environmental domains and can become cost-effective or even economically advantageous (Elmqvist et al. 2015; Lafortezza et al. 2018).

However, actions must balance the trade-offs between costs and benefits, rather than focusing solely on one type of domain (i.e., purely societal benefits without accounting for environmental benefits) (Sowińska-Świerkosz and García 2022). In this section, we will initially define these concepts and approaches and then provide examples and applications.

F. Orlandi (✉) · L. Bonciarelli · D. Muscas · M. Fornaciari
Department of Civil and Environmental Engineering, University of Perugia, Perugia, Italy
e-mail: fabio.orlandi@unipg.it

F. Bretzel · V. Lazzeri · A. Scartazza
National Research Council (CNR), Research Institute on Terrestrial Ecosystems (IRET), Pisa, Italy

N. Cannone
Department of Theoretical and Applied Sciences, University of Insubria, Varese, Italy

F. Malfasi
Department of Science and High Technology, University of Insubria, Como, Italy

S. Ceschin · L. D'Amato
Department of Sciences, Roma Tre University, Rome, Italy

V. Savo
Department of Education Sciences, Roma Tre University, Rome, Italy

G. Caneva et al. (eds.), *Urban Botany*, Future City 22,
https://doi.org/10.1007/978-3-032-28986-5_5

5.1.1 Nature-based Solutions

The concept of 'Nature-based Solution' (NbS) was introduced by the World Bank in 2008 (Sowińska-Świerkosz and García 2022) to outline adaptation strategies to climate change inspired by nature. In 2015, the European Commission has defined NbS as "*living solutions inspired by, continuously supported by and using nature, which are designed to address various societal challenges in a resource-efficient and adaptable manner and to provide simultaneously economic, social and environmental benefits*" (Lafortezza et al. 2018). More recently, the European Commission has reformulated the concept of NbS as "*new ways to approach socio-ecological adaptation and resilience, with equal reliance upon social, environmental and economic domains*" (Dumitru and Wendling 2021).

Since 2008, the concept and its application have received considerable attention from researchers and policymakers; however, it remains unclear what strategies can be classified as NbS and why (Sowińska-Świerkosz and García 2022). To provide global standards for the classification and identification of NbSs the IUCN has set eight criteria that actions must fulfill to be considered as a NbS: "*(1) address societal challenges; (2) landscape scale of intervention; (3) biodiversity gain; (4) economic viability; (5) governance capability; (6) equitably balance trade-offs; (7) adaptive management; (8) mainstreamed within an appropriate jurisdictional context*" (Sowińska-Świerkosz and García 2022).

Most literature on NbSs focuses on adaptation to climate change and urban sustainability (Sowińska-Świerkosz and García 2022), often overlapping in these areas. Indeed, urban environments are ideal settings for developing NbSs to increase the sustainability and livability of cities while also enhancing urban biodiversity, mitigating the heat island effect, reducing pollution, and so on (Pineda-Pinto et al. 2022). In fact, NbSs often provide multiple ecosystem services (Pineda-Pinto et al. 2022) in the long term (Lafortezza et al. 2018). In urban contexts, a NbS approach should have four main goals (Laforteza et al. 2018): (1) improve livability of cities for citizens and, at the same time, improve the environment (2) restore degraded urban lands to increase their ecological functionality and resilience and their provision of ecosystem services; (3) use NbSs to increase climate change adaptability and resilience; (4) leverage on the NbS multifunctionality to reduce multiple risks and provide multiple benefits (Fig. 5.1).

5.1.1.1 Nature-based Solutions in Practical Applications

Based on the four goals listed above, we provide four examples of NbSs in urban contexts.

1. <u>*Create community gardens on abandoned land*</u>. Community gardens are open spaces often managed by citizens or associations to cultivate vegetables, fruits, or flowers. These gardens offer multiple benefits, ranging from enhanced community cohesion and vitality to increased urban biodiversity. More specifically,

Fig. 5.1 Urban Parks at Aasee in Münster (North Germany), showing a strict relation among Art and Nature and multi-benefits for the citizens. (Photo by G. Caneva)

they provide food or medicinal plants, support urban biodiversity and maintain genetic diversity, improve soil conditions, mitigate urban heat and promote social cohesion (Caneva et al. 2020).

2. *Requalify urban brownfields by creating green spaces.* Brownfields are sites previously used for industrial or commercial purposes that are currently abandoned and may be contaminated by pollutants. This type of requalification provides multiple benefits, from economic to social and environmental positive effects (Atkinson et al. 2014). The environmental benefits include sequestration of carbon dioxide and pollutants, climate mitigation, stormwater regulation, noise attenuation, and restoration of natural habitats (Song et al. 2019). They also enhance the aesthetic appeal of cities and property values; they can improve citizens' well-being by providing places to relax, exercise, and socialize (Atkinson et al. 2014; Song et al. 2019).
3. *Implement green roof installations for climate change mitigation and adaptation*. Green roofs, also named "eco-roofs", "living roofs" or "roof gardens", are roofs with a vegetated surface. Green roofs can improve the thermal performance of buildings mostly by enhancing insulation and shading (Nguyen et al. 2021). Green roofs can also improve air quality by increasing green surfaces and can ultimately reduce the heat island effect. Depending on the layer's thickness, they can have noise-reduction properties, mitigate water runoff, and reduce stress on urban hydrology during heavy rainfall events. Green roofs enhance a building's appearance and beautify the urban landscape. Exposure to green spaces can have positive effects on both human mental health (i.e., restoration) and physical health (i.e., by reducing air and noise pollution). Green roofs can provide a habitat for urban wildlife, increasing the biodiversity of cities (Nguyen et al. 2021).
4. *Develop rain gardens for stormwater regulation and purification*. Rain gardens are depressions filled with coarse, porous material that can store and slowly release stormwater, and accommodate a vegetation layer. They can improve stormwater management, help remediate pollution, enhance nutrient cycling,

and reduce stormwater treatment costs. Ultimately, they possess a pleasing aesthetic appeal and can also enhance urban biodiversity (Ishimatsu et al. 2017).

5.1.2 Ecosystem Services and Disservices and a New Paradigm in Urban Design

It has been argued that the NbS concept embraces those of ecosystem-based adaptation, green infrastructure (GI), and ecosystem services. In the urban environment, green infrastructure includes public and private parks, community gardens, green roofs, and street vegetation, all of which provide crucial ecological functions and social and economic benefits. These benefits provided by nature are defined as ecosystem services (Costanza et al. 1997).

The ecosystem services specific to urban ecosystems are described in the TEEB report (TEEB 2011) and grouped into four main categories: Provisioning Services, Regulating Services, Supporting Services, and Cultural Services. The list includes the 17 ecosystem services considered by the TEEB report; to this list, we added noise reduction, as suggested by Gómez-Baggethun and Barton (2013), since this is a relevant ecosystem service provided by several urban green spaces (Table 5.1, Fig. 5.2).

Ecosystem services are often impacted or reduced in urban environments. One way to reverse this negative process is to create or redesign urban areas to provide or support ecosystem services (Pedersen Zari 2015), while simultaneously

Table 5.1 Urban Ecosystem Services. General description and contextualization in urban ecosystems (TEEB 2011; Gómez-Baggethun and Barton 2013)

PROVISIONING SERVICES: The benefits that pertain food and materials provided by ecosystems.	
Food	Urban green spaces such as community gardens, allotments, backyards, and green roofs can support local food production
Raw materials	Although limited in cities, some innovations using plants show how urban ecosystems can contribute to producing energy or materials
Fresh water	UGI helps store, filter, and regulate water, reducing the negative impacts of impermeable surfaces on the hydrological cycle
Medicinal resources	Urban residents often grow medicinal plants in gardens or balconies, sometimes preserving cultural traditions of migrant communities
REGULATING SERVICES: The benefits that pertain the regulation of ecosystem functions.	
Local climate & air quality regulation	Trees and vegetation cool urban areas, moderate rainfall runoff, and filter air pollutants
Carbon sequestration and storage	Mainly through trees, which absorb and store atmospheric CO_2 depending on growth rate, size, and leaf persistence
Moderation of extreme events	Green areas reduce urban heat-island effects and can mitigate damage from heatwaves or heavy rain
Waste-water treatment	Wetlands and aquatic plants can filter pollutants recycling nutrients

(continued)

Table 5.1 (continued)

Erosion prevention & soil fertility	Vegetation stabilizes soil, reduces erosion and landslide risks, and maintains organic matter
Pollination	Urban habitats that support bees and other pollinators contribute to plant reproduction
Biological control	Predators and parasites regulate pests; vegetation complexity in gardens enhances natural pest control
Noise reduction	Trees and shrubs act as barriers that reduce traffic and urban noise
SUPPORTING SERVICES: Ecosystems provide habitat for living beings and maintain their diversity.	
Habitats for species	UGI provides food, shelter, and breeding sites. Habitat quality depends on size, shape, connectivity, human pressure, and management
Maintenance of genetic diversity	Connectivity among urban green patches allows gene flow between populations of plants and animals
CULTURAL SERVICES: The non-material benefits people obtain from ecosystems.	
Recreation & health	Parks, green roofs, and walking paths support physical activity, relaxation, and mental well-being
Tourism	Urban green areas and iconic parks attract visitors and generate economic benefits
Aesthetic & artistic inspiration	Natural elements inspire art, architecture, and design
Spiritual experience & sense of place	Urban nature can hold symbolic, cultural, or spiritual meaning, contributing to belonging and identity

mitigating or removing ecosystem disservices (EDSs). Ecosystem disservices are negative effects (nuisances or even harm) that species or ecosystems can cause on humans (von Döhren and Haase 2019). The ecosystem dis-services that pertain to urban ecosystems are showed in Fig. 5.3.

5.2 Plants and Soil (F.B., A.S.)

5.2.1 Introduction

To ensure effective greening solutions, it is crucial to monitor and manage plant–soil interactions in urban environments during design and maintenance. These strategies promote soil health, plant growth, and ecosystem functionality by conserving soil properties. When human impact occurs, interventions such as improving soil structure, adding nutrients, controlling erosion, and using plants to restore degraded soils may be required (Minixhofer and Stangl 2021). Understanding soil–plant–microbiota interactions is essential for assessing soil health and its influence on urban vegetation (Giovannetti et al. 2023). Plants rely on soil for nutrients, water, and support, while root exudates and organic matter maintain soil structure and fertility (Fig. 5.4).

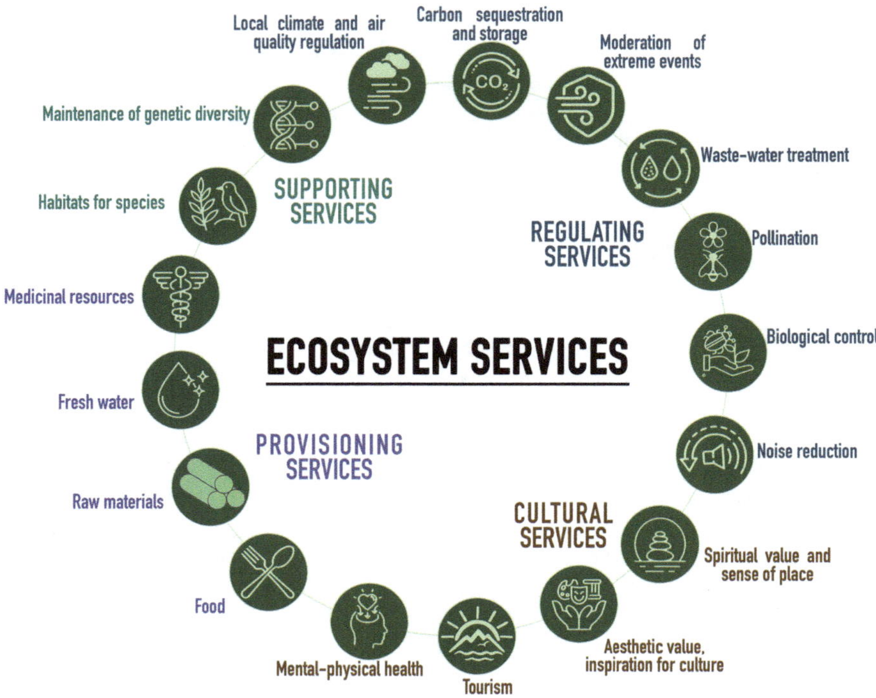

Fig. 5.2 Overview of the four categories of ecosystem services. Illustrating the diverse benefits that ecosystems provide to humans (Elaboration by F. Bartoli)

Soil microbiome activity forms stable aggregates that enhance carbon storage and water infiltration, and regulate nutrient cycling and air/water movement, improving plant resilience to stress.

Green infrastructure and soil conservation reduce degradation, sequester carbon, improve water quality, limit runoff, mitigate heat islands, enhance air quality, and provide psychological benefits, creating resilient and pleasant green urban spaces (Minixhofer and Stangl 2021).

Soil also interacts with the atmosphere, influencing air composition directly through dust and gaseous emissions from microbial and chemical processes, and indirectly by supporting plants that intercept particles and absorb pollutants (Giltrap et al. 2021).

However, soil compaction and alien materials (waste, rubble, contaminants) disrupt biogeochemical cycles and plant growth, impairing ecosystem services dependent on a healthy soil–plant system. Urban soils differ from natural ones due to human impact: modified profiles, altered structure, alkaline pH, compaction, and contamination. They often lack well-defined horizons, show mixed layers of topsoil and compacted subsoil, and alkaline pH from building materials (Fig. 5.5) (Macci et al. 2025).

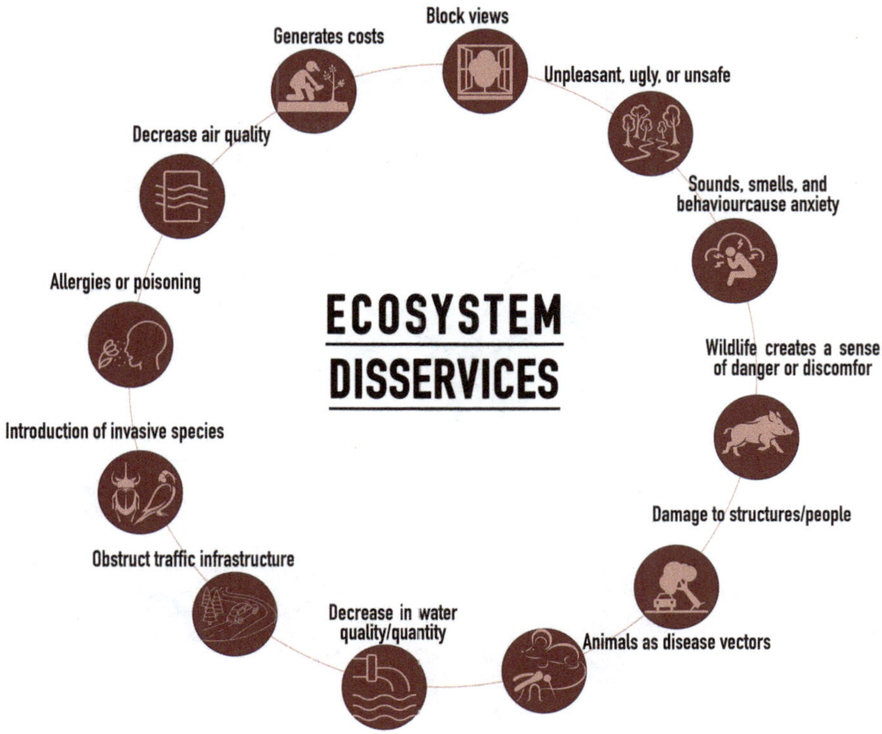

Fig. 5.3 Urban ecosystem disservices (Elaboration by F. Bartoli)

Soil compaction reduces pore space, limiting water and air flow, while rubble, alien materials and services restrict root expansion, threatening tree stability and causing falls (Craul 1992).

Urban soil heterogeneity, poor quality soil and harsh conditions limit suitable plant taxa, yet stress-adapted species thrive where some ornamentals fail. Herbaceous vegetation rich in therophytes, geophytes, and hemicryptophytes can form meadows that protect soil and enhance biodiversity (Bretzel et al. 2024). Leguminosae, as pioneer plants, efficiently colonize gaps: *Trifolium*, *Medicago*, *Lotus*, and other genera of the family enrich lawns, adding color and pollinator resources. Other colonizers of degraded urban soils include therophytes. Research on tree species focuses on tolerance to pollution and multi-stress conditions (drought, heat, compaction), species ability to mitigate air and soil pollution (Moura et al. 2025), and the effectiveness of multi species stands to improve soil organic matter.

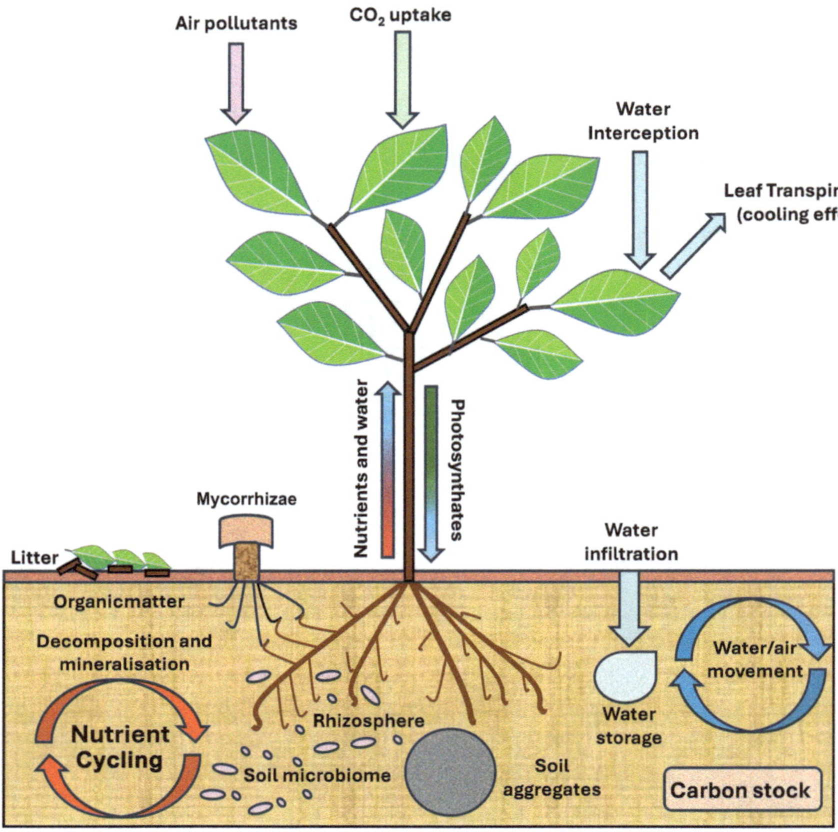

Fig. 5.4 The complex interplay between plant and soil in an urban environment. Plants contribute to CO_2 and air pollutants removal from the atmosphere and regulate water cycle and air temperature; rely on soil for essential nutrients, water and physical support, contribute to maintain soil structure and fertility, regulating nutrient cycling and water/air movement, supporting soil microbiome and promoting carbon stock, water infiltration/storage and soil aggregate formation (Elaboration by A. Scartazza)

5.2.2 The Role of Plants for Promoting Soil Health and Functionality in Urban Environment

Urban plants significantly improve soil physical and chemical properties. Roots penetrate and modify soil structure, enhancing porosity, aeration, and drainage. Ornamental species, depending on root systems, stabilize soil and slopes, reducing erosion. Root exudates bind particles, preventing compaction, while microbial decomposition releases organic matter that forms aggregates (Francini et al. 2021). The chemical and physical quality of plant residues determines decomposition rate, influencing aggregate stability and carbon and nitrogen cycles (Minixhofer and Stangl 2021). Thus, urban plants regulate nutrient cycling and hydrological

Fig. 5.5 Urban soil profiles in different cities, the absence of a defined layering and the materials not belonging to soil (bricks and rubble) are evident (Photos by R. Pini and L. Gardin)

dynamics. Green infrastructure can mimic natural hydrological processes, improving water retention, reducing runoff and erosion, and enhancing infiltration and sustainable drainage systems (McGrane 2016).

Urbanization also alters microbial diversity, a key soil health indicator. Plants and soil biota interact beneficially, but urbanization disrupts these relationships, potentially causing negative consequences for both plants and soil ecosystem. Understanding them is vital for managing technosol microbiomes to boost plant health and biodiversity (Zhao et al. 2013). Urbanization strongly affects soil biota richness: protists and meso/macrofauna are richer in urban than suburban soils (Yao et al. 2023), while bacterial and fungal richness shows little variation. Functionally, urbanization increases bacterivores, plant pathogens, and animal parasites. Soil biota also responds to plant functional types and park age: older deciduous trees increase microbes and earthworms (Francini et al. 2018). Plant types influence the quantity and quality of organic inputs via litter and root exudates, shaping microbial composition. Overall, targeted vegetation management can mitigate urbanization's negative effects on soil biodiversity and ecosystem services.

5.2.3 *The Role of Soil in Promoting Plant Health in Urban Environments*

Urban soils support plant growth, playing a crucial role in ecosystem function. Maintaining soil multifunctionality in an urban environment is essential for plant health, as it promotes nutrient supply, water retention, physical support, and microbial habitat. However, degraded urban soils are often compacted, contaminated, and nutrient-deficient, hindering plant growth and overall ecosystem health. The benefits of improved soil stem from its ability to support more vigorous plant growth, fostering the potential ecosystem services it can provide. In this approach, "good" soils are characterized by a diverse and balanced microbiome, optimal physical and chemical properties, and adequate nutrient availability and water/air circulation; "bad" soils exhibit a dysbiotic microbiome, poor nutrient availability, and suboptimal physical and chemical properties, negatively impacting plant health and physiological performance (Giovannetti et al. 2023). During heavy works, construction site operations damage soil and destroy the vegetation with the fertile layer (Fig. 5.6); to avoid soil degradation and plants' damage, it is useful to protect soil with structures (wooden boards or mats) and plant roots with jute or other materials kept humid, especially in summer.

The urban soil seed bank plays a role in vegetation regeneration after human damage to soil and plants. The soil seed bank is rich in early successional species, herbaceous pioneer plants, which quickly germinate and are then replaced by long-lived species (Albrecht et al. 2011). Therophytes are stimulated by soil disturbance, which breaks the seed dormancy, providing quick soil coverage. Subsequently,

Fig. 5.6 Construction sites are often placed on green areas, compromising soil quality and plant life (Photo by F. Bretzel)

stress-adapted perennials tend to take over in poor soils, ruderals in productive soils (Bretzel et al. 2024).

5.2.4 *Strategies for the Improvement of the Soil-Plant System in Urban Environments*

Soil policy and urban forestry are intrinsically linked because the policies, aimed at improving soil quality, are essential for creating a healthy urban environment that can support thriving urban forests, while urban forests contribute to soil health and provide numerous ecosystem services that benefit both people and the environment (Espinosa-Fuentes et al. 2023). Soil protection initiatives in cities implement greening benefits and sustain plants, microbiomes, and human health. These actions promote de-sealing and soil permeability, reducing runoff and preserving free soil. Strategies to improve fertility, structure, and water retention include adding compost and biosolids, as well as the use of constructed soils (Li et al. 2018). Furthermore, soil carbon stock depends on urban forest type, diversity, and structure, with higher soil organic carbon observed in coniferous stands compared to broadleaf stands. A meta-analysis examining the role of soil in enhancing urban resilience and biodiversity found that among potential urban vegetation strategies, wildflower meadows and community gardens can play a fundamental role in mitigating soil degradation and enhancing ecological stability (Halecki and Gąsiorek 2025). Urban lawns and other herbaceous vegetation help prevent soil erosion and compaction, reduce summer soil temperatures, and facilitate water penetration, thereby reducing the risk of flooding. Urban lawns, if mowing is reduced, are efficient at capturing pollutants and particulate matter (Przybysz et al. 2021) and also provide habitat for small fauna and pollinators. The soil underneath is also protected by excess trampling. Sometimes, good practices for safeguarding soil in urban areas are insufficient or unfeasible. If the soil is really of bad quality or contaminated, plants can be grown in raised beds to avoid pollutants and crop failure, especially for edibles. Furthermore, although soil rehabilitation in urban contexts has the potential to improve soil quality, the long-term effectiveness and impacts on tree growth remain poorly documented. Investigating soil–plant interactions in the urban environment is essential for understanding soil's capacity to support healthy plants and ecosystem services. Novel indicators capturing these complex interactions are needed to assess urban forestry functionality. To this end, plant health traits have been proposed as effective proxies for key physical, chemical, and microbiological soil features (Giovannetti et al. 2023). Furthermore, suitable soil–plant system indicators linked to nutrient cycles, such as ecological stoichiometry, enzyme activity, and carbon/nitrogen isotopes, have been proposed as valuable tools for planning and assessing the effectiveness of urban forestry interventions (Macci et al. 2025). This integrated approach, applied along urbanization gradients, showed that isotope signatures and enzymatic stoichiometry are reliable indicators of soil carbon and nutrient cycles, as well as microbial energy and nutrient demands. By tracking these indicators, urban planners and environmental managers can evaluate the success of greening initiatives and better understand how they contribute to a healthier, more resilient urban ecosystem.

The challenges that soil presents in urban areas, compaction, pollution, and loss of organic matter, can be faced by implementing Nature-based Solutions (NbS). All the following NbS are based on the conservation, use and protection of urban soil: green parking lots, vegetated grid paver, green tram tracks, community gardens, green cemeteries, urban forest, infiltration basin, pocket parks, tiny forests, tree pits, urban meadows, urban orchards, filter strips, soil de-sealing, spontaneous revegetation, soil fertility improvement (adding of compost and other bio-amendments), which help to maintain the functions and ecosystem services provided by the healthy soil in urban sites. On the other hand, raised beds, green roofs, green walls, mobile vertical greening, planters, engineered soils, rain gardens, and vertical forest are solutions to the loss of soil caused by widespread urban sealing. The second group of NbS employs techniques to reconstruct soil with substrates and waste materials, mineral and organic. All the solutions mentioned are in line with the conservation of biodiversity and the circular economy.

5.3 Plants and Water (Urban River and Water Bodies, Water Filtration) (S.C., L.B., M.F.)

5.3.1 Urban Rivers

Rivers are rich and complex ecosystems that can provide multiple ecosystem services, especially in urban settings (Everard and Moggridge 2012). In fact, urban rivers play important ecological roles, such as providing different habitats within a highly anthropic matrix, supplying water-related resources and transportation routes, increasing urban landscape diversity and providing recreational space for citizens (Baschak and Brown 1995). Rivers also act as ecological corridors, reducing isolation effects, promoting species colonisation dynamics and facilitating intraspecific exchanges, often playing a key role in the conservation of urban biodiversity (Johansson et al. 1996). The presence of river ecosystems within an urban context significantly enhances the floristic richness of the city they traverse. These ecosystems are characterized by a rich and distinctive flora associated with aquatic and riparian habitats, often supplemented by that from adjacent environments (Ceschin and Salerno 2021).

Anthropic activities often significantly alter the structure and function of urban river landscapes, negatively affecting their sustainability, ecological role, naturalness, and biodiversity (Baschak and Brown 1995). The most common alterations to natural river conditions in urban ecosystems include the construction of dams and various infrastructures, hydro-morphological modifications of watercourses, the replacement of natural banks with artificial embankments, water pollution, and illegal waste dumping (Nilsson and Berggren 2000). These anthropic pressures can cause significant changes in rivers' biodiversity and ecological functionality, through modifications in community composition, introduction of alien species (i.e., biological pollution), and loss of riparian habitats (Battisti et al. 2016).

5.3.2 Urban River and "River Contracts"

Considering the central ecological and landscape role of river ecosystems, during the second World Water Council forum (The Hague, 2000), the so-called "River Contracts" were officially introduced in Europe. River contracts are participatory agreements closely related to several European water and landscape regulations (e.g., Directive 2000/60/EC, the Landscape Convention), aimed at supporting the integrated planning and management of rivers and their complexity (Rouillard et al. 2018).

As part of the implementation of River Contracts in Italy, the Tiber River has recently been the subject of a River Contract, which also includes its stretch within the city of Rome. This River Contract, which also focuses on ensuring a proper planning and management of the urban river ecosystem from both an environmental and social perspective, must consider a range of actions, including: characterization of river biodiversity, restoration of deteriorated riparian habitats, recovery and enhancement of degraded areas along the river for public use, promotion of environmental education activities, development of historical-environmental projects, aimed at bringing citizens closer to the river ecosystem (Caneva et al. 2021). The presence of a river in a city, such as the Tiber within Rome, can be of great social importance, providing opportunities for citizens to engage in sporting activities, environmental education, and historical and natural events. In this sense, the river provides a real service to citizens and other users of the river ecosystem. Interest in a "social" restoration of rivers, especially in urban contexts, has been growing in recent years in many European cities (e.g., Stockholm, Vienna, London), where "social" benefits such as physical and mental well-being, recreation, social cohesion, and enhanced aesthetic, educational, and cultural values are prominent (Everard and Moggridge 2012).

In Italy, a recent example of environmental recovery of degraded riparian habitats in an urban context is represented by the "five parks overlooking the Tiber River", realized on the occasion of the Jubilee Year 2025 along the urban stretch of the Tiber River within Rome. The project was mainly based on an environmental requalification policy combining the recovery of the river environment from the ecological (re-naturalization activities, riverbank reconstitution, eradication of invasive plants, plantation of native bank species) and social perspective (creation of recreational areas with nature trails, cycling, pedestrian paths for citizens to closely enjoy the river environment). Another successful restoration example was the creation of the "Parco Tevere Magliana" (Rome) in 2014, aimed at recovering the river landscape and riverine biodiversity by encouraging the reinstatement of the original landscape. The Tiber riverbanks were turned into accessible green spaces through interventions in the most deteriorated areas, whereas the remaining floodplain was left in a semi-natural state. Further interventions, encouraging visitors to discover these newly restored areas and infrastructures, were installed, including stairs and ramps, a cycle route bank, an amphitheater, outdoor sports equipment, and a phytoremediation demonstration system (Roma Capitale, Progetti partecipati).

5.3.3 *The Urban Stretch of the Tiber Within Rome: A Botanical Case Study*

The final stretch of the Tiber River, which crosses the metropolitan area of Rome from north to south (Fig. 5.7), plays a crucial ecological role as biological corridor. It guarantees physical and functional connectivity between suburban and central urban areas, thereby supporting the enrichment and conservation of urban biodiversity (Ceschin et al. 2010a).

Indeed, the urban flora of the Tiber, comprising over 130 typical riverine species, contributes significantly to the city's spontaneous vascular flora, enriching and diversifying it (Ceschin and Salerno 2021). Of particular importance are the riparian tree populations with *Salix alba*, *Populus alba* and *Alnus glutinosa,* which can still be observed in the suburban sectors of the Tiber. However, comparing the current Tiber flora with that collected in previous studies, it emerges that richness has by more than 100 species, and its composition has partially changed over time (Ceschin et al. 2010a). Indeed, some species ecologically linked to good water quality (*Vallisneria spiralis*, *Hydrocharis morsus-ranae*, *Potamogeton pusillus*) or with riparian habitats (*Eleocharis palustris*, *Catabrosa aquatica*, *Cyperus badius*) have disappeared, while species tolerating water trophic pollution (*Myriophyllum spicatum*, *Sparganium erectum*), as well as ruderal, cosmopolitan, and alien species (*Acer negundo*, *Amorpha fruticosa*), were observed with increased frequency. The rise of these species suggests that anthropogenic pressures and the resulting environmental instability have intensified along this urban stretch of the river in recent decades, as anthropic disturbance generally limits native species while favoring tolerant, cosmopolitan, and alien species (Ceschin et al. 2010a).

Based on this botanical knowledge, appropriate planning and environmental management can be developed for this urban stretch of the Tiber, giving greater attention to rare and native river species of high naturalistic value, while promoting continuous monitoring of alien species to assess their invasiveness and eventually control their expansion along the river.

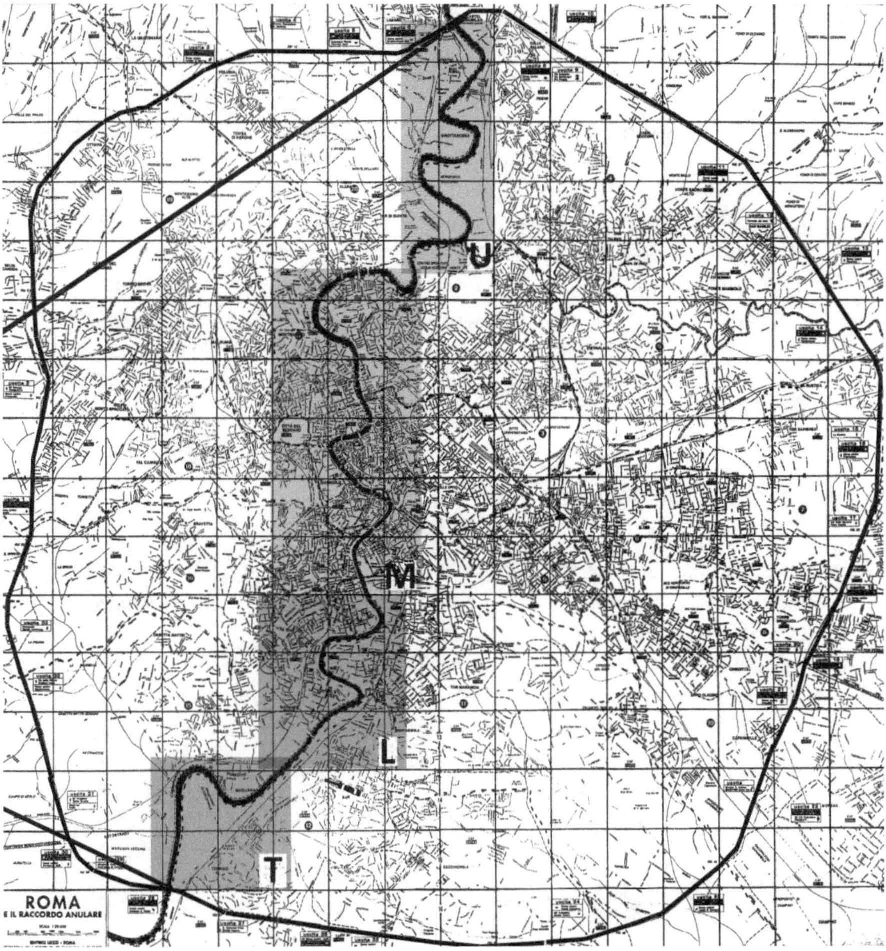

Fig. 5.7 In dark grey, the urban stretch of the Tiber River crossing the city of Rome is bounded by a ring motorway. From above, upper (U), middle (M), lower (L), and terminal (T) sectors of the urban Tiber. (Modified by S. Ceschin)

5.3.4 Aquatic and Riparian Plants and Their Applicative Uses in Urban Rivers

The case of the urban Tiber within Rome provides an excellent opportunity to highlight the importance of aquatic and riparian plants in cities ecosystem services they can potentially provide. Among the applicative aspects there are: use of vascular plants' pollen for the reconstruction of past urban environments (Pepe et al. 2016), use of aquatic and riparian plants (named *macrophytes*) as bioindicators of urban water quality (Ceschin et al. 2010b) or as biological agents for the remediation of polluted and contaminated urban waters.

5.3.5 Macrophytes as Bioindicators of River Water Quality

Numerous studies highlighted how the composition and distribution of macrophyte communities are strongly influenced by hydro-morphological and water chemical-physical factors. In particular, urbanization causes water mineralization and eutrophication, thereby altering water quality. These alterations are reflected in the macrophyte spatial patterns, making water quality (Ceschin et al. 2010b). Within the urban stretch of the Tiber in Rome, the frequent occurrence of macrophytes, such as *Potamogeton pectinatus*, *P. nodosus* and *Myriophyllum spicatum* (Fig. 5.8), indicates qualitatively poor waters, characterized by pollution and eutrophication, since these species are known for tolerating different contaminants, high conductivity and high nutrient concentrations (Ceschin et al. 2010b).

5.3.6 Water Filtration by Plants

Plants can play a key role in water filtration, given their well-known ability to phytoremediate surface waters. Phytoremediation is defined as "*a set of natural processes treating polluted water, based on the exploitation of the soil-vegetation system as a natural filter*" (Borin 2003). This system is both biotic (plants and microorganisms) and abiotic (substrate), and its filtering action involves physical (particle sedimentation), chemical (degradation), and biological mechanisms.

Water treatment by plants can be expressed in natural (undisturbed wetlands), semi-natural (buffer zones along rivers, reconstructed wetlands) and artificial

Fig. 5.8 On the top, from left: the Tiber River in the upper (at Ponte Milvio), middle (at Ponte Castel Sant'Angelo) and lower (at Ponte Marconi) stretches within the city of Rome. Below, aquatic plant communities dominated by *Myriophyllum spicatum*, *Potamogeton pectinatus*, and *P. nodosus* in some stretches of the urban Tiber. (Photo by S. Ceschin)

systems (constructed plants) (Borin 2003). Artificial greening solutions for water filtration are mostly represented by Constructed Wetlands (CWs), engineered infrastructures mimicking the purification functions of natural wetlands (Austin and Yu 2016). Other sustainable urban drainage systems also contribute to water quality management, primarily through bioswales and rain gardens, while green roofs and green walls further support stormwater filtration. Riparian buffer strips-vegetated areas adjacent to water bodies-represent an important semi-natural filtration system that can also be applied to urban areas (Borin 2003).

In artificial systems, selecting the most appropriate species is crucial for effective phytoremediation. Species used in CWs are mainly fast-growing macrophytes with high biomass production and a strong capacity for pollutants' absorption and filtration (Borin 2003). In Italian CWs, the most used species are the native *Typha latifolia*, *Phragmites australis*, *Juncus effusus*, *Schoenoplectus lacustris* and *Phalaris arundinacea*. Subsurface flow systems are planted with riparian plants only, while surface flow systems also include aquatic plants, such as *Lemna minor*, *Ceratophyllum demersum*, *Potamogeton nodosus,* and *Myriophyllum spicatum* (Table 5.1). Species' selection in urban rain gardens is broader and more complex since they have intermittent stormwater presence, with two or three moisture zones variable in time and space, according to climatic conditions, designed geometry, and infiltration rates (Dunnett and Clayden 2007). Phytoremediation of urban and agricultural wastewater primarily targets organic matter, nutrients, pathogenic bacteria, and pharmaceuticals, whereas remediation of industrial effluents and urban stormwater run-off focuses on heavy metals, hydrocarbons, and solvents (Austin and Yu 2016). Mechanisms involved in heavy metal phytoremediation incude phytostabilization, rhizoremediation and plant absorption at the roots and rhizosphere level, phytoextraction and phytovolatilization in the aerial part (Pouresmaieli et al. 2022). Translocation to the aerial part is variable and enhanced by high evapotranspiration rates; contaminated plant biomass (with bioaccumulated heavy metals) is harvested and appropriately disposed of as special waste.

For nutrient removal, plants act both directly, through root absorption, and indirectly, creating the conditions for microbial life and activity. Aquatic plants release excess oxygen-transported to the submerged parts for respiration-in the rhizosphere, creating aerobic sites that support aerobic bacterial adhesion to plant organs (Borin 2003). Organic contaminants (e.g., PCBs, PAHs), often present in urban wastewater, can be highly toxic. Plants can uptake, translocate and degrade PAHs, while plant-microbe interactions significantly enhance their biodegradation in soils (Tarigholizadeh et al. 2023). Some plant species can also absorb and metabolize pharmaceuticals, or promote microbial growth by providing surfaces for biofilm formation, releasing exudates and oxygen, and retaining suspended solids (Verlicchi and Zambello 2014) (Table 5.2).

Constructed wetlands planted with a mixture of *Typha latifolia* and *Phragmites australis* represent an optimal solution for removing heavy metals and nutrients. Both species exhibit a high metal bioaccumulation capacity in their tissues and can store elevated amounts of organic nitrogen in the vigorous rhizomes or aerial parts (Austin and Yu 2016). *Phragmites* successfully removes PAHs and, together with

Table 5.2 List of the main plant species for phytoremediation in Italy

Riparian plants	*Alismataceae*	*Alisma plantago-aquatica, Sagittaria sagittifolia*
	Apiaceae	*Apium nodiflorum, Berula erecta*
	Cyperaceae	*Carex acuta, C. elata, C. riparia, C. acutiformis, Cyperus longus, C. glomeratus, Schoenoplectus lacustris*
	Iridaceae	*Iris pseudacorus, I. sibirica*
	Juncaceae	*Juncus effusus, J. conglomeratus, J. articulatus,*
	Poaceae	*Glyceria maxima, Phalaris arundinacea, Phragmites australis*
	Typhaceae	*Sparganium erectum, Typha latifolia, T. angustifolia*
	Other families	*Nasturtium officinale, Mentha aquatica, Lythrum salicaria, Veronica anagallis-aquatica, V. beccabunga*
Aquatic plants	*Hydrocharitaceae*	*Hydrocharis morsus-ranae, Vallisneria spiralis*
	Lemnaceae	*Lemna minor, L. gibba*
	Other families	*Ceratophyllum demersum, Myriophyllum spicatum, Potamogeton nodosus, P. crispus, Salvinia natans*

Typha, is also able to remediate pharmaceuticals (Verlicchi and Zambello 2014; Tarigholizadeh et al. 2023).

Juncus effusus and *Phalaris arundinacea* are both efficient nitrogen absorbers and particularly effective in metal removal, accumulating Ni and Zn in roots and aerial parts, respectively (Austin and Yu 2016; Bonciarelli et al. 2025). Additionally, native riparian tree species, such as *Salix alba* and *Populus alba,* have proven highly efficient in phytoremediation of heavy metals and nutrients. *Salix* spp. stand out for metals removal, due to their remarkable biomass productivity, rapid growth, good rooting ability, high evapotranspiration rates and pronounced clone-specific metal uptake abilities (Pulford 2003).

5.4 Plants and Climate Mitigation & Air Pollutant Removal (N.C., F.M., F.O., D.M.)

5.4.1 Climate Change Effects Mitigation

Cities are particularly vulnerable to the impacts of climate change (including extreme climatic events) and pollution and, in general, to the biotic consequences of the multiple stresses associated with the urban environment. In line with the EU Strategy on adaptation to climate change, cities must increasingly act by implementing urban investments and managerial strategies to enhance urban resilience to these stresses and to promote their mitigation. Urban green spaces and natural areas play a crucial role in this context, performing multiple functions and providing significant environmental and societal benefits. They positively impact local climate, air quality, noise levels, and soil stability. Additionally, urban green spaces support diverse flora and provide habitats for numerous animal species, contributing to biodiversity conservation. In cities, plants cope with and adapt to the stresses of urban environments by modulating physiological and biophysical functions, such as carbon assimilation and transpiration rates, stomatal conductance, water use efficiency, carboxylation efficiency, mesophyll efficiency, and leaf thickness.

Depending to the physical features and the ecological requirements of the species, plants are able to mitigate the adverse impacts of urban climate (and urban climate change) by modifying air temperature, relative humidity and to contrast the UHI effect thanks to several processes, such as shading, evapotranspiration and albedo modification, that redistribute heat and encourage natural airflow, improving the thermal comfort within urban areas (Lin and Li 2025). Shading decreases the absorption of solar radiation and heat accumulation on urban surfaces, while evapotranspiration reduces temperature by releasing latent heat. A study conducted in Germany reported a reduction in air temperature (0.77–2.22 °C) and surface temperatures (5.5–15.2 °C) and an increase in relative humidity (1.11–6.48%) on shaded compared to unshaded surfaces, with higher UHI mitigation in species with high leaf area density and transpiration rates (Gillner et al. 2015). Shade can be strategically planned either for the whole year with evergreen trees or just for the warmest season with deciduous trees.

However, drought events and species' water stress behaviour can have consequences on the effectiveness of transpiration and cooling capacities (Moser-Reischl et al. 2019). High water-using species with diffuse-porous wood anatomy are recommended only for sites with high soil water availability, where they can buffer drought events and sustain transpiration. Differently, water-saving species with ring-porous wood anatomy have higher resilience and recovery ability from drought and provide cooling through stored water and latent heat exchange from the soil.

Albedo modification is the increase of surface reflectivity to minimize heat absorption. In an urban environment, most impervious surfaces have a lower albedo and higher heat storage than green spaces. Thus, green spaces as well as the application of good practice, such as the reduction of annual mowing, are two significant tools to increase albedo and water retention capacity and achieve climate mitigation (Ponti and Guglielmin 2023) (Fig. 5.9).

Generally, urban trees demonstrate a more substantial cooling effect than grasslands thanks to their higher evapotranspiration, canopy height, and moisture retention, especially under conditions of drought and summer heat, while rather open grass-dominated areas provide cool air during night/early morning. Thus, many plant traits influence the urban climate. In general, small trees, pyramidal crown shape, and high foliage density provide better cooling by shading, whereas tall trees with diffuse porous wood, high diameter and height increments, and leaf hairiness provide better cooling by transpiration (Kim et al. 2024).

Therefore, species must be selected according to local climate and environment, but also to several targeted adaptation traits, such as heat, drought, waterlogging, and pollution tolerance; and temperature, soil moisture, and light plasticity. The implementation of classified plant databases is fundamental to this selection and is of great interest for tackling climate change in cities.

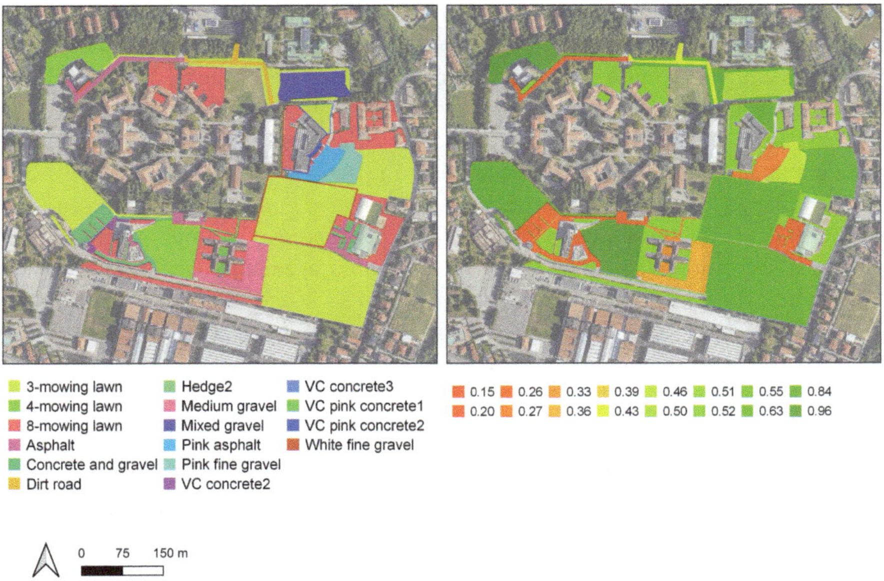

Fig. 5.9 List of the surface types (**a**) and their corresponding air temperature mitigation index (ATMI; 0 = min; 1 = max) (**b**): example from Bizzozzero University Campus in Italy. VC = vibro-compressed. (Modified by Ponti and Guglielmin 2023)

Other parameters of green spaces, such as composition and diversity, can have a higher relative contribution to the local climate than the single species traits themselves. As reported by Wang et al. (2021), a positive correlation between the diversity index and tree species richness, and the magnitude of cooling in China, with a spatial cooling range up to 600 m outside the green space edge. Similarly, Tran et al. (2019) found that species-diverse green roofs in three Canadian cities provide better fitness and functioning than monoculture green roofs. This provides a buffer against climate-related stresses and increases the magnitude of climate mitigation action.

Urban green spaces can strongly affect the energy balance, depending on density, canopy coverage, stratification, and plant community biomass. Indeed, the increase in tree density, canopy coverage, and canopy layers with different vertical structures and species composition directly influences sensible heat fluxes and lowers the ambient temperature. Indeed, a global meta-analysis confirmed that the shading and evapotranspiration of a dense tree canopy with high specific leaf area are more efficient at urban cooling than those of a less stratified, denser canopy of grass or shrubs (Kim et al. 2024).

The shape of the green spaces plays a significant role in climate mitigation in cities. A positive correlation of aggregation (i.e., spatial cohesion, connectivity, and spatial subdivision of patches) and area of the green spaces with the surface temperature highlighted the importance of the spatial interactions between green spaces, with specific morphological factors (i.e., largest patch index, mean patch size, edge density) more important in summer than in winter seasons. Few studies have examined the correlations among green space size, temperature variation, and cooling intensity. One example is Park et al. (2017) which reported the highest cooling effect (3.5 °C) with a 1500 m^2 mixed-type small green space. Also, the location is relevant, with strategically placed green spaces in heat-prone areas being more effective than the simple increase of green cover (Stojakovic et al. 2020).

Regional climate setting is another background parameter that affects the urban climate and the UGI's mitigation potential. At the global scale, the cooling effect of trees and grasslands is not spatially uniform and is affected by vapour pressure deficit, precipitation, and solar radiation patterns (Kim et al. 2024). Thermal mitigation efficiency is higher in the high than in the low latitudes, with evapotranspiration prevailing in the less water-limited northern cities and shading prevailing in more heat and drought-exposed southern cities. However, a wide variability in climate mitigation magnitude and idiosyncratic responses can't be excluded over this large spatial scale, highlighting the need to optimize urban greening strategies according to geographic and climate-specific context.

Many urban plant species with high cooling potential may also demand substantial irrigation, which can strain urban water supplies. Increase in pervious surfaces, native and drought-tolerant species, and soil water reservoirs ensure the temporal continuity of the mitigating effect of urban plants through physiological and biophysical processes, in particular in cities with dry and hot climates, for high water-using species, and in a climate change scenario with intense rainfall events and prolonged dry periods.

Alongside climate mitigation, urban green plays a significant role in removing air pollutants and limiting CO_2 emissions. CO_2 uptake via stomata and energy savings by tree presence is more significant in urban settings compared to forests. Carbon dioxide sequestration involves the annual accumulation of CO_2 in plant biomass during photosynthesis.

Studies have provided estimates of CO2 control by urban trees, highlighting the significant contribution of energy savings, particularly in regions with hot, dry summers, such as the Mediterranean. Planting trees for shading and reducing CO_2 emissions in urban areas is more effective than in forests, as it leads to substantial energy savings from reduced heating and air conditioning use. The amount of carbon absorbed and stored in urban areas varies with land use, tree species, tree size, and local climatic conditions.

5.4.2 Air Pollutant Removal

Moreover, urban green spaces and natural areas play a vital role, offering a wide range of environmental benefits that directly support human health and well-being. Among these benefits, vegetation's removal of air pollutants is particularly important in densely populated urban areas. Plants help improve air quality by capturing airborne pollutants such as nitrogen dioxide (NO_2), sulfur dioxide (SO_2), ozone (O_3), carbon monoxide (CO), Polycyclic aromatic hydrocarbons (PAHs), and particulate matter (PM_{10} and $PM_{2.5}$). This is achieved through both dry deposition, in which gaseous and particulate pollutants adhere to leaf surfaces, and gas exchange through stomata, where pollutants are absorbed and metabolized.

Air pollution is a serious threat to the health and quality of life of the urban population by reducing the life expectancy of residents in heavily polluted areas, particularly considering children and people with lung and heart diseases. The harmful PM in air pollution is mainly of anthropogenic origin and comprises a mixture of heavy metals, polycyclic aromatic hydrocarbons, and other substances suspended in the atmosphere. Natural sources have shown a higher contribution to PM10 than to PM2.5; the "unspecified source of human origin", industry, and traffic have shown a higher contribution to PM2.5 than to PM_{10}. However, we should also consider ultrafine particles, smaller than 100 nm (PM_1), produced in large quantities by vehicular combustion and their role in causing several adverse human health effects (Knibbs et al. 2011).

To reduce health risks for urban populations, pollution sources must be controlled, but vegetation remediation also has considerable potential. As found by Escobedo et al. (2008), planting vegetation in urban areas of the city of Santiago was a cost-effective measure for capturing air pollutants, while Cavanagh et al. (2009) observed significant effects of a forest patch on local-scale air quality in Christchurch, New Zealand. Moreover, plants remove polycyclic aromatic hydrocarbons (PAHs) and PM-associated heavy metals from the air in urban areas. PAHs still represent an environmental and human health issue. One of the strategies that could help reduce and mitigate the concentrations of these pollutants in the atmosphere is their uptake by trees, known as the "filter effect" (Terzaghi et al. 2013). Thus, the choice of tree and shrub species for urban environments should not be based solely on ornamental criteria but on their air-filtering potential.

Research conducted in Beijing, China, showed that trees in the city centre removed 772 tonnes of PM10 in one year, and across the US, urban trees and shrubs remove approximately 215,000 tonnes of PM10 per year (Nowak et al. 2006). Based on UK city studies, McDonald et al. (2007) concluded that planting trees in a quarter of the available urban area can reduce PM10 concentrations by between 2% and 10%. Therefore, the benefits of urban vegetation can have a direct positive effect on human health. Trees, with their large total leaf area, are considered the most effective type of vegetation for this purpose. This is mainly due to the size and structure of tree crowns, which create turbulent airflow, increasing PM deposition on leaves.

Plants vary in their ability to capture airborne pollutants depending on leaf traits such as structure, surface roughness, and the presence of fine hairs. These features enhance particle retention, and trees are particularly efficient because of their large leaf area and canopy architecture (Fig. 5.10).

Species selection is important because different plants exhibit different efficiencies in pollutant capture. For instance, broadleaf species with rough leaf surfaces trap particulate matter (PM) more effectively than species with smooth leaves (Hwang et al. 2011). Evergreen and deciduous trees contribute differently across seasons. Evergreen conifers can accumulate pollutants year-round, and their aerodynamic properties, small needles, and complex shoot structures make them

Fig. 5.10 Different leaf characteristics determining plant particulate absorption performances of (**a**) *Quercus pubescens* L and (**b**) *Tilia cordata* L. (Photos by L. Ruga)

capable of capturing larger amounts of PM than many broadleaf species (Przybysz et al. 2019). Since PM concentrations are typically higher in colder months, conifers provide key benefits in autumn and winter. Recent studies have also highlighted their ability to capture polycyclic aromatic hydrocarbons (PAHs) and filter both PAHs and PM, even during heavily polluted periods (Giraldez et al. 2022). These findings support the development of new equations for environmental models to better assess PAH mass balance in urban settings and refine air quality improvement strategies. Because many hydrophobic PAHs are associated with PM, a substantial portion of PAHs removed by trees comes from PM-bound PAHs. Urban trees also influence atmospheric chemistry beyond particle capture: integrated research on temperature, deposition, and emissions indicates that they can significantly help reduce ozone concentrations (Nowak et al. 2006). Although the percentage improvement in air-quality from pollutant removal may seem modest, the overall contribution of trees is substantial enough to support strategic urban tree cover management. Furthermore, biomonitoring of pollutants can be effectively conducted by analyzing leaves and bark, as sampling is simple and inexpensive (Aničić et al. 2011).

5.5 Plants and Animals (Pollinators, Food Webs, Unwanted Animal Species) (F.B., V.L.)

5.5.1 Insects

Among animals that benefit biodiversity, pollinators play a central role. However, pollinators have experienced significant declines over the last few decades (Potts et al. 2016). As expected, the greener the cities, the richer the pollinator communities. Among these animals, wild bees are generally considered the best pollinators, as they can pollinate almost 80% of flowering plants globally. Moreover, the biodiversity of wild bees has been beneficial for pollination and fruit and seed set in urban environments (Lowenstein et al. 2015). On the other hand, plant diversity is

Fig. 5.11 Entomogamous plants with visiting pollinators (top) and bee hotels (bottom) (photos by F. Bretzel, A. Hobelik)

linked to a richer overall bee diversity. Wild plant species and artificial nests enhance the presence of wild bees, with effects on urban agriculture, such as ornamental plants, which also play a role in attracting pollinators (Poole et al. 2025). As regards certain botanical families, such as Fabaceae, Asteraceae, Solanaceae, Rosaceae, and Lamiaceae, these are particularly favored by both wild and domestic bees (da Silva Mouga et al. 2015). It is well conceivable that certain management practices of urban green spaces (UGS), such as planting wildflower meadows, can be effective in fostering bee abundance and biodiversity (Anderson et al. 2023). On the other hand, other management practices, such as mowing, pruning, and insecticide treatments, have detrimental effects on pollinators (Poole et al. 2025). Also, the design of UGS is fundamental to maximizing the presence of insects, specifically pollinators. Different pollinator groups, such as bees, wasps, butterflies, diurnal or nocturnal moths, and flies, are more attracted by flowers displaying different shapes and colours, and the longer the flowering period, the higher the number of pollinator visits (Fig. 5.11).

Plants can also act as attractors of undesirable insects, some are well-known pests such as *Cydalima perspectalis*, *Paysandisia archon*, and *Rhynchophorus ferrugineus* for their respective target plants, *Buxus* spp., Argentinian ornamental palms, and *Phoenix* spp. Other insects, like mosquitoes, can be vectors of diseases. The presence of different mosquito genera in cities can follow complex models with several drivers, related to plants or the surrounding urban design, promoting one or another genus. UGS, especially if water bodies are available, can attract more mosquitoes than other urban contexts, while green roofs are possibly less visited by these insects; however, the complexity of the food web enables the presence of predators (birds, chiroptera).

5.5.2 Amphibians

Amphibians, both terrestrial and aquatic, are tightly linked to water, at least for reproduction, and they are often components of cities' faunas, sometimes abundant (Fig. 5.12).

Moreover, it is well-known that plants deeply affect amphibian populations. Burrow and Maerz (2022) reviewed the relevant literature and pinpointed three themes relative to plants' influence on amphibians: a) the abiotic conditions of their environment; b) the aquatic life stages through resource quality and availability; c) modulating predator-prey interactions. Consequently, mowing interventions in

Fig. 5.12 Urban open spaces and water bodies can be colonised by amphibians (photos by A. Somma, F. Bretzel)

natural vegetation that are often carried out in urban streams and wetlands are detrimental to amphibians. Different types of water vegetation can have differential effects on various anuran species. Floating-leaves vegetation (e.g., *Nymphaea* spp.) and emergent vegetation (e.g., *Nelumbo nucifera*) were effective in enhancing the presence of certain species, such as *Pelophylax nigromaculatus*, in urban, suburban, and rural areas of Shanghai, whereas submerged vegetation had a negative effect. Vegetation can also diminish the occupancy of fully aquatic amphibians in urban ponds, whereas, conversely, it can enhance that of terrestrial ones. A comparison of urban stormwater ponds and neighbouring ponds of other types in an urban area revealed strong similarities in amphibian communities and highlighted the role of vegetation in their assemblages (Holtmann et al. 2017). Indeed, in stormwater ponds, species richness increased with higher algal cover and decreased with higher tree cover, whereas in the other kinds of ponds, submerged vegetation cover proved to be the enhancing factor. Similar results were obtained in the comparison of natural and constructed ponds for native species breeding, with the artificial ones showing a higher presence of all the studied species.

5.5.3 Reptiles

Cities are commonly inhabited by reptiles that, in turn, have responded to the specific features of the ongoing process of urbanization. The presence of plants is, as expected, an important driver for reptiles in urban contexts. In cities, certain species may privilege vegetated areas while, on the other hand, tending to avoid more anthropized and more dangerous places like crossroads. As found out by Delaney et al. (2021), remnant fragmented green patches with wild vegetation could serve as micro-reserves for lizards in an urbanized area. From a conservation point of view, such micro-reserves should be as large as possible, even though certain small, vegetated patches can exhibit species richness comparable to that of larger patches. In a Californian urban area, it has been shown that an anthropized river had lower herpetofauna abundance and species richness than a rehabilitated one, while the latter had a similar abundance to that of a wild site (Banville and Bateman 2012). Moreover, certain features of the rehabilitated sites studied, such as vegetation complexity and tree species density, were positively correlated with the abundance of at least one lizard species. Revegetation actions in urban areas, aimed at increasing wildlife presence, can elicit different reactions among citizens depending on the taxonomic group. Birds and lizards were considered welcome, while the return of snakes could cause fear. Even though certain species can pose a serious threat to human safety, snakes can only be considered animals that provide ecosystem services in cities, thanks to their predation on unwanted mammals. In general, snakes tend to avoid highly urbanized areas and to prefer more natural sites characterized by groundcover vegetation, shrubs or sheltered by trees like vegetation patches or corridors (Lettoof et al. 2023). In this light, proper vegetation management in UGSs has already been carried out, for example, in peri-urban forests aiming to enhance the presence of brambles with the subsequent result of an increased number of detected individuals of the venomous *Vipera aspis* (Bonnet et al. 2016). Interestingly, this latter intervention and its results did not raise particular concern among citizens. Also, enhancing the presence of animals that snakes prey on can indirectly bolster snake populations; for example, the establishment of frog ponds can benefit anurophagous snakes.

5.5.4 Birds

Birds are naturally occurring animals in cities mainly due to their ability to fly. The species that can be found in cities, except for those that strictly avoid such contexts and are thus referred to as urban avoiders, fall into the categories of urban exploiters and urban adapters (Tryjanowski et al. 2020). Even though a reduction in avian richness has been observed along a gradient from the most to the least urbanized areas, vegetated spaces can enhance the presence of birds in cities (Batáry et al. 2018). UGSs can act as stepping stones for urban birds, for example, by providing nesting or resting sites, which is one of the most important factors influencing their presence in cities. Consequently, urban trees, either wild or planted, can serve this purpose, and street trees alone can be sufficient to positively contribute to the number of birds, especially if native plants and trees are more abundant. On the other hand, less natural sites like squares, gardens, and cultivated lawns were more visited by generalist species. The observations that the richness, diversity, and density of woody plant species were significant factors for bird species richness, abundance, and diversity suggest that efforts can be made to increase the number of plant species in UGSs. The availability of food is another important driver of avian communities in cities, and birds may exhibit varied diets, including insects, nectar, fruits, grains, and meat/fish, or be almost omnivorous. Consequently, plants and trees that produce flowers, fruits, and seeds can forage certain bird species, while shrubs can indirectly foster birds' presence through the attraction of insects. As for other animals, peculiar UGSs, such as green roofs and walls, together with private yards displaying both wild and cultivated plants, can serve as sites for birds to feed, shelter, and nest (Belcher et al. 2019). Also, for birds of prey, different types of urban vegetation, such as edges, hedges, and forested areas, are important factors driving their presence in cities, as observed along an urban-wildland gradient.

5.5.5 Mammals

Urban areas can also host mammal biodiversity and, in certain cases, endangered species (Fig. 5.13).

Mammals tend to be attracted to UGSs, where they move, forage, and nest more than in other urban areas. Their presence depends on body size, ecological

Fig. 5.13 *Martes martes*, *Hystrix cristata*, *Vulpes vulpes* (clockwise from top left) caught by night photo traps in cities. (photos by E. Mori)

preferences, and the characteristics and location of UGSs. Research in several Italian cities showed that synanthropic, mostly alien small mammals prefer managed UGSs, whereas native, non-synanthropic species are more associated with natural areas featuring shrubs and brambles (Dondina et al. 2025).

Similar patterns have been observed internationally: for example, rats are more abundant in residential gardens near forest patches or in areas with dense lower canopy, while urbanized surfaces may favor generalist species but reduce overall small-mammal diversity (e.g., Miller et al. 2022). Vegetation structure also plays an important role. High canopy cover can be associated with fewer specialist species, whereas tall human-modified vegetation near streams may benefit other small mammals.

For medium-sized mammals, vegetated patches such as natural areas, cemeteries, and golf courses tend to support higher biodiversity than typical city parks (Gallo et al. 2017). Residential gardens can also sustain native mammal populations, sometimes supporting abundance and reproductive levels comparable to those in nearby natural areas (Van Helden et al. 2020). Larger mammals show more variable responses: in some cases, vegetation does not influence their activity (George and Crooks 2006), while in others dense forest patches support the highest diversity (Borges et al. 2014).

Urban woodland area is generally positively correlated with mammal species richness, whereas long edge lengths may have negative effects. Large omnivores such as wild boars (Sus scrofa) are frequently observed in cities, often associated with green areas near natural habitats connected by ecological corridors, although their presence can lead to conflicts with humans.

Chiroptera are well-known urban dwellers whose presence is shaped by vegetation features. Trees can provide roosting sites (Corlett 2005), fruit-bearing species support frugivorous bats, and insect availability—often increased by UGS presence and plant diversity—attracts insectivorous species. Green roofs may also support bat activity when suitable habitats occur nearby, and overall structural complexity of UGS vegetation significantly influences urban bat communities (Suarez-Rubio et al. 2018).

References

Albrecht H, Eder E, Langbehn T, Tschiersch C (2011) The soil seed bank and its relationship to the established vegetation in urban wastelands. Landscap Urban Plann 100(1–2):87–97. https://doi.org/10.1016/j.landurbplan.2010.11.011

Anderson M, Crubaugh F, Greenslit C, Hill E, Kroth H, Stanislawski E, Ribbons R, Del Toro I (2023) BYO bees: managing Wild Bee biodiversity in urban Greenspaces. PLoS One 18:e0281468. https://doi.org/10.1371/journal.pone.0281468

Aničić M, Spasić T, Tomašević M, Rajšić S, Tasić M (2011) Trace elements accumulation and temporal trends in leaves of urban deciduous trees (Aesculus hippocastanum and Tilia spp.). Ecol Indic 11(3):824–830. https://doi.org/10.1016/j.ecolind.2010.10.009

Atkinson G, Doick KJ, Burningham K, France C (2014) Brownfield regeneration to greenspace: delivery of project objectives for social and environmental gain. Urban For Urban Green 13(3):586–594. https://doi.org/10.1016/j.ufug.2013.04.002

Austin G, Yu K (2016) Constructed wetlands and sustainable development. Routledge, T&F eBooks. https://doi.org/10.4324/9781315694221

Banville MJ, Bateman HL (2012) Urban and wildland herpetofauna communities and riparian microhabitats along the Salt River, Arizona. Urban Ecosyst 15(2):473–488. https://doi.org/10.1007/s11252-012-0228-5

Baschak LA, Brown RD (1995) An ecological framework for the planning, design and management of urban river greenways. Landsc Urban Plan 33(1–3):211–225. https://doi.org/10.1016/0169-2046(94)02019-C

Batáry P, Kurucz K, Suárez-Rubio M, Chamberlain DE (2018) Non-linearities in bird responses across urbanization gradients: a meta-analysis. Glob Chang Biol 24(3):1046–1054. https://doi.org/10.1111/gcb.13964

Battisti C, Poeta G, Fanelli G (2016) Role and effects of disturbances in natural systems. In: An introduction to disturbance ecology: a road map for wildlife management and conservation. Springer International Publishing, Cham, pp 13–29. https://doi.org/10.1007/978-3-319-32476-0_3

Belcher RN, Sadanandan KR, Goh ER, Chan JY, Menz S, Schroepfer T (2019) Vegetation on and around large-scale buildings positively influences native tropical bird abundance and bird species richness. Urban Ecosyst 22(2):213–225. https://doi.org/10.1007/s11252-018-0808-0

Bonciarelli L, Orlandi F, Muscas D, Fornaciari M (2025) Sustainable Stormwater management and bioretention: an overview of reviews of the last 10 years. Land 14(4):736. https://doi.org/10.3390/land14040736

Bonnet X, Lecq S, Lassay JL, Ballouard JM, Barbraud C et al (2016) Forest management bolsters native snake populations in urban parks. Biol Conserv 193:1–8. https://doi.org/10.1016/j.biocon.2015.11.001

Borges LH, Calouro A, Botelho AL, Silveira M (2014) Diversity and habitat preference of medium and large-sized mammals in an urban forest fragment of southwestern Amazon. Iheringia Série Zoologia 104:168–174. https://doi.org/10.1590/1678-476620141042168174

Borin M (2003) Fitodepurazione. Soluzioni per il trattamento dei reflui con le piante. Edagricole, Bologna, IT, pp 1–198

Bretzel F, Vannucchi F, Pezzarossa B, Paraskevopoulou A, Romano D (2024) Establishing wildflower meadows in anthropogenic landscapes. Front Horticulture 2:1248785. https://doi.org/10.3389/fhort.2023.1248785

Burrow A, Maerz J (2022) How plants affect amphibian populations. Biol Rev 97(5):1749–1767. https://doi.org/10.1111/brv.12861

Caneva G, Cicinelli E, Scolastri A, Bartoli F (2020) Guidelines for urban community gardening: proposal of preliminary indicators for several ecosystem services (Rome, Italy). Urban For Urban Green 56:126866. https://doi.org/10.1016/j.ufug.2020.126866

Caneva G, Ceschin S, Lucchese F, Scalici M, Battisti C et al (2021) Environmental management of waters and riparian areas to protect biodiversity through river contracts: the experience of Tiber River (Rome, Italy). River Res Appl 37(10):1510–1519. https://doi.org/10.1002/rra.3869

Cavanagh JAE, Zawar-Reza P, Wilson JG (2009) Spatial attenuation of ambient particulate matter air pollution within an urbanised native forest patch. Urban For Urban Green 8(1):21–30. https://doi.org/10.1016/j.ufug.2008.10.002

Ceschin S, Salerno G (2021) Exploring plant species richness along the Tiber River within the city of Rome. Phytotaxa 482(2):143–158. https://doi.org/10.11646/phytotaxa.482.2.3

Ceschin S, Salerno G, Bisceglie S, Kumbaric A (2010a) Temporal floristic variations as an indicator of environmental changes in the Tiber River in Rome. Aquat Ecol 44(1):93–100. https://doi.org/10.1007/s10452-009-9292-1

Ceschin S, Zuccarello V, Caneva G (2010b) Role of macrophyte communities as bioindicators of water quality: application on the Tiber River basin (Italy). Plant Biosyst 144(3):528–536. https://doi.org/10.1080/11263500903429221

Corlett RT (2005) Interactions between birds, fruit bats and exotic plants in urban Hong Kong. South China Urban Ecosystems 8(3):275–283. https://doi.org/10.1007/s11252-005-3260-x

Costanza R, d'Arge R, De Groot R, Farber S, Grasso et al (1997) The value of the world's ecosystem services and natural capital. Nature 387(6630):253–260. https://doi.org/10.1038/387253a0

Craul PJ (ed) (1992) Urban soil in landscape design. John Wiley & Sons

da Silva Mouga DMD, Nogueira-Neto P, Warkentin M, Ferett V, Dec E (2015) Comunidade de abelhas (Hymenoptera, Apidae) e plantas associadas em área de mata atlântica em São Francisco do Sul, Santa Catarina. Brasil Acta Biológica Catarinense 2(1):12–31. https://doi.org/10.21726/abc.v2i1.572

Delaney KS, Busteed G, Fisher RN, Riley SP (2021) Reptile and amphibian diversity and abundance in an urban landscape: impacts of fragmentation and the conservation value of small patches. Ichthyol Herpetol 109(2):424–435. https://doi.org/10.1643/h2019261

Dondina O, Tirozzi P, Viviano A, Mori E, Orioli V et al (2025) Spatial and habitat determinants of small-mammal biodiversity in urban green areas: lessons for nature-based solutions. Urban For Urban Green 104:128641. https://doi.org/10.1016/j.ufug.2024.128641

Dumitru A, Wendling L (2021) Evaluating the impact of nature-based solutions: a handbook for practitioners. European Commission: Directorate-General for Research and Innovation. https://data.europa.eu/doi/10.2777/11361

Dunnett N, Clayden A (2007) Rain gardens: managing water sustainably in the garden and designed landscape. Timber Press, Portland, Oregon

Elmqvist T, Setälä H, Handel SN, van der Ploeg S, Aronson J, Blignaut JN et al (2015) Benefits of restoring ecosystem services in urban areas. Curr Opin Environ Sustain 14:101–108. https://doi.org/10.1016/j.cosust.2015.05.001

Escobedo FJ, Wagner JE, Nowak DJ, De la Maza CL, Rodriguez M, Crane DE (2008) Analyzing the cost-effectiveness of Santiago, Chile's policy of using urban forests to improve air quality. J Environ Manag 86(1):148–157. https://doi.org/10.1016/j.jenvman.2006.11.029

Espinosa Fuentes MDLL, Peralta O, García R, González del Castillo E, Cerón Bretón RM, Cerón Bretón JG, Camal ET, Zavala García F (2023) Soil dynamics in an urban forest and its contribution as an ecosystem service. Land 12(12):2098. https://doi.org/10.3390/land12122098

Everard M, Moggridge HL (2012) Rediscovering the value of urban rivers. Urban Ecosyst 15(2):293–314. https://doi.org/10.1007/s11252-011-0174-7

Francini G, Hui N, Jumpponen A, Kotze DJ, Romantschuk M, Allen JA, Setälä H (2018) Soil biota in boreal urban greenspace: responses to plant type and age. Soil Biol Biochem 118:145–155. https://doi.org/10.1016/j.soilbio.2017.11.019

Francini A, Toscano S, Romano D, Ferrini F, Ferrante A (2021) Biological contribution of ornamental plants for improving slope stability along urban and suburban areas. Horticulturae 7(9):310. https://doi.org/10.3390/horticulturae7090310

Gallo T, Fidino M, Lehrer EW, Magle SB (2017) Mammal diversity and metacommunity dynamics in urban green spaces: implications for urban wildlife conservation. Ecol Appl 27(8):2330–2341. https://doi.org/10.1002/eap.1611

George SL, Crooks KR (2006) Recreation and large mammal activity in an urban nature reserve. Biol Conserv 133(1):107–117. https://doi.org/10.1016/j.biocon.2006.05.024

Gillner S, Vogt J, Tharang A, Dettmann S, Roloff A (2015) Role of street trees in mitigating effects of heat and drought at highly sealed urban sites. Landsc Urban Plan 143:33–42. https://doi.org/10.1016/j.landurbplan.2015.06.005

Giltrap D, Cavanagh J, Stevenson B, Ausseil AG (2021) The role of soils in the regulation of air quality. Philos Trans R Soc B 376(1834):20200172. https://doi.org/10.1098/rstb.2020.0172

Giovannetti M, Salvioli di Fossalunga A, Stringlis IA, Proietti S, Fiorilli V (2023) Unearthing soil-plant-microbiota crosstalk: looking back to move forward. Front Plant Sci 13:1082752. https://doi.org/10.3389/fpls.2022.1082752

Giráldez P, Aboal JR, Fernández JÁ, Di Guardo A, Terzaghi E (2022) Plant-air partition coefficients for thirteen urban conifer tree species: estimating the best gas and particulate matter associated PAH removers. Environ Pollut 315:120409. https://doi.org/10.1016/j.envpol.2022.120409

Gómez-Baggethun E, Barton DN (2013) Classifying and valuing ecosystem services for urban planning. Ecol Econ 86:235–245. https://doi.org/10.1016/j.ecolecon.2012.08.019

Halecki W, Gąsiorek M (2025) Urban soil ecosystem services as assets for green spaces in Europe: a meta-analysis of soil threat and potential biodiversity indices. Appl Soil Ecol 214:106337. https://doi.org/10.1016/j.apsoil.2025.106337

Holtmann L, Philipp K, Becke C, Fartmann T (2017) Effects of habitat and landscape quality on amphibian assemblages of urban stormwater ponds. Urban Ecosyst 20(6):1249–1259. https://doi.org/10.1007/s11252-017-0677-y

Hwang HJ, Yook SJ, Ahn KH (2011) Experimental investigation of submicron and ultrafine soot particle removal by tree leaves. Atmos Environ 45(38):6987–6994. https://doi.org/10.1016/j.atmosenv.2011.09.019

Ishimatsu K, Ito K, Mitani Y, Tanaka Y, Sugahara T, Naka Y (2017) Use of rain gardens for stormwater management in urban design and planning. Landsc Ecol Eng 13(1):205–212. https://doi.org/10.1007/s11355-016-0309-3

Johansson ME, Nilsson C, Nilsson E (1996) Do rivers function as corridors for plant dispersal? J Veg Sci 7(4):593–598. https://doi.org/10.2307/3236309

Kim J, Khouakhi A, Corstanje R, Johnston AS (2024) Greater local cooling effects of trees across globally distributed urban green spaces. Sci Total Environ 911:168494. https://doi.org/10.1016/j.scitotenv.2023.168494

Knibbs LD, Cole-Hunter T, Morawska L (2011) A review of commuter exposure to ultrafine particles and its health effects. Atmos Environ 45(16):2611–2622. https://doi.org/10.1016/j.atmosenv.2011.02.065

Lafortezza R, Chen J, Van Den Bosch CK, Randrup TB (2018) Nature-based solutions for resilient landscapes and cities. Environ Res 165:431–441. https://doi.org/10.1016/j.envres.2017.11.038

Lettoof DC, Parkin T, Jolly CJ, de Laive A, Von Takach B (2023) Snake life history traits and their association with urban habitat use in a tropical city. Urban Ecosyst 26(2):433–445. https://doi.org/10.1007/s11252-023-01327-x

Li G, Sun GX, Ren Y, Luo XS, Zhu YG (2018) Urban soil and human health: a review. Eur J Soil Sci 69(1):196–215. https://doi.org/10.1111/ejss.12518

Lin H, Li X (2025) The role of urban green spaces in mitigating the urban Heat Island effect: a systematic review from the perspective of types and mechanisms. Sustainability 17(13):6132. https://doi.org/10.3390/su17136132

Lowenstein DM, Matteson KC, Minor ES (2015) Diversity of wild bees supports pollination services in an urbanized landscape. Oecologia 179:811–821. https://doi.org/10.1007/s00442-015-3389-0

Macci C, Vannucchi F, Scartazza A, Masciandaro G, Doni S, Peruzzi E (2025) Soil–plant indicators for assessing nutrient cycling and ecosystem functionality in urban forestry. Urban Sci 9(3):82. https://doi.org/10.3390/urbansci9030082

McDonald AG, Bealey WJ, Fowler D, Dragosits U, Skiba U, Smith RI et al (2007) Quantifying the effect of urban tree planting on concentrations and depositions of PM10 in two UK conurbations. Atmos Environ 41(38):8455–8467. https://doi.org/10.1016/j.atmosenv.2007.07.025

McGrane SJ (2016) Impacts of urbanisation on hydrological and water quality dynamics, and urban water management: a review. Hydrol Sci J 61(13):2295–2311. https://doi.org/10.1080/02626667.2015.1128084

Miller KF, Wilson DJ, Hartley S, Innes JG, Fitzgerald NB, Miller P, van Heezik Y (2022) Invasive urban mammalian predators: distribution and multi-scale habitat selection. Biology 11(10):1527. https://doi.org/10.3390/biology11101527

Minixhofer P, Stangl R (2021) Green infrastructures and the consideration of their soil-related ecosystem services in urban areas—a systematic literature review. Sustainability 13(6):3322. https://doi.org/10.3390/su13063322

Moser-Reischl A, Rahman MA, Pauleit S, Pretzsch H, Rötzer T (2019) Growth patterns and effects of urban micro-climate on two physiologically contrasting urban tree species. Landsc Urban Plan 183:88–99. https://doi.org/10.1016/j.landurbplan.2018.11.004

Moura BB, Nissim WG, Manzini J, Scartazza A, Labra M, Hoshika Y et al (2025) The role of urban forests in tackling air and soil pollution in Italian cities. Urban For Urban Green 129066. https://doi.org/10.1016/j.ufug.2025.129066

Nguyen CN, Muttil N, Tariq MAUR, Ng AW (2021) Quantifying the benefits and ecosystem services provided by green roofs—a review. Water 14(1):68. https://doi.org/10.3390/w14010068

Nilsson C, Berggren K (2000) Alterations of riparian ecosystems caused by river regulation: Dam operations have caused global-scale ecological changes in riparian ecosystems. BioScience 50(9):783–792. https://doi.org/10.1641/0006-3568(2000)050[0783:AORECB]2.0.CO;2

Nowak DJ, Crane DE, Stevens JC (2006) Air pollution removal by urban trees and shrubs in the United States. Urban For Urban Green 4(3–4):115–123. https://doi.org/10.1016/j.ufug.2006.01.007

Park J, Kim JH, Lee DK, Park CY, Jeong SG (2017) The influence of small green space type and structure at the street level on urban heat Island mitigation. Urban For Urban Green 21:203–212. https://doi.org/10.1016/j.ufug.2016.12.005

Pepe C, Sadori L, Andrieu-Ponel V, Salomon F, Goiran JP (2016) Late Holocene pollen record from fiume morto (dead river), a palaeomeander of tiber river near ancient ostia (Central Italy). J Paleolimnol 56(2):173–187. https://doi.org/10.1007/s10933-016-9903-5

Pineda-Pinto M, Frantzeskaki N, Nygaard CA (2022) The potential of nature-based solutions to deliver ecologically just cities: lessons for research and urban planning from a systematic literature review. Ambio 51(1):167–182. https://doi.org/10.1007/s13280-021-01553-7

Ponti S, Guglielmin M (2023) How can the floor area types of a university campus mitigate the increase of urban air temperature? Landsc Ecol Eng 19(3):485–501. https://doi.org/10.1007/s11355-023-00553-x

Poole O, Costa A, Kaiser-Bunbury CN, Shaw RF (2025) Pollinators respond positively to urban green space enhancements using wild and ornamental flowers. Insect Conservation and Diversity 18(1):16–28. https://doi.org/10.1111/icad.12779

Potts SG, Ngo HT, Biesmeijer JC, Breeze TD, Dicks LV, Garibaldi LA et al (2016) The assessment report of the intergovernmental science-policy platform on biodiversity and ecosystem services on pollinators, pollination and food production. Intergovernmental Science-Policy Platform on Biodiversity and Ecosystem Services (IPBES)

Pouresmaieli M, Ataei M, Forouzandeh P, Azizollahi P, Mahmoudifard M (2022) Recent progress on sustainable phytoremediation of heavy metals from soil. J Environ Chem Eng 10(5):108482. https://doi.org/10.1016/j.jece.2022.108482

Przybysz A, Nersisyan G, Gawroński SW (2019) Removal of particulate matter and trace elements from ambient air by urban greenery in the winter season. Environ Sci Pollut Res 26(1):473–482. https://doi.org/10.1007/s11356-018-3628-0

Przybysz A, Popek R, Stankiewicz-Kosyl M, Zhu CY, Małecka-Przybysz M, Maulidyawati T et al (2021) Where trees cannot grow–particulate matter accumulation by urban meadows. Sci Total Environ 785:147310. https://doi.org/10.1016/j.scitotenv.2021.147310

Pulford ID, Watson C (2003) Phytoremediation of heavy metal-contaminated land by trees—a review. Environ Int 29(4):529–540. https://doi.org/10.1016/S0160-4120(02)00152-6

Rouillard J, Lago M, Abhold K, Roeschel L, Kafyeke T, Klimmek H, Mattheiß V (2018) Protecting and restoring biodiversity across the freshwater, coastal and marine realms: is the existing EU policy framework fit for purpose? Environ Policy Gov 28(2):114–128. https://doi.org/10.1002/eet.1793

Song Y, Kirkwood N, Maksimović Č, Zheng X, O'Connor D, Jin Y, Hou D (2019) Nature based solutions for contaminated land remediation and brownfield redevelopment in cities: a review. Sci Total Environ 663:568–579. https://doi.org/10.1016/j.scitotenv.2019.01.347

Sowińska-Świerkosz B, García J (2022) What are nature-based solutions (NBS)? Setting core ideas for concept clarification. Nature-Based Solutions 2:100009. https://doi.org/10.1016/j.nbsj.2022.100009

Stojakovic V, Bajsanski I, Savic S, Milosevic D, Tepavcevic B (2020) The influence of changing location of trees in urban green spaces on insolation mitigation. Urban For Urban Green 53:126721. https://doi.org/10.1016/j.ufug.2020.126721

Suarez-Rubio M, Ille C, Bruckner A (2018) Insectivorous bats respond to vegetation complexity in urban green spaces. Ecol Evol 8(6):3240–3253. https://doi.org/10.1002/ece3.3897

Tarigholizadeh S, Sushkova S, Rajput VD, Ranjan A, Arora J, Dudnikova T et al (2023) Transfer and degradation of PAHs in the soil–plant system: a review. J Agric Food Chem 72(1):46–64. https://doi.org/10.1021/acs.jafc.3c05589

TEEB (The Economics of Ecosystems and Biodiversity) (2011) TEEB Manual for Cities: Ecosystem Services in Urban Management (www.teebweb.org)

Terzaghi E, Wild E, Zacchello G, Cerabolini BE, Jones KC, Di Guardo A (2013) Forest filter effect: role of leaves in capturing/releasing air particulate matter and its associated PAHs. Atmos Environ 74:378–384. https://doi.org/10.1016/j.atmosenv.2013.04.013

Tran S, Lundholm JT, Staniec M, Robinson CE, Smart CC, Voogt JA, O'Carroll DM (2019) Plant survival and growth on extensive green roofs: a distributed experiment in three climate regions. Ecol Eng 127:494–503. https://doi.org/10.1016/j.ecoleng.2018.09.027

Tryjanowski P, Morelli F, Møller AP (2020) Urban birds: urban avoiders, urban adapters, and urban exploiters. In: Douglas I, Anderson PML, Goode D, Houck MC, Maddox D, Nagendra H, Tan PY (eds) The Routledge handbook of urban ecology. Routledge, London, pp 399–411

Van Helden BE, Close PG, Steven R (2020) Mammal conservation in a changing world: can urban gardens play a role? Urban Ecosyst 23(3):555–567. https://doi.org/10.1007/s11252-020-00935-1

Verlicchi P, Zambello E (2014) How efficient are constructed wetlands in removing pharmaceuticals from untreated and treated urban wastewaters? A review. Sci Total Environ 470:1281–1306. https://doi.org/10.1016/j.scitotenv.2013.10.085

von Döhren P, Haase D (2019) Risk assessment concerning urban ecosystem disservices: the example of street trees in Berlin. Germany. Ecosystem Services 40:101031. https://doi.org/10.1016/j.ecoser.2019.101031
Wang X, Dallimer M, Scott CE, Shi W, Gao J (2021) Tree species richness and diversity predicts the magnitude of urban heat Island mitigation effects of greenspaces. Sci Total Environ 770:145211. https://doi.org/10.1016/j.scitotenv.2021.145211
Yao H, Li Z, Geisen S, Qiao Z, Breed MF, Sun X (2023) Degree of urbanization and vegetation type shape soil biodiversity in city parks. Sci Total Environ 899:166437. https://doi.org/10.1016/j.scitotenv.2023.166437
Zari MP (2015) Ecosystem services analysis: mimicking ecosystem services for regenerative urban design. Int J Sustain Built Environ 4(1):145–157. https://doi.org/10.1016/j.ijsbe.2015.02.004
Zhao D, Li F, Yang Q, Wang R, Song Y, Tao Y (2013) The influence of different types of urban land use on soil microbial biomass and functional diversity in Beijing. China. Soil Use and Management 29(2):230–239. https://doi.org/10.1111/sum.12034

Part III
Ecological Integrated Approaches in Urban Green Interventions

Chapter 6
Botany in an Integrated View and in One Health Perspective

Giulia Caneva, Francesca Bretzel, Paolo Belardi, Marco Fornaciari, Carmelo Maria Musarella, Valeria Menchetelli, and Daniela Romano

Abstract This chapter analyses strategies to improve urban sustainability, enhancing green areas and green infrastructure in a Nature-based Solution approach. The integration across disciplinary knowledge often remains a weak point in complex systems, such as the urban areas.

The ecological view has the priority of an enhancement of autochthonous flora to protect biodiversity. This approach is also focussed on a true naturalistic enhancement and on an historical reconnection, rather than on aesthetic or decorative aspects. The methods used in selecting ecologically suitable plants, which are applied in interventions for ecological restoration, is explained. The fundamental need of a not superficial knowledge on the local vegetation, with its bioclimatic, biogeographical traits, such as on historical landscape transformations, to perform urban projects in a One Health perspective, is also stressed.

Urban horticulture for the cultivation of fruits, vegetables, herbs, and ornamentals in urban environments, can also provide diverse inclusive and equitable food access in urban food deserts, and offer environmental benefits. In a changing climate, and in the hard climatic conditions of cities, the reasons of using some non-native widely tolerant species (not invasive), that have a greater resilience respect

G. Caneva (✉)
Department of Sciences, University Roma Tre, Rome, Italy
e-mail: giulia.caneva@uniroma3.it

F. Bretzel
National Research Council (CNR), Institute of Research Terrestrial Ecosystems (IRET), Pisa, Italy

P. Belardi · M. Fornaciari · V. Menchetelli
Department of Civil and Environmental Engineering, University of Perugia, Perugia, Italy

C. M. Musarella
Department of AGRARIA, Mediterranean University of Reggio Calabria, Reggio Calabria, Italy

D. Romano
Department of Agriculture, Food and Environment (Di3A), University of Catania, Catania, Italy

G. Caneva et al. (eds.), *Urban Botany*, Future City 22,
https://doi.org/10.1007/978-3-032-28986-5_6

the native ones, are here enhanced. The most appropriate and sustainable cultivation techniques and some cases studies having this approach are also explained.

The view of technology and design in using plant material is finally analyzed. The awareness on the complex role of plants in the ecological, historical, cultural and aesthetic point of view, suggests the role of vegetation as a true building material, capable of determining the quality of the inhabited space, influencing its perception, improving its microclimate and contributing to human well-being. Design and Living Models, with solutions of Living in Spaces Immersed / Covered / Enveloped by Greenery, are also shown, giving examples of modern installations in cities.

Keywords Ecology · Cytotoxins · Historical reconnection · Urban Horticulture · Sustainable cultivation techniques · Technology and design · Urban spaces quality

6.1 Introduction (G.C.)

Cities face complex challenges, including climate change, population growth, declining human well-being, and the need to implement the One Health paradigm. Consequently, many city administrators and residents are looking for new and innovative strategies to improve urban sustainability (Elmqvist et al. 2019). Among these strategies, a priority is to enhance green areas (GA) and green infrastructure (GI) in a Nature-based Solution (NbS) approach.

Approaches to GI planning can vary by professional perspective. Therefore, an integrated vision is needed to achieve results that meet multiple objectives, and any project should be developed by a multidisciplinary team.

6.2 The Ecological View: Enhancement of Autochthonous Flora and Biodiversity (G.C., C.M.M.)

The urgent actions needed to protect biodiversity and slow down climate change-debated at the international level and enshrined in various conferences (see Chaps. 1 and 5)—have been formalized in the European Green Deal and the Next Generation EU (NGEU), developed after the Covid emergency (implemented through the Recovery and Resilience Facility, i.e., PNRR). At the European level, the National Biodiversity Strategy 2020–2030 aims to reduce ecological debt, promote the ecological transition, stimulate the economy and employment in line with the green economy and sustainable development, and protect biodiversity and the functionality of ecosystem services.

Today more than ever, botanists must assume a significant responsibility in what is commonly referred to as Green Infrastructure planning, particularly in urban contexts, though not exclusively. In the past, this aspect was often neglected in favour of the botanist's more "natural" interests, or, more precisely, strictly naturalistic interests. This situation may also stem from the limited recognition accorded to this key professional figure, whose unique and irreplaceable expertise could greatly contribute to planning activities that incorporate plants in various ways.

In the past, perhaps due to the widespread use of cultivated plant species—most of which have an allochthonous origin—in the creation of green areas or in ecological restoration projects, it was common practice for architects and engineers to involve only agronomists or specialized nursery professionals in plant selection. This approach was certainly influenced by the general perception that botanists and agronomists—both working with plants—possess highly overlapping competencies. Most people find it difficult to recognize the specific characteristics and differences in expertise among the various specialists who study plants and, when questioned on the subject, often fail to understand how a botanist may not address certain practical and applied aspects of botany, or how an agronomist may lack familiarity with issues related to natural flora and vegetation.

But how can a botanist be professionally distinguished from an agronomist if, ultimately, both deal with plants?

At a simplistic level, one might argue—as already noted at the beginning of this chapter—that the botanist is concerned with spontaneous (wild) plants, whereas the agronomist focuses on cultivated (domesticated) ones. This distinction, however, is not entirely accurate, or at least only partially so.

The agronomist applies scientific knowledge to the practical management of cultivated plants, possesses expertise in their requirements under non-natural or controlled conditions, and is capable of diagnosing and managing plant diseases. The botanist, by contrast, investigates and understands plants across a wide range of perspectives, including their classification, structure, physiology, genetics, ecology, evolution, and interactions with the environment.

Furthermore, the botanist is knowledgeable about phytogeography and the historical dimensions of plant life, both ancient—through archaeobotany—and more recent, such as plant introductions associated with trade networks, territorial expansions by certain civilizations, or geographical explorations. In essence, the botanist is an expert in plant biodiversity: not merely someone who knows the scientific names of plants, but a specialist capable of interpreting plants within an evolutionary and adaptive framework. In some cases, botanists also engage with topics that intersect with human culture, such as ethnobotany, which examines the relationships between humans and the plant world, particularly in regions where ancient traditions and long-standing cultural practices persist (e.g., Zocchi et al. 2022; Musarella et al. 2024). Other areas include historical botany and phytoiconology, which explores human-plant cultural relationships, as reflected in artistic representations (Caneva et al. 2005), as well as the botany of historic gardens and (Fig. 6.1) archaeological sites, which provide evidence of the beliefs and attitudes toward the plant world held by specific civilizations in different historical periods (Raimondo et al. 2022a, b).

Fig. 6.1 Ancient monumental individuals of planes; (**a**) *Platanus orientalis* at the Botanical Garden of Sapienza University of Rome; (**b**) *Platanus x hispanica* at the Rectorate of the University of Münster. (Photo by G. Caneva)

In all these contexts, the distinction between an agronomist and a botanist becomes particularly evident, as the latter possesses in-depth knowledge of the theoretical and cognitive aspects of plant science—namely, the ability to analyse the specific characteristics and peculiarities of plants. The agronomist, by contrast, is an expert in cultivation techniques, cultivar selection, acclimatization, agroforestry applications, plant utilization, and plant pathology—competencies that are essential for the practical care and management of plants.

In this sense, the botanist should play a more prominent role during the initial phases of plant selection and species choice, whereas the agronomist should be primarily responsible for overseeing plants during cultivation. Clearly, these roles and phases of intervention cannot be rigidly separated; rather, they must be closely interconnected to maximize the effectiveness of any intervention, considering different fundamental considerations (Fig. 6.2).

In fact, it raises the key question: *which plants should be selected for a project, and on what criteria*?

Every choice must be guided by well-defined criteria; thus, the answer appears, at least in principle, straightforward: namely, those plants that are floristically and ecologically suitable and functionally aligned with the established objectives. In practice, however, this answer is far less easy to implement, as it immediately raises the need to clearly define both suitability and functionality.

What, then, does floristic and ecological suitability mean?

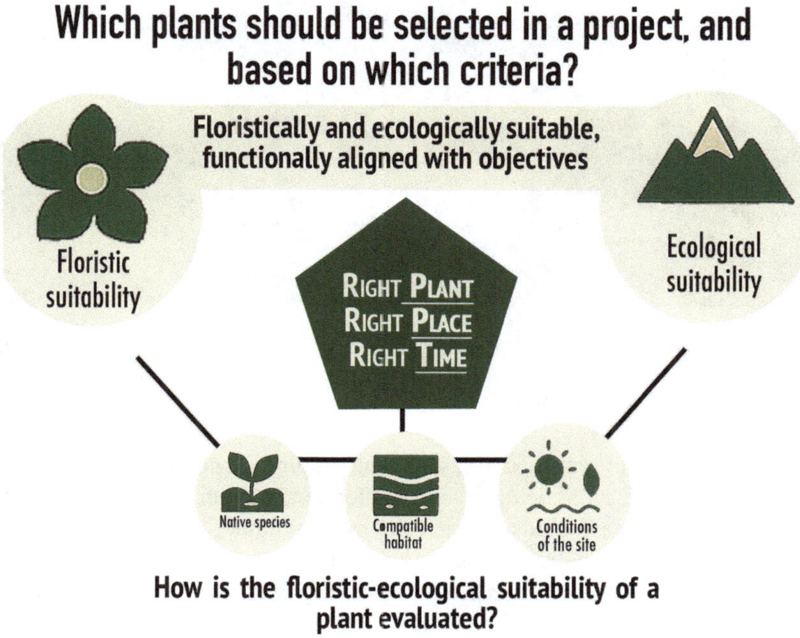

Fig. 6.2 Selection criteria in an ecological view. (Elaboration by the Authors)

A botanist would first argue that the most appropriate plants to select are native species from compatible habitats, that is, species that occur naturally within the geographical context under consideration and that simultaneously exhibit ecological suitability, meaning that they are well adapted to the conditions of the site where they are intended to be established. These conditions include soil characteristics, solar exposure, precipitation regime, daily and seasonal temperature fluctuations, wind exposure, and related factors.

As will be emphasized later, while this represents the baseline option, altenative choices cannot be ruled out when ecological or functional constraints make such an approach difficult to pursue. In a simplified formulation, one might say *that plant selection should rely on choosing the right plant, in the right place, at the right time* (see also Chap. 7).

But how can these three fundamental criteria be identified and applied?

In more technical terms, how is the floristic-ecological suitability of a plant evaluated?

This issue is well established in the scientific domain but remains less widely addressed in applied practice. Indeed, for more than a century, the development of vegetation science and the recognition of recurrent plant associations in nature whenever specific environmental conditions are present —a cornerstone of phytosociology—have provided the foundation for identifying which species are in equilibrium with given environmental settings and which, by contrast, should be considered unsuitable (Von Humboldt 1807; Braun-Blanquet 2013).

Fig. 6.3 Two side-by-side views compare a steep, rocky embankment under a bridge. (**a**) Work for the renaturalisation of a construction site along the Autostrada A2—Mediterranean motorway in Basilicata, Italy (**b**) Natural Forest cover in the surrounding area. (Photo by C.M. Musarella)

At the level of individual species, sufficient empirical knowledge has also accumulated to define the ecological ranges of different taxa with respect to major environmental factors. This has led to the development of numerical indicator scales that express species' ecological preferences (e.g., Ellenberg, Landolt, etc.).

In practical terms, the task consists of observing what nature presents and translating it into a design-oriented framework. Botanists are often involved late in the process, typically asked to resolve issues that arise when projects fail to obtain approval at various decision-making levels.

For example, in vegetation restoration interventions within construction sites, it is crucial to design re-naturalisation activities based on a prior study of the area, thereby enabling the designer to achieve a correct and ecologically appropriate restoration (Fig. 6.3).

However, this does not always occur, and the botanist is instead required to investigate what remains of the pre-existing vegetation by observing the surrounding areas and attempting to reconstruct the site's botanical history, often during a season unsuitable for this type of survey. At this point, the botanist's expertise reaches its limits, and all aspects related to potential aesthetic considerations—which are perceived differently depending on individual sensibilities and are often far from straightforward—as well as those more strictly tied to technical and agronomic issues, necessarily require the integration of additional expertise provided by other highly qualified professionals.

In pursuing their scientific objectives, botanists, both in relation to the interventions discussed above and to other types of projects, generally prioritize the use of native plant species. The reasons for favouring native plants are manifold. These species have undergone long-term genetic processes of natural selection that have enabled them to become finely adapted to their environments. As a result, they can establish and develop with minimal human intervention, exhibiting a high degree of resistance to limiting factors and various stressors. This leads to savings in both costs and maintenance requirements, while also ensuring good establishment success precisely because of this inherent adaptability.

Fig. 6.4 Typically Mediterranean species commonly used in green city plans. (**a**) *Cupressus sempervirens* trees bordering the perimeter of the Fira de Barcelona exhibition centre in Spain, (**b**) *Arbutus unedo* L.; (**c**) *Erica arborea* L.; (**d**) Pistacia lentiscus L.; (**e**) *Quercus ilex* L. (Photo by C.M. Musarella)

A further advantage lies in the ability of native plant systems to establish particularly effective trophic and functional relationships with the surrounding biological microcosm, ranging from soil microflora to pollinator communities, frugivorous birds, and other ecosystem components. This, in turn, helps maintain and strengthen the ecological balance of the environments into which they are introduced.

A very important aspect to consider is plants' capacity to serve as identity-forming and cultural elements for a specific place or for a broader geographical area, making such areas recognizable and distinctive. This holds true regardless of whether the species involved are native or long- established alien taxa. When thinking of the Mediterranean region, for instance, people immediately associate it with holm oaks (*Quercus ilex* L.) and olive trees (*Olea europaea* L.), but also with Mediterranean cypresses (*Cupressus sempervirens* L.) and stone pines (*Pinus pinea* L.), the latter two being species introduced in certain Mediterranean countries in ancient times (Fig. 6.4a–e).

Similarly, avenues lined with date palms (*Phoenix dactylifera* L.) evoke North Africa and the Near East, while araucarias (*Araucaria* sppl.) recall more distant American landscapes. In contrast, contemporary practice increasingly and repeatedly favours a limited set of cultivated species and ornamental cultivars—such as petunias, African marigolds, ornamental salvias, and pansies (*Petunia* sppl., *Tagetes erecta* L., *Salvia splendens* Sellow ex Nees, *Viola tricolor* L.)—which are used ubiquitously to embellish flowerbeds and parterres. These choices are often driven more by chromatic effects than by genuinely informed or context-specific selection.

It is evident, therefore, that all these factors must be carefully considered when selecting a species for the creation of a garden or for *ecological restoration*. In the first case, it may be acceptable to include a few alien species if they are particularly well-suited to the decorative or symbolic objectives of a specific site, which may be regarded as "artificial" or "human-made," and within a limited context. In contrast, ecological restoration should employ exclusively native species, as they are the only ones capable of sustaining the long-term ecological balance of a natural site.

Indeed, the persistence of numerous other plant, animal, bacterial, and lichen species depends on these native plants, as they have established intricate ecological relationships throughout the area's evolutionary and adaptive history. Failure to respect and properly re-establish this balance risks compromising the site's entire biodiversity, whether extensive or limited, and may also affect neighbouring ecosystems.

Thus, the goal is not to categorically exclude alien plant species for ideological or "patriotic" reasons; this is not science. Scientific reasoning inevitably intersects with history, particularly the history of migrations and biological introductions that have occurred across different periods.

Moreover, the principle of preferential use cannot be applied indiscriminately to all cultivated alien species, which have clearly defined origins and, in some cases, possess historical and cultural significance within a given area. Many of these cultivated species are also subject to periodic replacement due to economic, political, or environmental considerations. For example, in various Italian regions, the gradual replacement of citrus orchards or vineyards with kiwi plantations —originating from China, though most widely cultivated in New Zealand (*Actinidia chinensis* var. *deliciosa* A.Chev.), illustrates such shifts (Testolin and Ferguson 2009; Gullo and Dattola 2025).

Nevertheless, the role of plants in conveying *landscape identity* has already been emphasized, and certain limitations must be considered—particularly in the design of public areas, especially those with strong historical significance, such as archaeological sites and historic gardens (see Chap. 7). In these contexts, design decisions should follow more careful and deliberate criteria than those an individual citizen might reasonably when making personal choices, which may be influenced by intrinsic characteristics of the species or by the specific features of the site to be designed.

For instance, the potential invasiveness of alien species and the threat they pose to the conservation of natural communities must be considered (see also Chap. 3). Once acclimated, these species can become highly competitive with local flora, leading to their rapid decline and, consequently, affecting other species that depend on them. Furthermore, alien species originating from climates that differ substantially from those of the target site often require intensive cultivation and management, as they are unable to readily adapt to the new climatic and environmental conditions. This can result in unsuccessful plantings, increased economic costs due to ongoing maintenance or plant mortality, and the need to redesign or reschedule the planting intervention.

The issue of the competitiveness of alien species represents one of the most critical challenges: history demonstrates how plants introduced for specific purposes can, over time, prove to be disastrous choices (Ispra n.d.; Hu 1979), as in the case of the tree of heaven (*Ailanthus altissima* (Mill.) Swingle) (see Chap. 3), which was at times used in urban projects (Fig. 6.5).

Fig. 6.5 Planting of *Ailanthus altissima* (Mill.) Swingle in the Macro Museum in Rome (**a**, **b**) as an example of neglecting the role of br/alien invasive species. (Photo by in exhibit, 2015)

Thus, when selecting an alien species for a conservation or restoration objective, the short-, medium-, and especially long-term costs of the intervention must be carefully considered. What may initially appear to be a simple or low-cost introduction can later require substantial public expenditure for eradication measures, which themselves follow earlier public investments.

It is, of course, acceptable in certain cases to use alien species, if they are not classified as invasive or potentially harmful to natural habitats and do not pose risks to infrastructure, such as species whose seeds readily germinate in cracks in walls, pavements, or roads.

Furthermore, caution must be exercised when selecting alien plants that are typically favoured for their flowers, fruiting, or overall phenology, particularly when their aesthetic appeal may come at the expense of cultural or landscape integrity. It is therefore not surprising that an urban avenue in southern Italy, Spain, or Portugal might be lined with *Citrus* × *aurantium* L., a taxon traditionally used as a rootstock for various citrus species historically cultivated in these countries. (*Citrus* × *bergamia* (Risso) Risso & Poit., *Citrus* × *sinensis* (L.) Osbeck, etc.) (Continella et al. 2018). In addition to being highly resilient and requiring minimal maintenance, it also produces a striking aesthetic effect, with its fruits remaining attached to the canopy for an extended period.

The need and the benefits of protecting and enhancing characteristic elements from both naturalistic and cultural perspectives are increasingly recognized at the international level, particularly in a globalized world ever more shaped by human activities. Consequently, all planning and design processes should acknowledge and incorporate these considerations.

Finally, what actions should be taken beyond the design phase?

Transposing these concepts—general relevance—to the field of horticulture, it is essential to emphasize, ideally as one of the primary objectives of a national horticultural plan, the importance of ensuring that both production and subsequent design activities take into account the specific botanical, ecological, and historical characteristics of plant species. This approach aims to enhance the most significant elements of our landscape so that they contribute to the perception of identity within a given environmental context, as articulated in the manifestos of stakeholders committed to promoting collaborative and forward-looking design strategies (Caddeo et al. 2020). This entails guiding the selection of plant species in a broader and more informed manner, considering their multiple values and functions. Such selection should account for their contemporary and historical significance, as well as cultural, maintenance, or technical requirements relevant to their intended role.

All of this must be integrated into a philosophy of green design—particularly urban design, though not limited to it—that emphasizes true naturalistic enhancement and historical reconnection, rather than focusing solely on aesthetic or decorative aspects. Unfortunately, integration across disciplinary knowledge often remains a weak point in complex systems.

From this perspective, the process of green design presupposes a preliminary phase of acquiring knowledge about the distinctive characteristics of the local vegetative landscape, considering its defining bioclimatic and biogeographical traits, as well as those revealed through a historical analysis of landscape transformations. Where necessary, this should also include archaeobotanical and ethnobotanical investigations.

6.3 The Agronomical Approach: Cultivated Plants in Urban Ecosystems (D.R., F.B.)

Urban greenery—recreational or ornamental and generally accessible to the public— is a category of interventions applicable to parks within settlements (with or without public access), ornamental gardens, the green grounds of mansions, and botanical or zoological gardens located inside the urban fabric or within the peripheral contact zone of a settlement. It also includes city squares with greenery, inner courtyards of city blocks, cemeteries with vegetation located within or directly adjacent to settlements, and vegetated areas that can potentially be used for recreational purposes, even if this is not their primary function, such as wooded areas within urban fabric.

Urban horticulture, including rooftop gardens, has become increasingly popular both for productive purposes and as a tool for social cohesion. Urban horticulture is defined as *the cultivation of fruits, vegetables, herbs, and ornamentals in urban environments* in any form. It can provide diverse, nutritious foods, promote inclusive and equitable access to food in urban food deserts, and offer environmental benefits (Congreves 2022). Urban food forests complement urban horticulture in a more sustainable and natural context. The innovative aspects of these edible landscapes are encapsulated in two key concepts: *forest and permanence*. The term "*forest*" refers to the ability to emulate the successional functions and dynamics of natural and semi-natural forests by combining different species across multiple layers to maximize synergies and optimize the use of soil and light resources. The term *permanence* relates to time and to cultivation systems based on permaculture principles, resulting in consciously designed landscapes that mimic the patterns and relationships found in nature (Salbitano et al. 2019).

In any case, all areas covered by plants can provide benefits known as ecosystem services (ESs), which are fundamental to improving the sustainability of cities (Francini et al. 2022). Therefore, studying the GA and GI, as well as the agronomic strategies used to enhance their characteristics, has become a research priority in cities worldwide (Vila-Ruiz et al. 2014). The focus is on cultivated plants—those intentionally introduced by humans to perform various functions (aesthetic, nutritional, and environmental improvement). Growing plants in urban environments is necessary because we live in an anthropic setting, where the space and time needed for spontaneous evolutionary processes of plant communities to reach ecological balance are lacking. Cities are dynamic spaces where human use is generally the priority; as a result, nature is often vulnerable to damage or destruction, and risks posed by plants and animals must be controlled.

6.3.1 Choosing Criteria for Plants in the Urban Environment

The choice of plant species in relation to the type of green areas is linked to its intended use, even though NbS encourages a multifunctional approach in which plants simultaneously provide ornamental, mitigating, edible, and biodiversity-related functions. The selection criteria—or rather, the idiotype of plant species or plant assemblages—vary across typologies of GA or GI (Lundholm and Richardson 2010). However, when ecological conditions are not limiting, it is possible—and often recommended—to use a wide range of species, thereby increasing biodiversity and enhancing the capacity of plant cover to reduce pollution and improve air quality (Garbisu et al. 2020).

Botanists and ecologists generally recommend favouring native plant species due to their high adaptability to local climatic conditions. However, the urban environment differs significantly from the surrounding rural areas because of factors such as the heat-island effect and pollution, and historically it has served as a crossroads for plant introductions. In the review by Russo et al. (2025), several studies provided evidence that, in a constantly changing climate, the use of non-native species may be advisable due to the greater resilience of some widely tolerant species. Nevertheless, these species can also pose significant ecological risks, including competition with native flora, disruption of ecosystem dynamics, and potential invasiveness (see Chap. 3).

Some non-native plants—especially in regions characterized by high temperatures, low rainfall, or extreme weather events, conditions often found in the Mediterranean region—may exhibit greater resilience (Figs. 6.6 and 6.7).

Therefore, further research—through both monitoring and field studies—is needed to assess the long-term ecological consequences of introducing non-native species into urban environments. Understanding where and when non-native

Fig. 6.6 Examples of succulent plants, very resilient species to a drought environment. From top left: *Yucca rostrata* Engelm. ex Trel., *Myrtillocactus geometrizans* (Mart. ex Pfeiff.) Console, *Graptopetalum paraguayense* (N.E.Br.) E.Walther, *Echeveria setosa* Rose & Purpus, *Aeonium undulatum* Webb & Berthel., and *Portulacaria afra* Jacq. (Photo by D. Romano)

Fig. 6.7 Non-native species adapted to the Mediterranean climate: From left *Ficus macrophylla* Pers., *Archontophoenix alexandrae* (F. Muell.) H. Wendl. & Drude, and *Bougainvillea* Comm. ex Juss. sp. (Photo by F. Bretzel and D. Romano)

species can be used will help ensure that their potential benefits do not come at the expense of ecological integrity. At the same time, conservation strategies, such as biodiversity preservation and protection of remaining native vegetation, offer valuable means of supporting biodiversity and maintaining ecosystem services in urban areas.

To achieve this, raising public awareness is essential, as people are often attracted to more striking exotic species. Increasing awareness can help ensure that urban

Fig. 6.8 Some native Mediterranean shrubs with ornamental potential, hardly present in nursery catalogues. From top left: *Erica multiflora* L., *Prasium majus* L., *Euphorbia dendroides* L., *Cistus salviifolius* L., *Thymbra capitata* (L.) Cav., *Rosa canina* L., *Genista aetnensis* (Raf.) DC. (Photo by D. Romano)

green spaces function as conservation tools while maintaining the visual and recreational qualities that engage the public. To strengthen ecological resilience in cities facing climate challenges, participatory planning and public education must be actively promoted. Future research should also focus on identifying alternative species better suited to warmer, drier climates, thereby providing a concrete response to the growing need for functional green spaces in urban environments. Another crucial aspect concerns the commercial availability of native plants and the knowledge of their cultivation requirements, which are often lacking. In a review of Mediterranean native shrubs suitable for use in green areas, 369 out of 544 identified taxa (67.8%) were found to be unavailable in commercial nurseries (Fig. 6.8) (Leotta et al. 2025).

As noted by Capotorti et al. (2020), it is essential to place "*the right plant in the right place*". The selection of species for urban parks can be based on their adaptability to environmental conditions, their aesthetic and ecological value, their suitability for low-maintenance management, and the benefits they can provide in terms of ecosystem services (Fig. 6.9).

The selection of appropriate ornamental plant species is a key factor in creating a sustainable green space—one that reflects the local natural environment (Ghafari et al. 2020). Plant adaptation depends not only on genotype characteristics but also on environmental conditions and management practices (Van der Veken et al. 2008), given that the effects of climate change can be more rapid and extreme in cities than in rural areas.

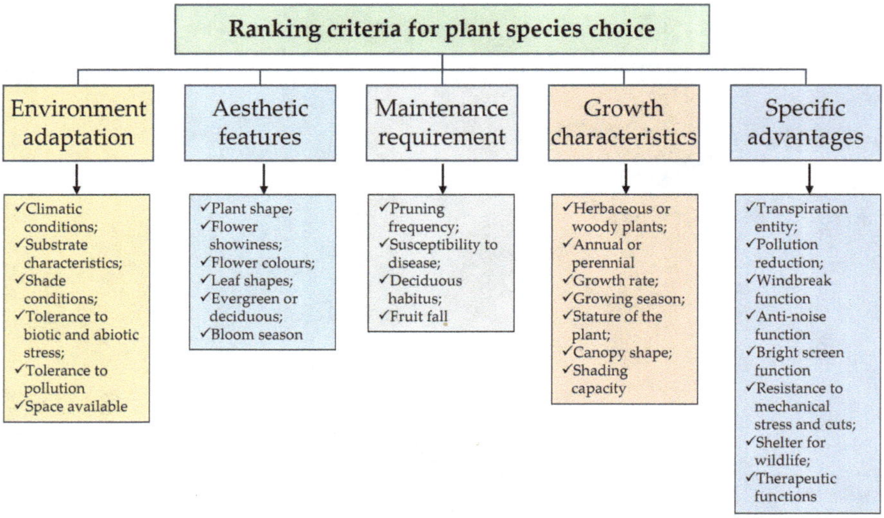

Fig. 6.9 Ranking criteria for plant species choice for sustainable green areas. (Courtesy by Leotta et al. 2023)

Miller (1997) proposed a species-selection scheme that considers site factors (i.e., environmental and cultural constraints), social factors (i.e., aesthetics, functions, and disruptions), and economic factors (i.e., planting and maintenance costs). Drought tolerance and cold hardiness are critical for the future survival of trees in a changing climate, and these traits can be assessed based on the climatic conditions of the species' places of origin (Roloff et al. 2009).

In addition to the characteristics of their environment of origin, plants may exhibit physiological plasticity, which enables them to adapt to stress and disturbance in multiple ways—sometimes even amplifying the severity of the effects (Toscano et al. 2021).

Higher temperatures, for example, increase the evapotranspiration demand and therefore intensify drought stress, making plants more susceptible to pest attacks (Dale and Frank 2014). Similarly, salinity resulting from recycled irrigation water or coastal flooding can negatively affect soil health and plant growth. Species with narrow tolerance ranges are likely to be the most adversely affected. Herbaceous and shrub life forms, owing to their greater plasticity, may be more suitable for such conditions than trees, which often have limited genetic capacity for adaptation due to their long lifespan.

The complexity involved in selecting the appropriate plant species often results in a reduced number of species being used—typically those considered more "reliable". However, maintaining biodiversity in urban green spaces helps reduce risks associated with pests, diseases, and climate change, thereby improving the resilience of ecosystem services provided by green infrastructure (Toscano et al. 2025).

Santamour proposed the 10/20/30 'rule of thumb', which states that urban forests should comprise no more than 10% of any plant species, 20% of any genus, or 30% of any single botanical family (Kendal et al. 2014). An important contribution to urban forest biodiversity can be made by shrubs, whose small size allows many individuals to coexist within a green area and whose resilience enables them to withstand numerous biotic and abiotic stresses. Wildflower meadows—which include a mix of flowering dicotyledons and monocotyledons and are generally established by sowing— can also enhance the biodiversity and resilience of GA. They are managed with low cultivation inputs, such as annual mowing, and typically without irrigation or fertilization. As a result, flowering follows a seasonal pattern determined by the site's climatic conditions and species coexistence. Other plants cultivated in urban spaces include those in private areas—gardens, balconies, windowsills, and vegetable gardens. These plants contribute to the provision of ES and should be considered integral to the urban green heritage.

Urban environments can sustain plant species biodiversity due to environmental and land cover heterogeneity (Corine classification 2025), socioeconomic factors, and the introduction of new species. Urban green spaces include many different types of habitats, ranging from intact patches of native vegetation to highly engineered systems such as green roofs (Aronson et al. 2017). In addition to the pedo-climatic conditions, the level of care and management is a key factor in determining which species can be cultivated in a GA (Fig. 6.10).

Fig. 6.10 The pedo-climatic conditions and the degree of maintenance drive the plant choice for urban gardens. (**a**) The Mediterranean maquis at the Stavros Niarchos Park in Athens, (**b**, **c**) flowering borders in Santiago de Compostela (Spain), (**d**) public park in Trieste (Italy). (Photo by F. Bretzel)

Improving our understanding of the contribution of urban biodiversity will enable GI to deliver the ecosystem services needed to support an increasingly urban global population.

6.3.2 The Most Appropriate and Sustainable Cultivation Techniques

Traditional ornamental plants require substantial cultivation inputs, including fertilization, pruning, watering, high-quality soil, and skilled gardeners who can distinguish weeds from cultivated species and apply appropriate horticultural techniques. In the absence of these optimal conditions, plants deteriorate and lose their aesthetic value, failing to provide the ES. In some cases, however, sustainable cultivation techniques can increase the plant tolerance to abiotic and biotic stresses, for example, through the use of *biostimulants* that induce or enhance stress-response mechanisms (Toscano et al. 2018).

Landscape plants are grown in nurseries under controlled, optimal conditions of nutrient availability and humidity. Once transplanted into GA, they are exposed to post-transplant stressors, including root loss, water deficit, pests and diseases, and poor soil fertility (Koller 1977). Therefore, in addition to choosing the appropriate planting season, it is essential during the initial establishment period to minimise stress by compensating with higher cultivation inputs. Transplanting methods play a crucial role in ensuring full establishment. The timing of planting is particularly important: in Mediterranean climates, autumn transplanting—before the onset of the rainy season—promotes better plant establishment. However, the limited ability of container-grown seedlings to rapidly develop deep and well-structured root systems after planting (Vallejo et al. 2000) can hinder survival. Arbuscular mycorrhizal (AM) symbioses can enhance root system function by improving water uptake and transport efficiency (Caravaca et al. 2003; Green et al. 2005).

In recent decades, high temperature, drought, salinity, and pollutants have increasingly been recognized worldwide as major stresses that negatively affect plant growth and survival. To cope with these adverse conditions, plants activate a range of morphological, physiological, and biochemical mechanisms. In addition, *biostimulants* are widely used to support plant resilience and survival. According to EC Regulation 2019/1009, *biostimulants* are products that stimulate plant nutrition processes independently of their nutrient content, with the aim of improving one or more of the following plant characteristics: (a) nutrient use efficiency; (b) tolerance to abiotic stress; (c) quality traits; (d) availability of confined nutrients in soil or rhizosphere.

Biostimulants are products derived from natural raw materials, such as hydrolysed proteins and amino acids from animal and plant by-products, micro-algae and seaweed extracts, humic substances, plant extracts, and microorganisms (La Torre et al. 2016). They are typically applied as foliar spray and can enhance plant growth, stress tolerance, and yield and productivity of many crops. Plants treated with commercial biostimulants—including mycorrhizae, hydrogel, and similar products—may show increased resistance to constraints such as poor soil fertility, drought, disease, pests, and urban pollutants (Newman and Davies Jr 1987).

Arbuscular mycorrhizal (AMF) and humic substances are two of the seven main categories of biostimulants that have been used separately in previous studies to improve plant nutrient uptake, growth, and development, as well as to enhance tolerance and resistance to abiotic stress and to promote soil structural stability (Aalipour et al. 2020). Numerous studies indicate that the symbiotic relationship between plants and AMF is a key factor in helping plants tolerate and/or resist abiotic stress.

Mulching serves multiple functions, including weed control and reducing competition between ornamental plants and spontaneous species. It can also be effective in reducing water loss and improving the water-use efficiency of ornamental plants. Mulching materials may be organic—such as pine bark, biodegradable films, crushed plant waste, straw, or compost—or inorganic, such as crushed shells (Fig. 6.11).

Plant assemblages can help reduce management costs and prevent agronomic problems by taking advantage of complementarity among species at multiple levels. Such a combination must be planned to ensure at least no negative interference at the root and canopy levels; however, synergistic effects can also be achieved. Plant density and species distribution should be arranged by placing plants with different root expansion types—both shallow and deep—so they can exploit nutrients at different soil depths.

Plants belonging to the Fabaceae family, which are able to fix atmospheric nitrogen through symbiosis with *Rhizobium*, are considered pioneer species capable of colonising poor soils and supplying nitrogen to neighbouring plants. This can reduce the need for nitrogen fertilizers and, consequently, lower management costs. It is important to note that although chemical fertilization is used to promote plant growth, it can make plants more vulnerable to pests and drought and increase maintenance requirements (such as mowing and pruning). In contrast, the addition of organic amendments (such as compost and similar materials) improves long-term soil fertility and helps counteract soil degradation.

Fig. 6.11 Different types of mulching (pine bark on the right; crushed shells on the left). (Photo by the Authors)

Positive interactions may also arise through improvements in soil structure: for example, tree roots can enhance water infiltration, reducing flooding and benefiting species that struggle in heavy or compacted soils.

Interactions among plants—both *interspecific* (between different species) and *intraspecific* (within the same species)—also include antagonistic relationships. Species that produce allelopathic compounds should be planted at a sufficient distance from each other to avoid negative effects. All situations of antagonism should be avoided, as they can lead to losses in ornamental value, such as leaf yellowing and stunted growth.

The tree canopy should be managed to prevent branch crossing, which can restrict air circulation, increase the accumulation of gas, and raise the incidence of plant diseases or pest infestations. High planting density leads to elevated maintenance costs—particularly for pruning—and increases the likelihood of plant failure due to excessive shading. Combining plants with canopies of different heights can provide several positive benefits, including the possibility of planting shade-lover species underneath, which often develop darker leaves when grown under reduced light. The Japanese agronomist Miyawaki developed a precise method for establishing small forests (100–2000 m^2) in urban gardens by mimicking natural succession. The method involves improving soil fertility with mycorrhizae and mulching, densely planting young individuals of native trees and shrubs (1–3 m^2) and allowing them grow with minimal maintenance. Examples of Miyawaki-style tiny forests can be found in Rome at the San Lorenzo neighbourhood and Botanical garden (Schirone et al. 2025).

Drought stress imposed during the nursery phase can enable seedlings to develop greater root biomass and branching (Green et al. 2005). Alpha grass (*Stipa tenacissima* L.) has been cultivated in nurseries to facilitate the establishment of several native Mediterranean shrubs (*Medicago arborea* L., *Quercus coccifera* L., and *Pistacia lentiscus* L.). In this context, the grass significantly reduced photosynthetically active radiation and soil temperature, thereby improving shrub survival near *S. tenacissima* compared with open areas. Predawn water potentials measured before and after the summer were significantly higher in the area with the alpha grass (Maestre et al. 2001).

The micro-catchment water harvesting (Reddy 2016) can provide an effective solution even in very dry climates, as it captures rainwater and improves soil moisture and vegetation establishment. This technique promotes deep root development and reduces shrub mortality rates (Ali and Yazar 2007). required for effective water capture can also be incorporated into the plant composition and design of GI.

6.3.3 Case Studies

Here, we report some examples of urban parks (see also Chap. 7) that have been successful not only in terms of planting and management but also in attracting visitors.

The Stavros Niarchos Park covers an area of 21 hectares in the peripheral district of Kallithea —meaning "beautiful view"— and hosts a rich variety of flora, including trees (such as carobs, laurels, cypresses), evergreen shrubs, and an extensive selection of native aromatic plants (*https://www.snfcc.org/en/*). Each species is highlighted monthly on the park's website. In 2017, a Vegetable Garden was created as a dedicated area for both urban cultivation of organic vegetables and educational activities for children and adults (Fig. 6.12).

The BAM park is developed around the green building "Bosco verticale", designed by Stefano Boeri (see also Chap. 9). It includes 100 plant species, 500 individual trees, and 135,000 aromatic plants, bulbs, herbaceous species, shrubs, climbers, and aquatic plants (https://bam.milano.it/-). The Dutch landscape designer Piet Oudolf contributed to the design of the perennial garden (Fig. 6.13).

Gray to Green is a project carried out in the city centre of the former industrial city of Sheffield, Yorkshire, aimed at encouraging citizens to walk and cycle instead of driving in the city (https://www.greytogreen.org.uk/.). At the same time, the city council implemented NbS to manage the flooding and other challenges associated with climate change. Many pollinator-friendly species were planted to improve habitat quality (Fig. 6.14).

Fig. 6.12 Stavros Niarchos Park (Athens, Greece) (https://www.snfcc.org/en/)

Fig. 6.13 BAM (Milano Italy) (https://bam.milano.it/-)

Fig. 6.14 Gray to Green (Sheffield, UK) (https://www.greytogreen.org.uk/.)

6.4 The View of Technology and Design (M.F., V.M., P.B.)

6.4.1 Plant Material: Identity and Function

In landscape design, plant material plays a role that extends far beyond the ornamental dimension to which it was relegated for decades. Although greenery is profoundly connected to human nature, the construction of modern cities has often marginalized it, treating it as an accessory or a simple service element, lacking the design dignity attributed to architectural materials. In reality, vegetation carries deeply layered historical, cultural, and aesthetic meanings and represents an essential component of contemporary responses to the ecological crisis. This awareness requires that we recognize vegetation as a true building material—one capable of determining the quality of the inhabited spaces, shaping their perception, improving their microclimate, and contributing to human well-being.

From this perspective, green space should be conceived as a continuum between nature and architecture, a fusion in which traditional disciplinary boundaries blur and give way to an integrated design language—not two separate worlds, but a single living structure in which plant and built elements work together to shape the environment. In this sense, a garden can be understood as an "open house", articulated across three fundamental levels: ground, vertical, and ceiling, where vegetation performs functions equivalent to those of the floor, walls, and roof.

Within this framework, plant material performs two fundamental functions. On the one hand, it can replace elements of the built environment in creating, delimiting, and organizing spaces, through hedges that mark out areas, trees that establish lines of visual separation or act as windbreaks, masses of vegetation that protect from solar radiation, and low plantings that guide pathways or soften edges. On the other hand, greenery can reinforce and enhance the architectural features of buildings: climbing plants that enrich façades, colors and forms that draw attention to specific details, and contrasting vegetation that mitigates rigid volumes or amplifies focal points.

In this sense, plant material acquires the status of an autonomous architectural component, endowed with identity, function, dynamism, and the ability to create a dialogue between the built environment and nature. Plants act primarily as elements of spatial organization, defining different areas and guiding perception through their shapes, sizes, and colors. For example, alternating shrubs and taller trees allows for the modulation of visual depth and helps guide movement; the intervals created between groups of shrubs introduce perceptual pauses and small clearings that reduce the density of the vegetation and invite visitors to stop and rest; and alternating lawns and groundcover plants responds to different needs, facilitating foot traffic in active areas while enhancing more difficult or less frequented zones.

The plant material, therefore, not only defines the appearance of the space but also implicitly communicates its function: a grassy surface invites people to pause, play and engage in physical activity, while groundcover plants, herbaceous species and low shrubs create decorative surfaces and borders that are unsuitable for walking on but excellent for controlling erosion, slowing the growth of weeds, and enhancing the transitions between paths and ornamental areas.

The structuring of space through vegetation unfolds across the three fundamental planes that characterize any inhabited environment: the ground plane, the vertical planes, and the ceiling plane.

On the ground level, different surfaces determine the character and the functionality of the space: durable stone, wood, or concrete paving is necessary in areas with heavy foot traffic, while grass or groundcover is more suitable for secondary pathways, relaxation areas, decorative zones, or surfaces that must remain permeable to water and cool from a microclimatic standpoint. Lawn offers soft, welcoming surfaces, whereas groundcover plants serve a decorative and ecological function, particularly in challenging or extensive areas. Ensuring proper transition between different surface types guarantees functionality, durability, and aesthetic coherence.

Vertical planes delimit and separate spaces, define their perimeter, and provide order, balance, and perceptual comfort. They include both built elements, such as walls and fences, and plant elements, such as trees, hedges, and groups of shrubs. Trees, thanks to their trunks, act as true vertical structural components. When planted densely and in alignment, they can form natural walls that create visual and acoustic barriers while guiding the gaze and movement through their spatial rhythm. The foliage of shrubs and trees, on the other hand, generates softer and more variable boundaries: the denser and more compact it is, the more enclosed and intimate the space appears; conversely, lighter, airier foliage allows light and views to filter outward. Evergreen shrubs, with vegetation beginning at ground level, are particularly effective in defining boundaries and articulating areas. They maintain their function throughout all seasons and often offer ornamental features such as flowers and fruits.

The upper closure, formed by tree canopies and any plant pergolas, plays a dual role: it visually defines the space from above, giving it a sense of completeness and protection, and it contributes to environmental comfort through shading, light filtration, wind protection, and noise attenuation. Broad tree canopies can create intimate, welcoming green rooms with a significantly more comfortable microclimate.

The strategic use of plant elements of different heights allows for graded levels of enclosure and diversified spatial perceptions. A layered vegetative wall can include tall trees, intermediate shrub rows, and herbaceous or groundcover species at the ground level, producing an articulated depth and spatial hierarchy that optimizes both functionality and perceived quality. Moreover, plant variety is not only an aesthetic asset: it enriches biodiversity, supports pollinators, provides shelter and food for birds and small mammals, ensures ecological continuity, and strengthens the wider food chain.

The choice of species requires precise criteria: floristic and ecological suitability, based on the plant's compatibility with the site's environmental conditions, and functional suitability, based on development, phenology, spatial role, cultivation requirements, and aesthetic value. Native species are generally preferable for their resilience and ecological importance. However, exotic species may be used with care when they enrich the composition without becoming invasive or requiring excessive maintenance. Ultimately, plant material is not a decorative complement but an architectural, perceptual, and ecological device that constructs living, dynamic spaces, articulates functions and perceptions, mediates between nature and the built environment, and contributes to a new way of conceiving the city as a habitable, complex, and interconnected ecosystem.

6.4.2 Surveying and Representation

Within the field of architectural surveying and representation, urban greenery is increasingly conceived as an autonomous design material, variable in form, dynamic, and dependent on environmental conditions. Vegetation, understood as a complex object of knowledge rather than a mere figurative complement to the built environment, requires specific investigative tools, measurement procedures, and representation techniques. The evolution of digital technologies (from digital photogrammetry to parametric modeling) and their integration with territorial information systems have, over the past two decades, laid the groundwork for the development of a true *discipline of vegetation surveying* within architectural and landscape sciences (Bianchini and Hewage 2012; Apollonio et al. 2013).

From a methodological standpoint, surveying vegetation does not merely involve the morphological and metric recording of species and plant volumes; it requires an understanding of a synergistic set of geometric, biological, and environmental properties. Tree canopies influence visual permeability, leaf density affects light distribution and spatial characterization, and seasonal cycles generate formal and perceptual variations. Representation must therefore also convey the dynamic transformation of vegetation and its relationships with built space (Forman 2014). Today's methodologies allow greenery to be treated, at different levels of accuracy and for varying purposes, as a component that is objectively measurable and parametrically modellable. Urban vegetation surveying can be articulated into three main methodological strands, reflecting its intrinsic complexity and derived from architectural and urban surveying, yet characterized by thematic distinctions specific to vegetation.

The traditional approach is direct surveying, suited to specific sites or *in situ* assessments. It involves metric measurements carried out with distance meters or topographic instruments that document vegetation masses and their relationships with built structures. Botanical survey forms (MATTM 2017) are typically compiled alongside these operations to record taxonomic identification, phytosanitary conditions, and maintenance requirements. Photographic and chromatic surveys document textures, phenological states, and potential perceptual alterations.

Three-dimensional digital surveying—based on data acquisition through active and/or passive sensors and complementary to direct surveying—is now central to both research and practice. Digital photogrammetry (terrestrial and aerial via UAV) generates high-resolution 3D models for volumetric and phenological assessments; laser scanning (terrestrial TLS or airborne LiDAR) captures the internal geometry of canopies, extracting quantitative parameters such as height, trunk diameter, and leaf area density. These data are then integrated into parametric models and Landscape BIM environments (Landscape Institute 2016), forming an interoperable information model suitable for advanced design simulations.

Cataloguing and the development of information systems integrate the two methods, providing a systemic framework for managing vegetative assets. Geographic Information Systems (GIS) enable geolocation, the study of ecological correlations, and the modeling of environmental scenarios. Botanical databases and digital repositories (e.g., i-Tree or PlantNet) support quantitative analyses and taxonomic comparisons, while digital atlases of vegetation unify surveying, monitoring, and design within a single digital environment.

The methodological synthesis depends on the desired level of analysis. For morphological–spatial investigations, photogrammetry and laser scanning provide geometric and volumetric data for 3D modeling and architectural integration. Biological–physiological analysis employs botanical forms and sensors to collect phytosanitary and phenological data, supporting species selection and maintenance strategies. Environmental and territorial analysis, conducted using GIS and climate datasets, generates ecological and performance information to improve sustainability and ecosystem services. Finally, perceptual–sensorial surveying– based on observation and the production of photoplanes from photogrammetry– captures chromatic, luminous, and tactile data that underpin the design of spatial quality and its aesthetic representation (Disney 2019).

Representing vegetation poses a distinctive challenge for architectural drawing, as it requires accounting for its temporal variability—growth, seasonal changes, and volumetric variations—as well as its density and visual permeability. Parametric models and BIM/IFC systems for landscape (Landscape BIM) enable the simulation of different vegetation development scenarios, offering operational support for evaluating microclimate, shading, and interactions with built space. At the same time, traditional two-dimensional representation is adopting more systematic forms: in surveying and design, greenery is expressed through graphic conventions that reflect vegetation categories, phenological states, density, and spatial functions. Drawing thus reveals its nature as a language of translation and critical mediation between natural processes and the activities of documentation and design.

6.4.3 Design and Living Models

In Franco Purini's drawing (1984), the archetype of the house appears: a hut made of branches forming a structural frame, covered with foliage, and placed at the center of a clearing to observe and control its surroundings. The drawing, intentionally theoretical, underscores the complex relationship between nature and the city—ambiguous yet also synergistic and dialectical—since the latter originates precisely from the natural matter from which it simultaneously seeks protection (Purini 2022). It is no coincidence that, since ancient treatise writers on architecture—most notably Marcus Vitruvius Pollio—the forest has been regarded not only as the antithesis but also the foundation of architecture. The primordial dwelling type of the hut-in-the-clearing addressed the need for shelter from climate and predators while preserving a nostalgic bond with nature. Likewise, Laugier, in his *Essai sur l'architecture* (1752), saw in the primitive hut the precursor of both the classical temple and of the Corinthian order.

Urban history—through its millennial evolution, which has never renounced the *natura naturans* explored by Giordano Bruno and Baruch Spinoza—confirms this relationship by incorporating within itself "disciplined forests": from the Hanging Gardens of Babylon to Benedictine enclosed gardens, from complexes such as Villa d'Este and the *Saline royale* of Arc-et-Senans to Regent's Park and Central Park, up to modern utopias by Le Corbusier (*Ville Radieuse,* 1930) and Wright (*Broadacre City,* 1932). Today, continuing the ecological sensibility developed over the twentieth century, experiments such as green roofs and vertical gardens no longer treat nature as mere "scenography" or building material but as a design principle (Bellini and Daglio 2009). From this shift emerge three living models—dwelling immersed in greenery, beneath greenery, or enveloped by greenery—characterized by low ecological impact and aimed at improving quality of life through sustainability. This contemporary tendency, which seeks not only to reforest cities (Young 2011) but also to integrate vegetation and buildings, does not constitute an expressive renunciation; rather, it marks the beginning of a "multinaturalist" design that subverts anthropocentrism: "*a design that does not merely shape ephemeral objects - let alone ephemeral dwellings - but that rises to become a guardian of life in all its forms, a design that, with humility and respect, engages with the complexity of the natural world, drawing inspiration and instruction from it*" (Staid 2025).

6.4.3.1 Living in a Space Immersed in Greenery

The idea of living immersed in greenery (Fig. 6.15) is tied to the concept of afforestation, which has been revitalized today in the wake of the pandemic and climate change (Clark et al. 1997), though rooted in antiquity. In Mesopotamian, Egyptian, and Greek cities, trees served communicative and symbolic functions (Sander and Randrup 1997), while in Roman cities they provided shade and comfort (Warden and Romano 1994). In later centuries, urban greenery assumed various roles: food production in the Middle Ages, ornamentation in the Renaissance, and a means of improving environmental health in the industrial era (Helphand, 2006). Only recently, however, has afforestation become a foundational component of design, both in urban planning and as an architectural material. Notable examples that merge artifice and nature include Condominio 25 Verde by Luciano Pia (Turin, 2014), a true inhabitable forest supported by a metal structure; the Circle Wood

Fig. 6.15 Living in a Space Immersed in Greenery. From the left: Luciano Pia, Condominio 25 Verde (Torino, 2014); Mobius Architekci, Circle Wood House (Warszawa, 2021); Stefano Boeri Architetti, Liuzhou Forest City (Liuzhou, 2024)

Fig. 6.16 Living in a Space Covered with Greenery. From the left: Roberto Gabetti and Aimaro Isola, Olivetti West Residential Unit (Ivrea, 1971); Gunnar Birkerts, Law Library Extension (Ann Arbor, 1981); William Morgan, Hilltop House (Brooksville, 1975)

House (Mobius Architekci, 2021), where an oval building intersected by trees integrates seamlessly with the surrounding natural landscape; Liuzhou Forest City (Stefano Boeri Architetti 2024), a forest-city with one million trees and shrubs belonging to over one hundred species, improving air quality, absorbing CO_2, producing oxygen, and mitigating temperatures.

6.4.3.2 Living in a Space Covered with Greenery

The idea of living in a space covered by greenery is tied to hypogeism, long regarded as utopian and later rejected by the Modern Movement, which based its principles on the biological necessity of natural light and attributed only negative connotations to its absence (Nicoletti et al. 1980) (Fig. 6.16).

However, early modern precedents are not devoid of visionary "subterranean" impulses: Expressionist architecture, inspired by the myth of the cosmic mountain, imagined ground-like cities in the form of earthen swells (Finsterlin, Taut) or rocky spires (Krayl). Only between the 1960s and 70 s, amid hippie culture and the energy crisis, did the prospect of living spaces covered by greenery reemerge: from Ralph Erskine's *Subarctic City* (1958) to Chanéac's *Ville-Taupe* (1963), and examples such as the Olivetti West Residential Unit in Ivrea (Gabetti and Isola 1971) and the Law Library Extension in Ann Arbor (Birkerts 1981). Crucial was the work of William Morgan, who defined nine *Earthform categories* (Burger 1987) and built hypogean houses such as the Dune House, Hilltop House (1975), and Forest House (1979). A new season of subterranean architecture emerged, later epitomized by Emilio Ambasz's slogan *"Green Over Grey."* Ambasz's works (the *Casa de Retiro Espiritual* in Seville, the Lucille Halsell Conservatory in San Antonio, the ACROS building in Fukuoka) transformed green roofing from a merely functional component into a central formal element.

6.4.3.3 Living in a Space Enveloped by Greenery

As illustrated by the wall frescoes from the Villa of Publio Fannius Sinistore at Pompeii, the idea of vegetal cladding was already present in antiquity. From ancient times onward, plants were used in ways that combined religious symbolism and representational meaning with decorative purposes (Caneva et al. 2025; Hosseini and Caneva 2022).

This tradition continued for centuries, from the pergolas depicted in the *Hypnerotomachia Poliphili* (1499) and the exceptional painted pergolas by Raphael in Agostino Chigi's Roman Villa Farnesina (Caneva et al. 2025) to the industrial age, when natural supports (willow, wicker, or wood) were gradually replaced by iron.

This material shift enabled the hydroponic experiments of Hector Horeau, who envisioned Parisian boulevards covered by iron-and-glass structures adorned with flower basins. Although never built, his proposal anticipated the vertical garden system later patented by Patrick Blanc, first implemented in 1986 at the *Cité des Sciences* in Paris and later reinterpreted by Gaetano Pesce in the Organic Building in Osaka (1993), with cantilevered, uniquely shaped pots that host more than 80 native plant species. The green wall later became an icon of sustainability with the Musée du quai Branly (2006) and Madrid's CaixaForum (2007). This line of research also led Blanc to design *La Robe végétale*, a wedding dress presented by Jean-Paul Gaultier in 2002, merging habitat and fashion into a single living envelope (Fig. 6.17).

Fig. 6.17 Living in a Space Enveloped by Greenery. From the left: Gaetano Pesce, Organic Building (Osaka, 1993); Jean Nouvel, Musée du quai Branly (Paris, 2006); Jacques Herzog and Pierre de Meuron, CaixaForum (Madrid, 2007)

References

Aalipour H, Nikbakht A, Ghasemi M, Amiri R (2020) Morpho-physiological and biochemical responses of two turfgrass species to arbuscular mycorrhizal fungi and humic acid under water stress conditions. J Soil Sci Plant Nutr 20(2):566–576. https://doi.org/10.1007/s42729-019-00146-4

Ali A, Yazar A (2007) Effect of micro-catchment water harvesting on soil-water storage and shrub establishment in the arid environment. Int J Agric Biol 9(2)

Apollonio FI, Gaiani M, Sun Z (2013) 3D modeling and data enrichment in the digital reconstruction of architectural heritage. Int Arch Photogramm Remote Sens Spat Inf Sci 40:43–48. https://doi.org/10.5194/isprsarchives-XL-5-W2-43-2013

Aronson MF, Lepczyk CA, Evans KL, Goddard MA, Lerman SB, MacIvor JS et al (2017) Biodiversity in the city: key challenges for urban green space management. Front Ecol Environ 15(4):189–196. https://doi.org/10.1002/fee.1480

Bellini OE, Daglio L (2009) Verde verticale. Aspetti figurativi, ragioni funzionali e soluzioni tecniche nella realizzazione di living walls e green façades. Maggioli Editore, Santarcangelo di Romagna (RN)

Bianchini F, Hewage K (2012) How "green" are the green roofs? Lifecycle analysis of green roof materials. Build Environ 48:57–65. https://doi.org/10.1016/j.buildenv.2011.08.019

Braun-Blanquet J (2013) Pflanzensoziologie: grundzüge der vegetationskunde. Springer

Burger E (1987) Geomorphic architecture: multifamily residential design solutions. Tunn Undergr Space Technol 2(3):287–297

Caddeo A, Cogoni A, Iiriti G, Loi MC, Marignani M, Podda L, et al (2020) Dai balconi ai parchi urbani: buone pratiche per un giardinaggio consapevole. Life ASAP. Pubblicazione realizzata nell'ambito dell'azione B4 del progetto LIFE15 GIE/IT/001039 "Alien Species Awareness Program" (ASAP)

Caneva G, Pacini E, Signorini MA (2005) Le rappresentazioni delle piante nel Medioevo e Rinascimento. In: Caneva G (ed) La biologia vegetale per i beni culturali, vol 2. Nardini Editore, pp 101–108

Caneva G, Hosseini Z, Bartoli F, Capotorti G, Attorre F, Blasi C (2025) Revaluating CUNA places (CUltural place of high relevance for NAture): Rome as a multifaced example of outstanding values and potentials. Land 14(2):226. https://doi.org/10.3390/land14020226

Capotorti G, Bonacquisti S, Abis L, Aloisi I, Attorre F, Bacaro G et al (2020) More nature in the city. Plant Biosyst 154(6):1003–1006. https://doi.org/10.1080/11263504.2020.1837285

Caravaca F, Barea JM, Palenzuela J, Figueroa D, Alguacil MM, Roldán A (2003) Establishment of shrub species in a degraded semiarid site after inoculation with native or allochthonous arbuscular mycorrhizal fungi. Appl Soil Ecol 22(2):103–111. https://doi.org/10.1016/S0929-1393(02)00136-1

Clark JR, Matheny NP, Cross G, Wake V (1997) A model of urban forest sustainability. Arboricult Urban For 23(1):17–30

Congreves KA (2022) Urban horticulture for sustainable food systems. Front Sustain Food Syst 6:974146. https://doi.org/10.3389/fsufs.2022.974146

Continella A, Pannitteri C, La Malfa S, Legua P, Distefano G, Nicolosi E, Gentile A (2018) Influence of different rootstocks on yield precocity and fruit quality of 'Tarocco Scirè'pigmented sweet orange. Sci Hortic 230:62–67. https://doi.org/10.1016/j.scienta.2017.11.006

Corine Classification (2025) Green urban areas. https://land.copernicus.eu/content/corine-land-cover-nomenclature-guidelines/html/index-clc-141.html

Dale AG, Frank SD (2014) Urban warming trumps natural enemy regulation of herbivorous pests. Ecol Appl 24(7):1596–1607. https://doi.org/10.1890/13-1961.1

Disney M (2019) Terrestrial Li DAR: a three-dimensional revolution in how we look at trees. New Phytol 222(4):1736–1741. https://doi.org/10.1111/nph.15517

EC (2019) Regulation (EU) 2019/1009 of the European Parliament and of the Council of 5 June 2019 Laying down Rules on the Making Available on the Market of EU Fertilising Products and

Amending Regulations (EC) No 1069/2009 and (EC) No 1107/2009 and Repealing Regulation (EC) No 2003/2003. Available online: https://eur-lex.europa.eu/legal-content

Elmqvist T, Andersson E, Frantzeskaki N, McPhearson T, Olsson P, Gaffney O, Takeuchi K, Folke C (2019) Sustainability and resilience for transformation in the urban century. Nat Sustain 2(4):267–273. https://doi.org/10.1038/s41893-019-0250-1

European Commission (2025) Green infrastructure. https://environment.ec.europa.eu/topics/nature-and-biodiversity/green-infrastructure_en. Accessed 3 Sept 2025

European Union (2014) Regolamento (UE) n. 1143/2014 del Parlamento europeo e del Consiglio, del 22 ottobre 2014, sulle specie esotiche invasive. https://eur-lex.europa.eu/legal-content

Forman RT (2014) Urban ecology: science of cities. Cambridge University Press

Francini A, Romano D, Toscano S, Ferrante A (2022) The contribution of ornamental plants to urban ecosystem services. Earth 3(4):1258–1274. https://doi.org/10.3390/earth3040071

Garbisu C, Alkorta I, Kidd P, Epelde L, Mench M (2020) Keep and promote biodiversity at polluted sites under phytomanagement. Environ Sci Pollut Res 27(36):44820–44834. https://doi.org/10.1007/s11356-020-10854-5

Ghafari S, Kaviani B, Sedaghathoor S, Allahyari MS (2020) Ecological potentials of trees, shrubs, and hedge species for urban green spaces by multi-criteria decision making. Urban For Urban Green 55:126824. https://doi.org/10.1016/j.ufug.2020.126824

Green JJ, Baddeley JA, Cortina J, Watson CA (2005) Root development in the Mediterranean shrub *Pistacia lentiscus* as affected by nursery treatments. J Arid Environ 61(1):1–12. https://doi.org/10.1016/j.jaridenv.2004.09.001

Gullo G, Dattola A (2025) Kiwi, una svolta positiva per la frutticoltura calabrese. Rivista di frutticultura 7:22–24

Hosseini Z, Caneva G (2022) Lost gardens: from knowledge to revitalization and cultural valorization of natural elements. Sustainability 14(5):2956. https://doi.org/10.3390/su14052956

Hu SY (1979) *Ailanthus*. Arnoldia 39(2):29–50. https://doi.org/10.5962/p.250166

ISPRA (n.d.) *Ailanthus altissima*. https://www.specieinvasive.isprambiente.it

Kendal D, Dobbs C, Lohr VI (2014) Global patterns of diversity in the urban forest: is there evidence to support the 10/20/30 rule? Urban For Urban Green 13(3):411–417. https://doi.org/10.1016/j.ufug.2014.04.004

Koller GL (1977) Transplanting stress-a view from the plant's perspective. Arnoldia 37(5):230–241. https://www.jstor.org/stable/42954151

La Torre A, Battaglia V, Caradonia F (2016) An overview of the current plant biostimulant legislations in different European Member States. J Sci Food Agric 96(3):727–734. https://doi.org/10.1002/jsfa.7358

Leotta L, Toscano S, Ferrante A, Romano D, Francini A (2023) New strategies to increase the abiotic stress tolerance in woody ornamental plants in Mediterranean climate. Plants 12(10):2022. https://doi.org/10.3390/plants12102022

Leotta L, Toscano S, Ferrante A, Romano D (2025) The use of Mediterranean native shrubs for improving the sustainability of urban environments. Front Horticult 4:1652517. https://doi.org/10.3389/fhort.2025.1652517

Lundholm JT, Richardson PJ (2010) MINI-REVIEW: habitat analogues for reconciliation ecology in urban and industrial environments. J Appl Ecol 47(5):966–975. https://doi.org/10.1111/j.1365-2664.2010.01857.x

Maestre FT, Bautista S, Cortina J, Bellot J (2001) Potential for using facilitation by grasses to establish shrubs on a semiarid degraded steppe. Ecol Appl 11(6):1641–1655. https://doi.org/10.1890/1051-0761(2001)011[1641:PFUFBG]2.0.CO;2

MATTM (2017) Linee guida per il governo sostenibile del verde urbano. Comitato per lo sviluppo del verde pubblico. Ministero dell'Ambiente e della Tutela del Territorio e del Mare. https://www.mase.gov.it

Miller RW (1997) Urban forestry: planning and managing urban greenspaces, 2nd edn. Prentice-Hall, Hoboken, p 560

Musarella CM, Patti M, Laface VL, Spampinato G (2024) An overview of ethnobotanical knowledge for the enhancement of typical plant food and the development of a local economy: the

case of Calabria region (Southern Italy). Vegetos 37(6):2230–2241. https://doi.org/10.1007/s42535-024-00975-4
Newman SE, Davies FT Jr (1987) High soil temperature and water relations of endomycorrhizal nursery crops. J Environ Hortic 5(2):93–96
Nicoletti M, Nicoletti M, Nicoletti M, Nicoletti M, Architect I (1980) L'architettura delle caverne. Laterza, Bari, p 16
Purini F (2022) Discorso sull'architettura: cinque itinerari nell'arte del costruire. Marsilio Editori spa
Raimondo FM, Mazzola P, Bartoli F (2022a) Analisi della componente vegetale di Villa Palmeri nella Città di Termini Imerese (Palermo). Quaderni di Botanica Ambientale e Applicata 33:21–28
Raimondo FM, Bajona E, Bartoli F (2022b) Analisi della componente vegetale dei Giardini storici di Palermo. Il Parco Piersanti Mattarella e la Villa Falcone-Morvillo su Via Libertà (già Giardino Inglese). Quaderni di Botanica Ambientale e Applicata 33:29–34
Reddy PP (2016) Micro-catchment rainwater harvesting. In: Sustainable intensification of crop production. Springer Singapore, Singapore, pp 209–222
Roloff A, Korn S, Gillner S (2009) The Climate-Species-Matrix to select tree species for urban habitats considering climate change. Urban For Urban Green 8(4):295–308. https://doi.org/10.1016/j.ufug.2009.08.002
Russo A, Esperon-Rodriguez M, St-Denis A, Tjoelker MG (2025) Native vs. non-native plants: public preferences, ecosystem services, and conservation strategies for climate-resilient urban green spaces. Land 14(5):954. https://doi.org/10.3390/land14050954
Salbitano F, Fini A, Borelli S, Konijnendijk CC (2019) Urban Food Forestry: current state and future perspectives. Urban For Urban Green 45:126482–126484. https://doi.org/10.1016/j.ufug.2019.126482
Sander H, Randrup TB (1997, December) Urban forestry in the Nordic and Baltic countries. In: Proceedings of a Nordic Workshop on Urban Forestry held in Tallinn, Estonia
Schirone B, Pica A, Fratini F, Menegoni P, Cianfaglione K (2025) Miyawaki and urban tiny forests in Italy. Earth 6(4):116. https://doi.org/10.3390/earth6040116
Staid A (2025) Dare forme al mondo. Per un design multinaturalista. UTET, Segrate (Mi)
Testolin R, Ferguson AR (2009) Kiwifruit (Actinidia spp.) production and marketing in Italy. N Z J Crop Hortic Sci 37(1):1–32. https://doi.org/10.1080/01140670909510246
The Landscape Institute (UK) (2016) BIM for landscape. Routledge, London
Toscano S, Romano D, Massa D, Bulgari R, Franzoni G, Ferrante A (2018) Biostimulant applications in low input horticultural cultivation systems. Italus Hortus 25(2):27–36. https://doi.org/10.26353/j.itahort/2018.1.2736
Toscano S, Ferrante A, Romano D, Tribulato A (2021) Interactive effects of drought and saline aerosol stress on morphological and physiological characteristics of two ornamental shrub species. Horticulturae 7(12):517. https://doi.org/10.3390/horticulturae7120517
Toscano S, Romano D, Lazzeri V, Leotta L, Bretzel F (2025) How can plants used for ornamental purposes contribute to urban biodiversity? Sustainability 17(9):4061. https://doi.org/10.3390/su17094061
Vallejo VR, Serrasolses I, Cortina J, Seva JP, Valdecantos A, Vilagrosa A (2000) Restoration strategies and actions in Mediterranean degraded lands. Desertification in Europe: mitigation strategies, land-use planning. European Commission, Luxembourg, pp 221–233
Van der Veken S, Hermy M, Vellend M, Knapen A, Verheyen K (2008) Garden plants get a head start on climate change. Front Ecol Environ 6(4):212–216. https://doi.org/10.1890/070063
Vila-Ruiz CP, Meléndez-Ackerman E, Santiago-Bartolomei R, García-Montiel D, Lastra L, Figuerola CE, Fumero-Caban J (2014) Plant species richness and abundance in residential yards across a tropical watershed: implications for urban sustainability. Ecol Soc 19(3). https://doi.org/10.5751/ES-06164-190322
Von Humboldt A (1807) Ideen zu einer Geographie der Pflanzen nebst einem Naturgemälde der Tropenländer: auf Beobachtungen und Messungen gegründet, welche vom 10ten Grade nördli-

cher bis zum 10ten Grade südlicher Breite, in den Jahren 1799, 1800, 1801, 1802 und 1803 angestellt worden sind (vol 1). Cotta

Warden PG, Romano DG (1994) At the villa of the papyri at Herculaneum. Art History 17(2):228–254. https://www.researchgate.net/profile/David-Romano-2/publication/263753360_The_Course_of_Glory_Greek_Art_in_a_Roman_Context_at_the_Villa_of_the_Papyri_at_Herculaneum/links/59f4c1fda6fdcc075ec4b84a/The-Course-of-Glory-Greek-Art-in-a-Roman-Context-at-the-Villa-of-the-Papyri-at-Herculaneum.pdf

Young RF (2011) Planting the living city: best practices in planning green infrastructure—results from major US cities. J Am Plan Assoc 77(4):368–381. https://doi.org/10.1080/01944363.2011.616996

Zocchi DM, Bondioli C, Hamzeh Hosseini S, Miara MD, Musarella CM, Mohammadi D et al (2022) Food security beyond cereals: a cross-geographical comparative study on Acorn Bread Heritage in the Mediterranean and the Middle East. Foods 11(23):3898. https://doi.org/10.3390/foods11233898

Chapter 7
Forests, Parks and Gardens

Carmelo Maria Musarella, Flavia Bartoli, Mara Bodesmo, Giulia Caneva, Eugenio Domínguez-Vilches, Marco Fornaciari, Carmen Galán, Zohreh Hosseini, Marco Maovaz, Fabio Orlandi, Aldo Ranfa, and Thekla Tsitsoni

Abstract Urban forests are now regarded as fully-fledged complex ecosystems, designed with native species and capable of replicating natural processes, thereby helping to improve the urban climate, reduce pollution and protect biodiversity and human well-being. Alongside their functional role, urban parks and gardens are taking on an increasingly significant educational role, acting as 'living laboratories' that promote learning, social inclusion and mental health, despite certain challenges related to resources and integration into educational systems. Historic gardens,

C. M. Musarella (✉)
Department of AGRARIA, Mediterranean University of Reggio Calabria, Reggio Calabria, Italy
e-mail: carmelo.musarella@unirc.it

F. Bartoli
Institute of Heritage Science (ISPC), National Research Council (CNR), Montelibretti, Rome, Italy

M. Bodesmo
Phytosanitary Inspector, Umbria Region, Perugia, Italy

G. Caneva · Z. Hosseini
Department of Sciences, University Roma Tre, Rome, Italy

E. Domínguez-Vilches · C. Galán
Department of Botany, Ecology and Plant Physiology, University of Cordoba, Cordoba, Spain

M. Fornaciari · F. Orlandi
Department of Civil and Environmental Engineering, University of Perugia, Perugia, Italy

M. Maovaz
Centro di Ateneo per i Musei Scientifici (CAMS), University of Perugia, Perugia, Italy

A. Ranfa
Former Professor of Environmental and Applied Botany—University of Perugia, Perugia, Italy

T. Tsitsoni
Department of Forestry and Natural Environment, Aristotle University of Thessaloniki, Thessaloniki, Greece

G. Caneva et al. (eds.), *Urban Botany*, Future City 22,
https://doi.org/10.1007/978-3-032-28986-5_7

archaeological sites and monumental areas are also of great importance as 'living monuments', where nature and culture intertwine, acting as reservoirs of biodiversity, yet requiring a balanced approach to management that combines conservation, sustainability and adaptation to climate change within a context often complicated by regulatory and managerial constraints.

Therapeutic gardens also play a significant role; drawing on a long tradition, they use contact with nature to promote physical and mental recovery, particularly in vulnerable individuals, through careful, multisensory and multidisciplinary design, although their widespread adoption is limited by costs and maintenance. At the same time, community gardens are shared spaces that combine food production, environmental education and social inclusion, contributing to environmental quality and collective well-being. However, they too face challenges relating to access to land, management and resources, making participatory and sustainable strategies essential.

The various forms of urban greenery, therefore, play a fundamental role in improving the quality of life and sustainability of cities. Integrated planning and adequate institutional support are essential to fully realise their potential.

Keywords Archaeological sites and monumental areas · Community gardens · Educational Spaces · Historical gardens · Parks · Therapeutic gardens · Urban Forests · Urban Gardens

7.1 Urban Forests (C.M.M., T.T.)

Urban forests can be defined "as a network of spaces within and around urban and peri-urban settlements, including woods, groups of trees, roadside trees, and isolated trees in parks, gardens, and other urban contexts, aimed at creating functional and resilient ecosystems, highlighting the link with natural forest dynamics" (Fig. 7.1) (see FAO meeting (2016) in Salbitano et al. 2016).

Subsequently, the National Committee for Public Green Areas of the Ministry of the Environment developed the National Urban Green Strategy, introducing elements that bring urban plant communities closer to real plant communities (MATTM 2018). The document requires the exclusive use of native species, an increase in the area devoted to plant communities, the removal of asphalt surfaces, the creation of urban forests, and the maintenance of planted trees for at least seven years.

While continuing to use the term "urban forests," the National Strategy emphasizes the potential of local vegetation, the use of native species, and long-term management. On this basis, the Committee has been tasked with managing €30 million in funding, provided under the so-called "Climate Decree," through a call for proposals for Italian Metropolitan Cities to create urban forests (IR 2019). This establishes a more strictly ecological vision, according to which urban forest structures must be real forests and not heterogeneous sets of green spaces with non-forest functions. Furthermore, the legal and administrative definition of a forest (2000 m^2

Fig. 7.1 Panoramic view of part of the Turia Urban Park in Valencia (Spain). (Photo by C.M. Musarella)

and 20% tree cover) (IR 2018) is not considered; instead, larger areas and a greater number of native woody species are required.

This gradual shift away from the concept of urban forests as simply 'urban greenery' has been superseded by a more modern, ecological approach adopted in implementing the Italian PNRR (National Recovery and Resilience Plan) measure dedicated to urban forestry (Blasi et al. 2025). According to the PNRR Plan, promoted by the Italian Ministry of the Environment and supported by the European Commission, an urban forest is an authentic forest ecosystem in an urban or peri-urban area, with the structural and functional characteristics of a wood (Gèhu 2006).

It has a *multi-level structure, a high species diversity, and adequate spatial dimensions, equal to at least 1 ha and preferably greater than 3 ha, to enable* ecological dynamics to be activated (Fig. 7.2).

The urban forest is therefore not a green decorative element, but a tool for ecological restoration that integrates biodiversity, education, environmental functions, and urban quality of life, contributing to the reduction of pollution, the mitigation of climate change, and the conservation of biodiversity, as well as improving the well-being and health of citizens (Cano-Ortiz et al. 2025a, b).

It is important that urban forests maintain an essential character that is known and loved by citizens. The Urban Climate Green Assets Strategy is based on a vision to ensure healthy, green infrastructure with a different structure and composition, capable of contributing most effectively to mitigating the impacts of climate change in the city (Tsitsoni et al. 2015).

The creation of new urban forests is based on the principle of "the right plant in the right place" (EC 2021), to which we can also add "at the right time," respecting the ecological and biogeographical characteristics of the area, as well as the biology

Fig. 7.2 The Central Park of New York, as an example of an Urban Forest. (Photo from Google Earth)

of the plants. In line with the provisions of the Nature Restoration Regulation, the aim is to bring nature back into human living spaces by restoring degraded habitats and ecosystems in line with local flora and vegetation characteristics (EU 2024).

The PNRR measure therefore develops a new reforestation model applicable to both urban areas and natural and semi-natural systems, based on the use of numerous native woody species consistent with the local vegetation series. The project, which involved 13 Metropolitan Cities, definitively overcomes the idea of the urban forest as an indistinct collection of urban plant types, instead providing for woods, shrubbery, bushes, and herbaceous communities planned according to the principles of synphytosociology and plant ecology, to protect biodiversity and enhance ecosystem services to improve the quality of urban life.

Their presence positively affects the mental and physical behavior of city residents. In fact, as shown in Chap. 5, one of the most important functional effects of urban forests is their contribution to improving the city's climate. The tree canopy controls solar radiation because the leaves retain, absorb, and reflect part of it, and only a small amount passes through the canopy. Furthermore, trees also contribute to wind protection, reducing wind speed by creating windbreaks on both sites, the windward and leeward and regulate wind movement by acting as a barrier (Ren et al. 2023; Yuan et al. 2017), such as to precipitation retention and to soil erosion of air pollution reduction and air renewal (the effect of the urban forest on the quality of the atmospheric air of cities) (Nowak and Dwyer 2007). The urban forest exhibits a dual action in its behavior against the large quantities of aerosols contained in the atmosphere of large cities. It functions as a biological barrier against pollutants and as a decontamination factor by physical-mechanical means, retaining or absorbing aerosols with the leaves (Nowak et al. 2006), and by biochemical means, depolluting the atmosphere and through the metabolism of trees (Samara and Tsitsoni 2014; Tsitsoni 2010; Diener and Mudu 2021). An urban forest of average productivity of 1 ha produces 4 tons of O_2 per year, with a net production of 2.5 tons. The total annual production of the forest in O_2 is ten times greater than that of any other terrestrial ecosystem (Dafis 2001).

Urban forests also help reduce noise by absorbing sound waves from tree leaves and branches (Samara and Tsitsoni 2007). This reduction depends on the forest species, and it was found that on the Thessaloniki ring road, the existing Calabrian pine (*Pinus brutia* Ten.) urban forest reduces traffic noise by 6 dB at a depth of 60 m from the road (Samara and Tsitsoni 2011). In Mediterranean cities, one of the most important problems for urban forests is the risk of forest fires. Most urban forests in Greece belong to Mediterranean ecosystems, with Mediterranean pines (*Pinus halepensis* Mill. and *P. brutia*) as the main species; therefore, the biggest problem they face today is fire. A significant number of peri-urban forests serve as "flint stores" because of the fuel and garbage they accumulate from human activities. After the great fire in the urban forest of Thessaloniki, "Seih-Sou," in July 1997 (Fig. 7.3a), the course of regeneration and reforestation was documented. Over 800 ha had been reforested in a short period after the fire, until March 1998.

Although many species were planted in the area where the study was conducted to create a mixed urban forest of conifers and broadleaf trees, very few survived. Twelve years after the fire, natural regeneration was very rich. However, the survival rate of the planted species was very low, and the dominant species, the Calabrian pine (*P. brutia*), reached 81% (Fig. 7.3b) (Tsitsoni et al. 2009).

At the European level, the National Biodiversity Strategy 2020–2030 dedicates a specific objective to forest conservation: "B.9 Achieve more connected, healthier, and more resilient forests" by actively contributing to the EU target of planting at least 3 billion trees in urban, peri-urban, and extra-urban areas, particularly in large metropolitan areas (Fig. 7.4).

Fig. 7.3 Two side-by-side views compare a forest area before and after recovery. (**a**) Devastating effect of fires occurring in the urban forest of Thessaloniki, in July 1997 and (**b**) following reforestation with pines. (Photos by T. Tsitsoni)

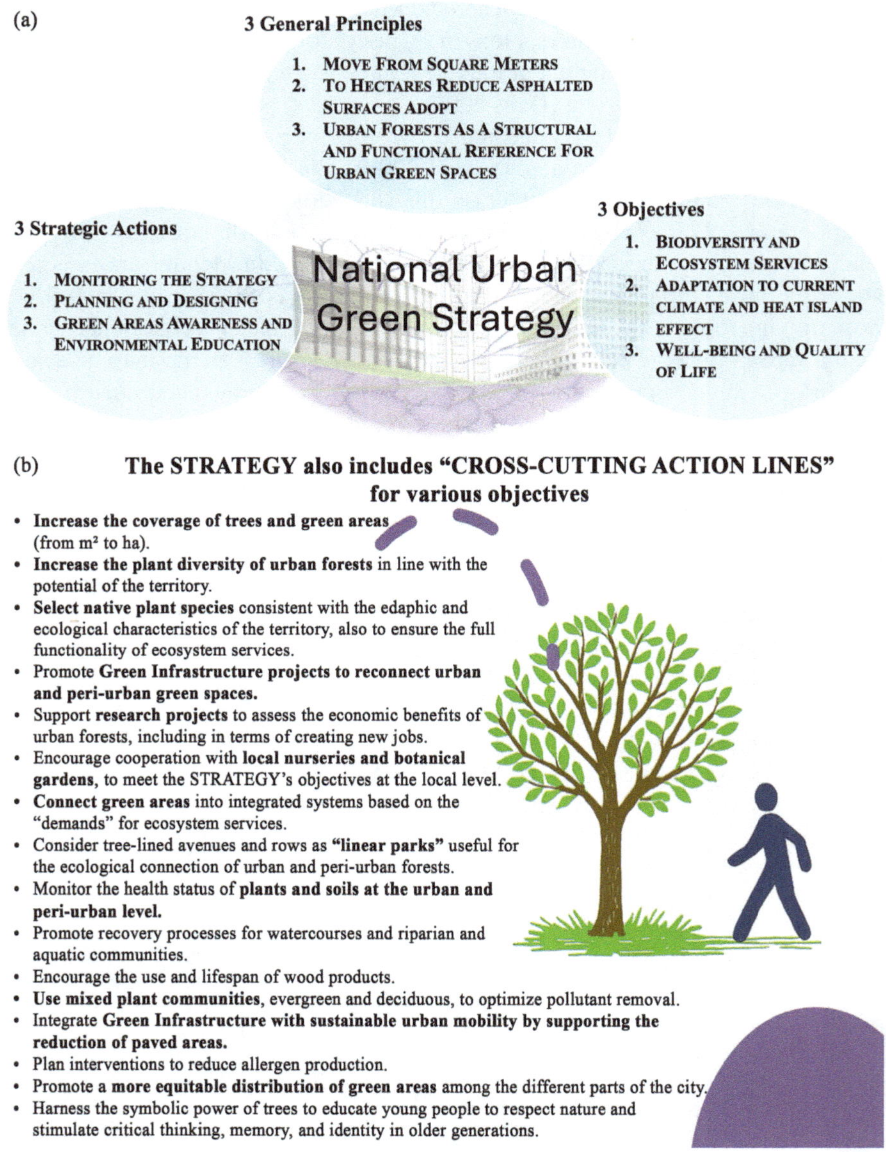

Fig. 7.4 National Urban Green Strategy from the PNRR (original elaboration from UE documents)

At the urban level, it is a tool that enables all national metropolitan cities to follow a common methodology, based on solid scientific references, to identify and plant the right tree in the right place (a tree consistent with biogeographical and ecological characteristics).

According to the current concept, sustainable forest resource management (both urban and non-urban) aims primarily to manage them to maintain and improve their biodiversity, improve their biodiversity, ecosystem services, and renewal capacity, to ensure their potential and related ecological, economic, and social functions indefinitely (Cano et al. 2024a, b). In Italy, the project identified 14 metropolitan cities, home to more than 21 million people, covering an area equal to 14.47% of the national territory.

7.1.1 Case Study

"Riforest@graria": the new urban forest of the Metropolitan City of Reggio Calabria (Italy)

In this context, the Metropolitan City of Reggio Calabria and the Department of Agriculture (AGRARIA) of the Mediterranean University of Reggio Calabria initiated a technical-scientific collaboration in 2020 to plan, finance, and design urban, peri-urban, and extra-urban afforestation activities. Among the presented projects within the the so-called "Climate Decree", the project "Riforest@graria" refers to the hill on which the department itself stands ("AGRARIA's hill"). This project, chosen as a "Third Mission" activity by the University, was based on the principles of restoration and natural enhancement as a strategic approach to restoring urban ecosystem functionality. It is part of a perspective on developing peri-urban spaces near the city center of Reggio Calabria to recover urban/rural boundary areas that are continuously affected by anthropogenic disturbances (grazing, fires, illegal landfills, etc.).

The interventions were carried out on eleven plots owned by the University, with a total area of 4.34 hectares, using native species, in accordance with Italian and EU forest strategies (Forest Europe 2008; ISPRA 2015): this allows to form a biotope with the typical characteristics of the Mediterranean environment and creating ecological continuity between the plain system and the hilly landscape lines. In fact, the short distance between this site and the SAC "Collina di Pentimele" (IT9350139) can create a valuable functional connection from an ecological point of view and contribute to landscape defragmentation (in accordance with the National Biodiversity Strategy, the European Strategy for Green Infrastructure, and the Habitat Directive) (Fig. 7.5). The forestation activity was developed using a vegetation-based approach and analyses of the area's pedological, bioclimatic, and morphological characteristics. Specifically, the structural and ecological functions of the existing plant communities were analyzed, as well as their dynamic succession relative to the long-term climax. This approach gave the project significant environmental value (including estimated benefits in terms of CO_2 absorption and air quality improvements) and socio-economic value (Spampinato et al. 2021). The reforestation project included the pre-existing plant communities to enhance their connectivity and ecological role with the local natural context. Specifically, in the intervention area, remnants of a *Quercus pubescens* Willd. subsp. *pubescens* [= *Q. virgiliana* (Ten.) Ten.] forest already occurring, mixed with steppe grasslands of *Ampelodesmos mauritanicus* (Poir.) T. Durand & Schinz. and *Spartium junceum* L., as well as anthropogenic vegetation characterized by *Arundo donax* L. and *Rubus ulmifolius* Schott., along with individuals of *Ceratonia siliqua* L. (Fig. 7.6).

Fig. 7.5 Panoramic view of the intervention area. The green patch at the top corresponds approximately to the SAC 'Collina di Pentimele' (IT9350139), while the green patch at the bottom corresponds to the Riforest@graria forest. Metropolitan City of Reggio Calabria (Italy). Photo from Google Earth (C) modified

Fig. 7.6 Residual oak forest with *Quercus pubescens* Willd. [= *Q. virgiliana* (Ten.) Ten.] (left) and grasslands with *Ampelodesmos mauritanicus* (Poir.) T. Durand & Schinz. (right). (Photos by Carmelo M. Musarella)

Thanks to the phytosociological analysis and bioclimatic study of the intervention area, two potential vegetation series were identified:

- Southern Apennine Thermo-Mediterranean Series of the Chestnut Oak (*Oleo-Querceto virgilianae sigmetum*). Climax association: *Oleo-Quercetum virgilianae* Brullo 1984 (forest of chestnut oak and olive tree). Habitat Code 91AA*: Eastern white oak woods.

- Southern Apennine Thermo-Meso-Mediterranean Xerophilous Series of the Tree Spurge and Olive Tree (*Oleo-Euphorbieto dendroidis sigmetum*). Edaphoclimax association: *Oleo-Euphorbietum dendroidis* (scrubland with tree spurge and olive tree).

The soil preparation was carried out using techniques typical of traditional agronomic practices, including digging holes measuring 40 × 40 × 40 cm. Starting with local germplasm, the seedlings used were produced in a nursery in polyethylene trays. Before planting, an adequate amount of fertilizer was applied at the bottom of the hole. In some plots, where stabilization was needed on excessively steep slopes, gabions, stakes, and living willow fences were constructed. Additionally, geogrids were placed for slope reinforcement, and geotextile was used for drainage, providing anti-erosive protection.

The reforestation was mixed, using various species (from a minimum of 4 to a maximum of 10 per plot), planted at an average spacing of 3 × 3 m, with a density not exceeding 1200 plants per hectare, based on ecological and silvicultural considerations. In fact, in a reforestation plan, plant density plays an important role in stand management from both ecological and cultural perspectives and influences evolutionary dynamics. Different planting densities have varying effects on individual plant growth and overall soil stability. The planting did not follow a regular pattern; to maintain natural conditions, a random arrangement was used, considering species associations and plant size, while always ensuring a uniform distribution and future coverage of the area. In total, 5019 plants, both tree and shrub species, were planted across the entire area.

The project followed guidelines based on natural environmental recovery processes, beginning with the introduction and/or maintenance of pioneer species that can quickly cover the soil and create a natural habitat for the establishment of more mature plant communities.

The selected native plant species help maximize the intervention's effectiveness while minimizing maintenance and plant-care costs. Along the slopes, individuals of *Quercus ilex* L. subsp. *ilex* were planted, alongside pioneer species such as *Cytisus infestus* (C. Presl) Guss. subsp. *infestus* and *S. junceum*, which contribute significantly to substrate enrichment. In sites with the best conditions, additional plants of *Q. ilex* subsp. *ilex* were planted, alternated in small quantities with *Q. pubescens* subsp. *pubescens*. On the warmer, South-facing slopes, specimens of *Pyrus communis* L. subsp. *pyraster* (L.) Ehrh. were also used. On shaded slopes with greater rocky outcrops, preference was given to the establishment of *Q. ilex* subsp. *ilex* and *Celtis australis* L. subsp. *australis*, with some shrubs of *Euphorbia dendroides* L. and a wild variety of *Olea europaea* L. In sites with lower substrates, species such as *Q. pubescens* subsp. *pubescens*, *Q. suber*, and *S. junceum* were planted, as they are expected to dominate over time.

Finally, urban furniture was installed to enrich the forest and make it more accessible to the local academic community and citizens who visit to enjoy its benefits. Specifically, picnic tables, trash bins, benches of various designs, two informational

Fig. 7.7 View of the Riforest@graria forest. Plot characterised by various tree and shrub species of different ages, planted at irregular distances, and spontaneous herbaceous vegetation in winter before mowing in late spring. (Photo by C.M. Musarella)

totems serving as signage and entrance markers, and a solar-powered LED lighting system were installed.

Over the next seven years, annual operations are scheduled to ensure the best possible success of the reforestation, including invasive species and weed control, replanting after failures, pruning, and emergency irrigation. Despite a fire in the summer of 2023 that destroyed a significant portion of the works, the 'Riforest@graria' project was completed and presented to the public on 1st October 2024 (Fig. 7.7).

7.2 Urban Gardens and Parks as Educational Spaces (C.G., E.D.-V.)

7.2.1 Definition and Classification

Urban gardens and parks are a diverse, multifunctional category of green infrastructure that plays a pivotal role in transforming urban environments into healthier, more sustainable, and inclusive living spaces. In the context of the 2030 Agenda for

Fig. 7.8 The new settlement of the Geobotanical Garden of the University Roma Tre, where students and citizens can learn about natural associations among plants and rocks of the Latium region. (Photos by G. Caneva)

Sustainable Development and the Sustainable Development Goals (SDGs), these spaces also contribute directly to SDG 4 (Quality Education), SDG 11 (Sustainable Cities and Communities), and serve as living laboratories and interactive classrooms where citizens—particularly children and youth—can experience environmental processes, biodiversity, and sustainability in action.

Greening solutions include a wide array of forms: from formal municipal parks and botanical gardens (Caneva et al. 2016) (Fig. 7.8) to informal community gardens, schoolyards, rooftop greenhouses, and edible corridors along sidewalks or transit routes.

In educational terms, these spaces are classified as "outdoor learning environments" or "living labs" (s.l.), and their incorporation into urban design represents a shift toward holistic sustainability education (Cano-Ortiz et al. 2025b). They can host interdisciplinary curricula, support inquiry-based science education, and promote emotional and civic engagement with environmental issues.

> *"Urban green spaces are more than lungs of the city—they are its memory, its classroom, and often, its conscience."*—Matthew Gandy, *The Botanical City.*

7.2.2 Methodological Approach for Plant Taxa Selection

Selecting plant species for educational gardens and parks requires balancing ecological resilience, didactic function, aesthetic appeal, and maintenance constraints. A sound methodology considers both the environmental conditions of the urban microclimate and the pedagogical goals of space.

Key criteria include:

- Native and climate-resilient species: these support local biodiversity and require fewer resources.
- Pollinator-friendly and layered plantings: to illustrate ecological interactions.

- Seasonality and diversity: for observing change over time and fostering long-term engagement.
- Cultural and historical value: plants with ethnobotanical significance, local culinary uses, or traditional medicinal roles.
- Edibility and food literacy potential: species like tomatoes, lettuce, beans, or fruit trees engage students with urban agriculture and sustainability.

7.2.3 Examples

7.2.3.1 Barcelona, Spain

In Barcelona's "Escoles + Sostenibles" program, school gardens incorporate native Mediterranean plants, e.g., *Thymus vulgaris* L. (thyme) and *Salvia rosmarinus* Spenn. (rosemary), drought-resistant perennials, and seasonal vegetables. The species are selected to support lessons in biology, climate change, and food systems. Teachers use observation diaries, QR-coded signage, and student-led planting calendars to structure the learning process.

7.2.3.2 Lisbon, Portugal

In Lisbon's "Hortas de Lisboa" initiative, planners combined local aromatic shrubs, e.g., *Lavandula stoechas* L. (lavender) and *Salvia rosmarinus* (rosemary), with fruiting trees and companion-planting schemes to teach students about pollination, soil regeneration, and edible landscapes (Dickinson et al. 2012). Plant labels include both scientific and local names, as well as QR codes linked to multilingual educational content.

7.2.4 Specific Benefits

Urban educational gardens and parks offer a rich spectrum of ecosystem services and human-centered benefits, making them uniquely positioned at the intersection of ecology, education, public health, and social cohesion.

7.2.4.1 Ecological Benefits

- Biodiversity Enhancement: by planting native and diverse flora, these spaces support urban wildlife, particularly pollinators like bees and butterflies, and contribute to biological connectivity.
- Microclimate Regulation: trees and shrubs reduce surface and air temperatures, mitigating urban heat island effects.

- Air and Water Quality: vegetation filters airborne pollutants and promotes stormwater infiltration through permeable soils.
- Soil Regeneration: the introduction of organic composting and non-invasive ground cover improves soil health.

7.2.4.2 Educational Benefits

- Hands-on Learning: students engage directly with ecological processes, enhancing understanding through observation, experimentation, and care tasks.
- Interdisciplinary Teaching: gardens provide platforms to teach sciences, mathematics, geography, literature, and arts.
- Citizen Science and Inquiry: learners monitor biodiversity, conduct soil or water tests, and contribute to long-term datasets.
- Civic Engagement: children and adults develop a sense of environmental responsibility and place-based identity.

7.2.4.3 Social and Health Benefits

- Mental Health: contact with greenery has been shown to reduce anxiety and increase concentration, especially in children.
- Social Inclusion: gardens promote interaction across generations, cultures, and socioeconomic groups.
- Food Access and Equity: edible gardens can mitigate urban food insecurity and promote nutritional literacy.

7.2.4.4 Cultural and Symbolic Benefits

- Cultural Memory and Heritage: traditional plant uses and local knowledge are preserved and shared.
- Celebration of Diversity: multicultural communities can share agricultural practices and festive rituals tied to seasonal crops or plants.
- Art and Expression: green spaces are often combined with murals, installations, or storytelling programs that anchor emotional learning.

Example—The Edible Schoolyard Project (USA). Founded by Alice Waters in Berkeley (California), this initiative transformed a school cafeteria garden into a cross-disciplinary learning hub. Students learn about plant biology, the history of agriculture, cooking, and the ethics of sustainability in an immersive environment. Academic performance and social interaction improved markedly in the participating schools (Waters 2017).

7.2.5 Maintenance and Monitoring

The success of urban educational gardens depends on consistent care, monitoring, and community engagement. Maintenance systems should be low-cost, ecologically sound, and participatory.

7.2.5.1 Maintenance Strategies

- Soil Health: composting, mulching, and seasonal soil amendments maintain fertility and structure.
- Water Efficiency: Use of drip irrigation, greywater systems, and rainwater harvesting reduces dependence on municipal sources.
- Integrated Pest Management (IPM): Natural pest control methods protect biodiversity and reduce chemical inputs.
- Rotational Planting: seasonal changes in crops support biodiversity and prevent soil depletion.

7.2.5.2 Monitoring Tools

- Citizen Science Platforms: tools like iNaturalist or eBird allow students and community members to document species.
- QR-coded Plant Labels: offer educational content and foster digital engagement.
- Biodiversity Scorecards: track indicators such as the number of flowering species, pollinator visits, or bird sightings.

7.2.5.3 Organizational Models

- School-Led: integrated into curriculum and managed by teachers, parents, and students.
- Community Co-Managed: shared by neighborhood associations, NGOs, and city staff.
- Hybrid Governance: partnerships between municipalities, universities, and private sponsors.

Example—Paris's "Main Verte" Program. This municipal initiative supports over **130** gardens run by residents and schools. It offers training, tools, and seasonal evaluations and requires biodiversity education components. Many gardens host weekly workshops on composting, herbal medicine, or pollinator conservation.

7.2.6 Challenges and Limitations

Despite their potential, educational green spaces face several structural and environmental limitations.

- Urban Space Scarcity: competing interests (e.g., parking, real estate) often outweigh green infrastructure in planning.
- Inequity in Access: marginalized communities usually lack public green space, perpetuating environmental injustice.
- Short-Term Funding: many gardens rely on limited grants or volunteer labor, hindering continuity.
- Curricular Disconnect: teachers may lack time or training to integrate outdoor learning into formal syllabi.
- Environmental Stressors: pollution, heatwaves, invasive species, or vandalism can compromise garden functionality.

7.2.7 Solutions

- Integrating gardens into city resilience and education strategies.
- Establishing legal protection for community-managed land.
- Offering professional development for educators.
- Encouraging interdisciplinary research and documentation.

7.2.8 Case Studies: Urban Corridors and Green Schoolyards

Cheonggyecheon Stream is an urban ecological restoration project that transformed a buried stream beneath an expressway into a green corridor. Features include interpretive signage, public art, biodiversity monitoring, and seasonal student tours. The project restored riparian habitats and improved air quality while serving as a prominent model of urban regeneration (Kim et al. 2019) (Fig. 7.9).

Landschaftspark Duisburg-Nord, a former steel plant, was converted into an urban park emphasizing ecological succession and minimal intervention. Native species were introduced gradually, and industrial ruins were repurposed as interpretive elements. It now hosts biodiversity education trails, ecological art exhibits, and community-led tours (Latz 2001) (Fig. 7.10).

The High Line is an elevated railway transformed into a biodiverse linear park with seasonal plantings curated by landscape architects and botanists. Visitors

Fig. 7.9 Cheonggyecheon Stream (Seoul, South Korea). (Photo by Kylen Ishioka)

access interactive digital guides and participate in volunteer-led ecology walks. The High Line emphasizes native and drought-tolerant species (Stalter and Kincaid 2021) (Fig. 7.11).

Barcelona's "Let's Transform Schoolyards" project reimagined over 200 asphalt schoolyards as green spaces. These now include native planting zones, shaded areas, and water retention features. Each redesign is participatory, involving students, teachers, and families. Biodiversity indicators and climate-resilience features are monitored annually.

As part of the "Corredores Verdes" initiative, the city planted hundreds of fruit-bearing and pollinator-supporting plants along major roads. The program involves schools in mapping plants, maintaining pollinator hosting, and harvesting edible species. Its goal is both food security and biodiversity education (Fig. 7.12).

In conclusion, urban gardens and parks, when designed and managed as educational spaces, serve as a powerful interface among ecology, pedagogy, and civic

Fig. 7.10 Landschaftspark Duisburg-Nord (Germany). (Photo by Malcolm Miles) (Creative Commons Attribution 4.0 International)

Fig. 7.11 The High Line (New York City, USA). (Photo by Kalina Prelik)

engagement (Cano-Ortiz et al. 2025a). Their integration into city landscapes contributes not only to biodiversity and climate resilience but also to fostering environmental literacy, community cohesion, and equitable access to green infrastructure. These living laboratories enable experiential learning that connects students and citizens to seasonal rhythms, ecological interactions, and sustainable food systems. Despite challenges, such as funding limitations, maintenance burdens, and spatial constraints, successful case studies from diverse cities demonstrate that multi-stakeholder governance and thoughtful plant selection can yield enduring educational and ecological benefits. As tools of nature-based education, these spaces also align with global policy agendas, such as the SDGs, reinforcing their relevance in building inclusive and sustainable urban futures (Cano-Ortiz et al. 2025b).

Fig. 7.12 Urban Edible Corridors (Elaboration by Alcaldía de Medellín, Colombia)

Strengthening their role in urban planning and curricula will be key to cultivating resilient, environmentally conscious generations (UNESCO 2021).

7.2.9 *Future Directions and Innovations*

To further enhance the role of urban gardens and parks as educational spaces, future initiatives could explore:

- Digital Integration: Augmented Reality (AR) apps to identify plants and explain ecological processes (Nassauer 1995).
- Climate-Resilient Designs: incorporating drought-tolerant species and smart irrigation systems (Tan et al. 2013).
- Global Collaboration: networks for sharing best practices across cities (e.g., UNESCO's Green Cities Initiative) (UNESCO 2021).
- Policy Advocacy: lobbying for urban agriculture and green education to be included in national curricula (Wolch et al. 2014).

7.3 Historical Gardens, Archaeological Sites and Monumental Areas (G.C., M.M., Z.H.)

7.3.1 *Definition and Classification: The Botanical Significance of Historical Heritage Sites*

The Florence Charter (ICOMOS 1982) defines historic gardens as "*an architectural and horticultural composition of public interest from historical or artistic points of view*" (Art. 1). Because gardens are primarily vegetal and subject to decay and regeneration, they were classified as living monuments that required special rules of protection (Arts. 2–3). These gardens are characterized by its plan and topography, vegetation (species, proportions, color schemes, spacing, heights), structural and decorative features; and water elements" (Art. 4), serving as a cosmic representation of an idealized world and a testament to a culture, a style, an age, and often to the originality of a creative artist (Art. 5).

UNESCO (1972) describes archaeological sites as areas containing significant evidence of past human activity, recognized for their outstanding universal value from historical, aesthetic, ethnological, or anthropological perspectives (Fig. 7.13).

Heritage parks encompass all these categories, and such terminology seems particularly appropriate when considering sites with a relevant stratigraphy of history, as in urban areas, where archaeological sites and gardens can have inextricable links.

In heritage sites, plant composition reflects aesthetic, religious, and functional philosophies, and natural elements are a critical component of identity, functioning as ecological infrastructure, cultural archive, and material witness of long-term

Fig. 7.13 *Horti Farnesiani* in the archaeological site of Palatine (Rome, Italy)—a historical garden and an archaeological area, as a combined example of a "CUNA place", conserving biodiversity and cultural values. (Photos by G. Caneva)

human–environment interactions (Hosseini et al. 2024; Romanato 2024). Then, they can become "CUltural place of high relevance for NAture" (CUNA places) (Caneva et al. 2025) (Fig. 7.14).

7.3.2 The Botanical Significance

Historical gardens, archaeological sites, and monumental areas often preserve a rich natural heritage, comprising interesting plant species, vegetation communities, and landscape patterns with naturalistic value, especially when they occur in urban areas. These sites frequently act as biodiversity hotspots, harboring plants coming from surrounding landscapes. In some cases, such as the Colosseum (Rome), different floristic studies have been conducted throughout history, and they showed the exceptional floristic richness of such a site, even if in an urban context (a total of 684 species identified over the past four centuries) (see Chap. 2). In that case, it was due to the great variety of habitats and the state of abandonment, with successions from pioneer to evolute stages corresponding to different vegetation series (Caneva et al. 2002).

In fact, microhabitat variations combined with limited disturbance, through restricted access and conservation regulations, create conditions for species persistence, supporting species richness and hosting rare or endemic taxa protected under the Habitats Directive (92/43/EEC) and CITES protocols, as observed in Mediterranean areas (Zangari et al. 2023; Panitsa et al. 2024).

Heritage parks, for instance, sustain rare woodland plants and protected species and often feature old-growth trees, heirloom cultivars, and long-established naturalized species that reflect historical horticultural practices and plant introductions (Fig. 7.15). In urban and peri-urban contexts, where intensive land use has

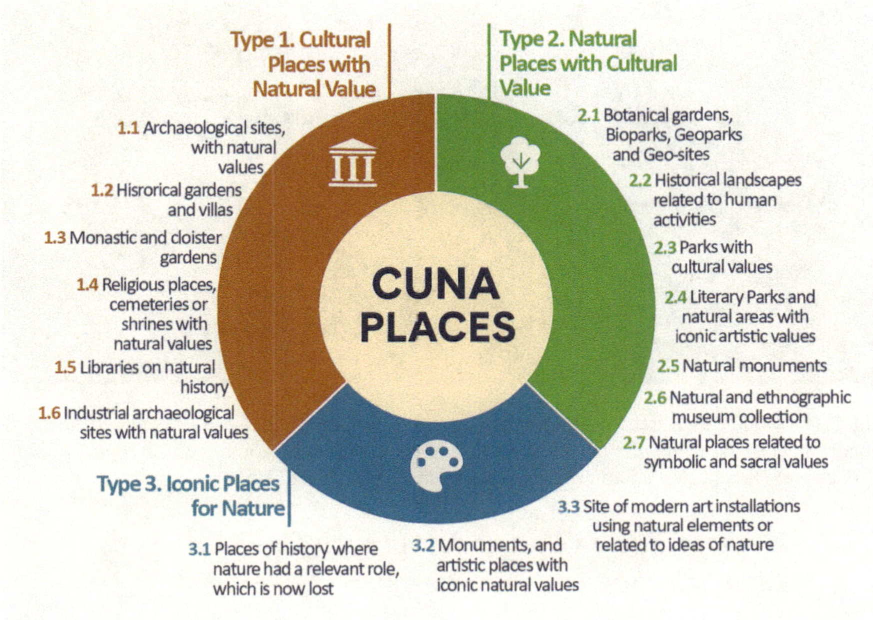

Fig. 7.14 Categories of cultural places with a natural interest and vice versa (CUNA places) in the international framework. (Modified from Caneva et al. 2025)

eliminated native vegetation, these sites serve as nature reserves and critical refugia for biodiversity (Báez and Torreblanca 2024).

Since the late twentieth century, there has been a paradigm shift toward viewing these sites as cultural landscapes that embody both cultural and natural values. UNESCO's 1992 recognition of cultural landscapes as World Heritage sites (including "clearly defined landscape[s], intentionally designed and created by man... such as garden and park landscapes") illustrates this integrative perspective. Individual gardens and park ensembles are now seen as part of larger landscape contexts, rather than isolated artifacts.

Recent studies advocate a "landscape approach" to historic gardens, examining how a garden's spatial evolution, ecological processes, and surrounding environment contribute to its heritage value (Wagenhofer et al. 2025). The creation of a corridor connecting historical parks with monumental trees was completed in Thessaloniki, (Stavrinou and Tsitsoni 2018) (Fig. 7.16).

Fig. 7.15 Monumental trees in Rome: (**a**) *Cupressus sempervirens* L.—XVI cent.—Chiostro di Michelangelo, (**b**) *Magnolia grandiflora* L.—XVII cent.—Lungara Street (**c**) *Cedrus libani* A. Rich.—XIX cent., Villa Farnesina. (Photos from Caneva et al. 2025)

7.3.3 Heritage Sites: Views and Approach for Plant Taxa Selection

Historic gardens, such as archaeological areas, are "living monuments" in which natural growth and change must be understood and managed, considering their cultural significance (Funsten et al. 2020). Increasingly, research addresses gardens' ecological functions—their flora and fauna, habitat value, and ecosystem services—linking biocultural diversity conservation with heritage management, stressing the importance of preserving them for future generations, with a high level of authenticity.

Beyond biodiversity, plants can act as "cultural indicators", revealing symbolic and functional values that shaped design and ritual practices, mirroring religious beliefs, social hierarchies, and aesthetic views (Caneva 1999). Their conservation thus demands interpretive approaches that acknowledge their repository's role of intangible heritage, as symbols of identity, memory, and artistic expression (Olivadese and Dindo 2022; Caneva et al. 2025). Heritage sites also provide ecological baselines for reconstructing historical vegetation patterns. Combining current botanical surveys with archaeobotanical evidence, such as pollen, seeds, and charcoal, reveals how ecosystems have changed over centuries, offering insights into long-term vegetation dynamics and resilience (Hosseini and Caneva 2022; Kanellou and Papafotiou 2025).

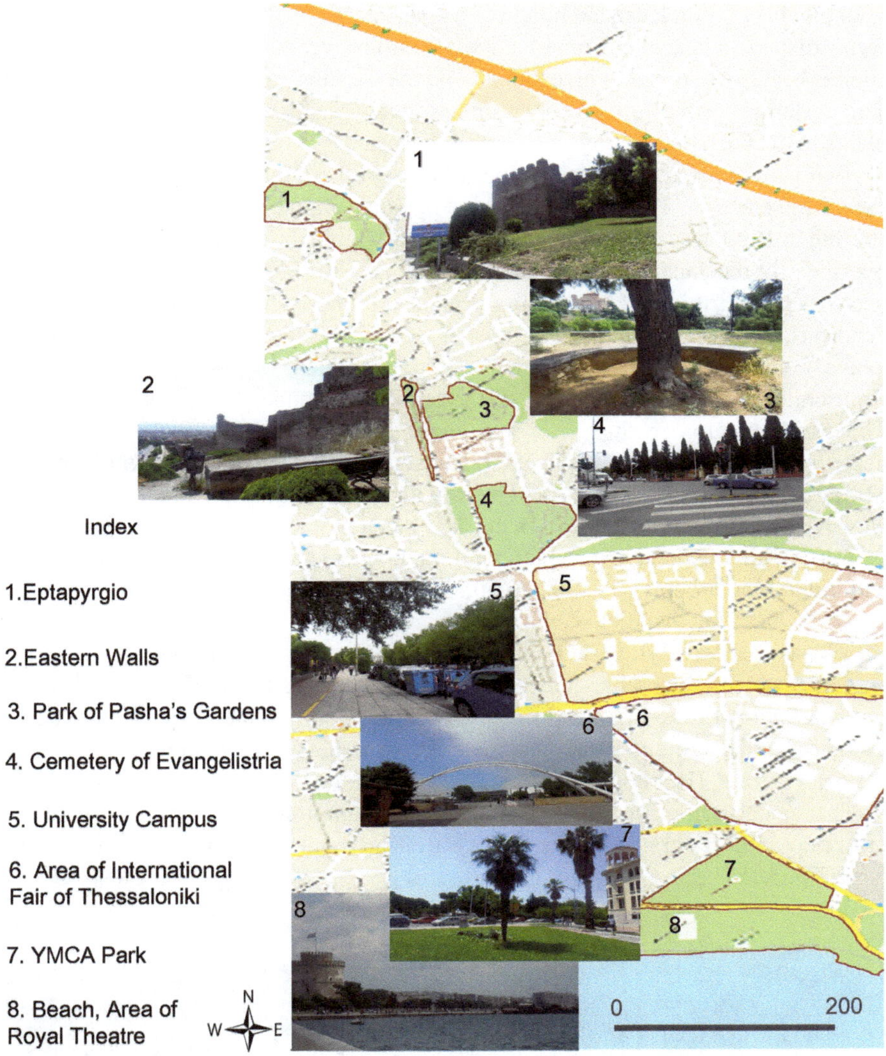

Fig. 7.16 The corridor connecting historical parks with monumental trees in Thessaloniki, Greece. (Photos by T. Tsitsoni)

7.3.4 Garden Styles and Varietal Choice

When considering garden restoration, Edouard André, in the second half of the nineteenth century, argued that the guiding principle was historical truth (verité historique). André was a French botanist and landscape architect who, in 1879, published the treatise "L'art des jardins"(André 1879), in which gardens were considered works of art worthy of preservation in their own right (Giusti 2004). André's scientific background also led him to cite plant species that characterized gardens of styles distant in space and time, and within a few years, this attention to varietal choice began to spread. For example, in the Japanese-inspired gardens created across Europe at the turn of the twentieth century, this tendency to use original plants to evoke exotic settings was put to the test.

Two gardens created by the Rothschilds were particularly noteworthy for their botanical selection. The first, built in the Parisian suburb of Boulogne-Billancourt starting in 1900, was commissioned by Edmond Benjamin James de Rothschild and overseen by the Japanese gardener Hata Wasuke. Wasuke worked for twenty-five years to perfect this park, importing ornamental cherry trees, hydrangeas, rhododendrons, aquatic irises, and the most prized chrysanthemum cultivars from his native country. The second garden was inaugurated in 1901 by Leopold de Rothschild, of the English branch of the family, in Gunnersbury, west of London. The park's creator, in this case, was an English gardener, James Hudson, who described the garden's genesis in 1907. Hudson built fences and gates with bamboo, a bridge and a summer pavilion, and paths with stones laid "following the Japanese". However, what interests us most is the care taken in choosing the ninety plant taxa native to China, Korea, and Japan: conifers, maples, bamboo, peonies, magnolias, and lilies (Maovaz and Legé 2024) (Fig. 7.17).

Throughout the twentieth century, scholars continued to focus on the plants typical of past garden styles: in 1961, Georgina Masson published her book Italian Gardens, which included an appendix listing the various types of plants cultivated from antiquity to the eighteenth century (Masson 1961). Penelope Hobhouse, a designer and historian, then published a comprehensive monograph "Plants in Garden History" (Hobhouse 2004).

That same year, a fire at the royal palace of Hampton Court sparked one of the most important restoration projects of a European royal palace in the twentieth century (Fig. 7.18).

Fig. 7.17 Two views of the Japanese Garden at Gunnesbury (from Hudson 1907)

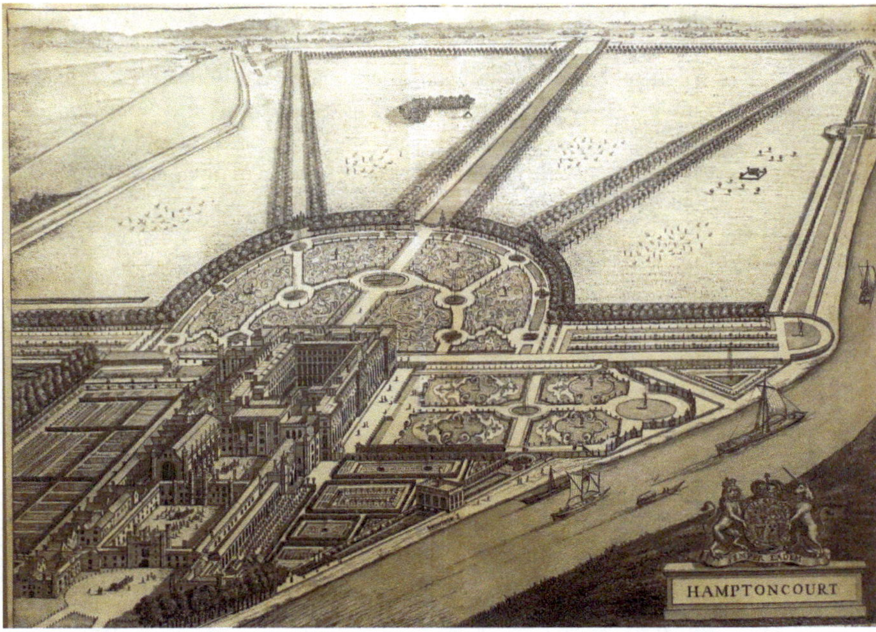

Fig. 7.18 Hampton Court Park (London, UK) with the Privy Garden overlooking the Thames (from Kip 1720)

Initially, the work focused on the building damaged by the fire, but it was soon decided also to tackle the renovation of the so-called 'Privy Garden', planted in front of the palace's south façade in 1702.

In the original design, the garden framed views of the Thames, but as the plants grew in the Victorian era, these views were obscured. The restoration of the original appearance was based on two lines of research: archaeological and documentary. The royal archives had preserved many testimonies of the past: plans, views, financial statements, descriptions of work carried out, and, above all, lists of planted plants. All the collected material enabled the botanical collections of the garden and the adjacent orangery to be recreated, restoring the clarity of the complex.

7.3.5 Maintenance and Monitoring

The vegetation of heritage sites is an essential character-defining element, and for a garden's survival, continuous maintenance, specialized horticultural knowledge, and the propagation of historic plant varieties are needed. As dynamic socio-ecological systems, gardens are always in flux, as plants grow and age, and their structure depends on maintenance, use, and management practices over time (Lian et al. 2024). Abandonment or inadequate care often triggers natural succession, transforming designed compositions into spontaneous plant communities (Gullino et al. 2020).

In the lack of a comprehensive legal instrument to uniformly protect nature–culture heritage worldwide, a systematic botanical assessment should be included in management plans (Table 7.1). Conserving such sites requires balancing the preservation of designed features with nature's dynamic cycles, while recognizing that authenticity itself evolves over time.

In recent decades, scholars and practitioners have approached historic gardens and heritage sites through an ecological lens, recognizing them as biodiversity refuges and integral components of urban green infrastructure. Organizations such as UNESCO and ICOMOS advocate integrated management strategies that link the conservation of cultural and natural heritage but excluding biodiversity as a core value can weaken decision-making and management plans. Even in the case of historic gardens, governance frameworks remain weak; only 7% of countries have explicit legal provisions for historic gardens, and botanical values are rarely included in designation criteria (Hosseini et al. 2024).

In these sites, the management of spontaneous and planted vegetation has traditionally prioritized the preservation of historical structures, often overlooking the botanical component. However, plants exert dual roles.

On one hand, they can cause structural damage and accelerate biodeterioration. Such damage occurs through chemical processes, including hydrogen ion exchange, colloidal particle interactions, and secretion of chemical exudates; however, the primary mechanism is mechanical root pressure, which enlarges fissures and cracks, reduces stone cohesion, and leads to erosion (Caneva 1999). Then, Caneva et al. (2023) proposed a comprehensive Risk assessment method, the RHV index, that evaluates Hazards linked to plant colonization and environmental conditions, as well as Vulnerability factors such as substrate bioreceptivity, microsite characteristics, and conservation status.

This multifactorial approach explains why the same species may exhibit different levels of aggressiveness depending on geographical, bioclimatic, and edaphic conditions (e.g., ivy (*Hedera helix* L.) showed a higher risk in Mediterranean climates than under cold-humid temperate conditions in England).

On the other hand, vegetation can enhance aesthetic value, support biodiversity, and contribute to site stability by reducing erosion and buffering the microclimate. These landscapes embody "cultural layering," in which present-day vegetation reflects interactions among potential natural vegetation, historical land-use legacies, soil modifications from occupation, and spontaneous colonization, with species distributions shaped by millennia of human influence (Cicinelli et al. 2018). Floristic survey sites consistently document plant communities structured by disturbance regimes, substrata, and monument microhabitats -distinct "floristic identities" emerging from the interplay of culture and nature (Kanellou and Papafotiou 2025).

Detailed floristic inventory, vegetation classification, assessment of conservation value, analysis of historical plant composition, and evaluation of ecosystem services provide the evidence base needed to justify management investments, navigate

Table 7.1 Assessment of vegetation in Heritage Sites

Assessment category	Methods/indicators	Note for interpretation
1. Floristic & Vegetation Survey	Systematic field survey, botanical keys and taxonomic guides, and herbarium verification	Foundation for all other assessments. For archaeological sites, distinct communities may indicate environmental gradients (moisture, substrate, disturbance) shaped by archaeological features. For gardens, note species documented in historical records vs. those occurring spontaneously.
	Classification of Raunkiaer life forms, biogeographical origins	
	Classify vegetation into plant communities using phytosociological classification	
	Distinguish planted, legacy, and spontaneous colonizers through archival sources + field patterns	Essential for interpreting garden identity or archaeological processes
2. Conservation value assessment	International (e.g., CITES, habitats directive); national (red lists, national protection laws	The presence of protected species increases site conservation priority and may trigger legal obligations.
	Endemics, relics, rare species	
	Dendrochronology; archival data (age, size, historical records)	Major contributors to identity and landscape continuity
	Species with religious, artistic, or historical significance through iconography, texts	Adds intangible heritage dimension
3. Vegetation–monument interaction	RHV index, microsite, root penetration, bioreceptivity	Essential for conservation planning
	Soil stabilization, microclimate buffering, pollinator support, vegetation cover mapping, species traits	Supports ecological sustainability and site resilience
4. Landscape and contextual factors	Microtopography & exposure data	Explains plant distribution patterns
	Past horticultural practices, agriculture, abandonment cycles, historical maps, excavation data	Reveals cultural layering and long-term dynamics
5. Cultural-historical analysis	Library & archive research	Essential for assessing authenticity
	Image analysis: Paintings, engravings, photographs	Identifies lost or altered plantings
	Interviews; ethnobotany	Connects community heritage with site management

conflicts between authenticity and biodiversity, and ensure botanical elements receive attention comparable to architectural features.

7.3.6 Challenges and Limitations

7.3.6.1 Fragmented Legal Instruments

Fragmented legal frameworks hinder integrated management of archaeological and historic sites. Sites may fall under multiple classifications—such as cultural landscapes, protected areas, or mixed properties—leading to conceptual overlaps and inconsistent policies (UNESCO-SCBD 2010). Heritage and nature conservation sectors often use different terminologies and prioritize different values, resulting in misalignments and a lack of shared language. This dual approach fosters conflicting competencies and fragmented agendas, weakening efforts to integrate cultural and ecological values. However, practitioners can partially compensate for weak legal frameworks by providing systematic documentation of botanical assessments.

7.3.6.2 Lack of Botanical Value in Identity

Defining the identity of historic gardens remains underdeveloped. While gardens are traditionally understood as cultivated, enclosed spaces for human use and pleasure, their identity is complex and shaped by evolving aesthetics and cultural contexts. Botanical elements—central to this identity—are often overlooked in conservation frameworks, with limited analysis, protection, or valorization. This gap creates uncertainty in cataloging and managing historic gardens (Dreija 2012; Hosseini et al. 2024).

7.3.6.3 Climate Impacts on Historic Plants

Climate change threatens the survival of traditional plant species historically associated with garden design. Adaptive restoration strategies must balance authenticity with resilience, selecting climate-tolerant cultivars that maintain visual continuity with historical compositions (Singh 2020; Hosseini and Caneva 2022). This requires evidence-based decisions rather than rigid adherence to obsolete species lists.

7.3.6.4 Authenticity vs. Biodiversity

Managing vegetation in historic gardens and archaeological landscapes involves reconciling authenticity with biodiversity. However, this is not necessarily a binary choice. An authenticity-focused approach prioritizes historically documented species and design fidelity but may reduce ecological resilience (ICOMOS 1982). Conversely, biodiversity-oriented strategies enhance habitat value and ecosystem services but risk compromising cultural significance (Kanellou and Papafotiou 2025; Báez and Torreblanca 2024).

7.4 Therapeutic Gardens (A.R., M.B.)

7.4.1 Definition and Classification

Since ancient times, many scholars have demonstrated the importance of Nature's spiritual and therapeutic role and its influence on human mental and psychophysical well-being. Suffice it to recall that all the world's mythologies and religions originated in a garden. The term "Paradise" derives from the two Persian words (*pairi* = around and *daeza* = wall) as a *hortus conclusus*.

Nature's positive influence on physical well-being was clear in the ancient Greek cultures, where students were trained walking in Lyceums (from holm oaks woods), such as in the Middle Ages, as evidenced by the emergence of cloisters, gardens, and orchards among various European monastic orders, where time was spent in contemplation. In the thirteenth century, Humbert, former Master of the Dominican Order, decreed that part of the land surrounding the monastery's infirmary should include fields and gardens for patients' recreation (Montford 2004). In 1700, Benjamin Rush, the father of American psychiatry, stated that "... working the land and cultivating plants has a beneficial effect on human mental health".

The first hospitals in the Western world were monastic communities where herbs and prayers were essential to healing the sick, and internal gardens were integral to the monastic complexes (Cooper Marcus and Barnes 1999). In the second half of the nineteenth century, some American hospitals began building greenhouses for therapeutic purposes. However, it was only in the 1920s that specific programs for the rehabilitation of patients with physical and cognitive disabilities were established. In 1817, Friends Hospital in Pennsylvania was one of the first hospitals to create woodland trails, where the natural environment was seen as a safe and peaceful place.

In 1919, C.F. Menninger and his son, Dr. Karl, founded the Menninger Foundation in Kansas, where gardening was central to patient care (Vigorelli 2005). In the 1950s, after both world wars, a strong need for rehabilitation for soldiers with mental and physical disabilities emerged; this encouraged the creation of rehabilitation centres that combined pharmacological treatments with contact with the natural environment (Söderström 2000).

Over the years, research has shown that spending time outdoors offers significant therapeutic benefits, particularly for people with disabilities and dementia, as contact with the natural world facilitates psychophysical recovery (Antonelli and Casavecchia 2005). These studies have led to the concept that the *"healing garden"* offers relief by stimulating and rewarding sensory interaction between patients and nature (Cooper Marcus and Barnes 1999).

In large psychiatric hospitals, patients tend vegetable and ornamental gardens, as daily gardening activities such as digging, planting, and harvesting help them perceive daily rhythms and the changing seasons. This idea was put into practice in Swedish psychiatric hospitals as early as the 1940s (Söderström 2000). Indeed, several studies have confirmed the close link between length of hospital stay, use of painkillers, and the ability to admire the natural world from the hospital window (Ulrich 1984).

In patients with dementia, and in particular Alzheimer's disease, the physical environment plays a therapeutic role, therefore the arrangement of outdoor spaces becomes an integral part of the treatment to overcome cognitive and functional deficits (Cooper et al. 1999; D'Andrea et al. 2008; Edwards 2013; Pomposini 2004; Rivasseau Jonveaux et al. 2013; Valla 2002). The *Healing gardens* for people with Alzheimer's disease use elements such as scented plants, simple, logical layouts with individual pathways, and safe surfaces to help reduce anxiety, encourage reminiscence, and provide a sense of accomplishment and calm (Ranfa et al. 2014, 2017).

7.4.2 Healing Garden in Italy

In Italy, the importance of *healing gardens* was only recognized in the 1980s. Since 1989, the most important gardens for Alzheimer's patients have been established. The Garden of the Grimani Butteri Foundation in Osimo (Marche) has primarily chosen aromatic herbs to stimulate memories of the past and reduce behavioural problems and the tendency to wander aimlessly (Valente and Cooper Marcus 2015). In the 1990s, the first "Alzheimer Experimental Project" was launched in Lombardy, followed by projects in Piedmont, Emilia-Romagna, and Tuscany. A garden designed by Giovanni Ingrao in 1996 in Monza (Lombardy) also included an experimental garden (Valla 2002). In a long-term care home in the province of Varese, artificial vegetation and wallpaper were used to create a life-size forest in an indoor greenhouse. At the Villa Serena nursing home in Montaione, near Florence (Tuscany), the region's first green space for those patients was specifically designed to include scented, seasonal, and non-poisonous plant species.

In Umbria, a project called "Therapeutic Parks" was launched in 2012, which has developed therapeutic activities in parks, including community gardens, musical and theatrical events, animal activities, and amateur sports such as mountain biking and Nordic walking. All these activities are aimed at people with disabilities, those experiencing financial hardship, the elderly, and those undergoing rehabilitation, to create spaces for psychomotor rehabilitation projects. Other gardens have also been created, such as the one for Alzheimer's patients attached to the Muzi Betti Nursing Home in Città di Castello (Perugia), conceived to show that the sensory garden represents a step forward in healthcare by promoting patient autonomy.

In 2013, a *healing garden* for Alzheimer's patients was created at the Fontenuovo Assisted Living Facility in Perugia (Ranfa et al. 2014, 2017).

7.4.3 Methodological Approach

These projects were carried out using a multidisciplinary approach by a team of professionals, including landscape architects, engineers, agronomists, botanists, physicians, therapists, and environmental psychologists. The garden's design adhered to three essential requirements based on guidelines established in previous studies (Cooper Marcus and Barnes 1999):

1. Natural mapping (Norman 1988).
2. Latent image elements (Lynch 1960).
3. Housing zones (Zeisel and Welch 1981).

According to Natural mapping (Norman 1988), the garden's layout was designed to be "natural," with a clearly defined entrance, main path, and exit. Based on image elements (Lynch 1960), five fundamental elements were introduced relating to how the brain processes information:

- Opportunity for choice, privacy, and a sense of control. It is important for a garden to encourage stress reduction and provide a sense of control. Physical exercise has been shown to provide significant mental and psychological benefits, increasing cardiovascular health and reducing depression in adults and children (Koniak-Griffin 1994; Brannon and Feist 1997).
- Visibility and direct contact with nature. The presence of a garden in a retirement home is essential, as it allows for a constant connection with the outside world. It is also important to have an abundant variety of plant species of different sizes, colours, and textures, including species that bloom seasonally; attract non-threatening animals, such as birds, squirrels, and butterflies; have leaves and grasses that sway in a light breeze (Ulrich 1984).
- Sense of Security. In addition to the garden's visibility and accessibility, it is essential that users have a sense of security, both physiological and psychological, or they will not benefit from it (Cooper Marcus and Barnes 1995).
- Psychological Comfort and Tranquillity. Many patients require medications that require avoiding direct exposure to the sun; others may fear the consequences of being outdoors. Therefore, a garden must provide psychological comfort by offering seating in the sun or in the shade (Cooper et al. 1995). If a garden is to have therapeutic value, quiet is essential, in complete contrast to the hustle and bustle of the public, the TV, and the interior noises.
- Clear and positive design. The garden should be designed considering the patients' impaired cognitive abilities and daily challenges. Since spending time outdoors has been shown to have a calming effect and help reduce anxiety and aggression, encouraging interest in the outdoors is obviously particularly impor-

tant (Zeisel 2001). These gardens use elements such as scented and textured plants, a simple, logical layout with single paths and safe surfaces, to help reduce anxiety, encourage reminiscence, and offer a sense of accomplishment and calm (Ranfa et al. 2014, 2017).

7.4.4 Specific Benefits

The creation of *healing gardens* has been shown to improve psychophysiological outcomes in people with dementia. Interest in this field has grown over the past decade. However, there remains a need to study the beneficial effects of therapeutic gardens in relation to the type and severity of dementia, as well as to develop guidelines for garden design (Murroni et al. 2021)."Dementia-friendly" therapeutic gardens provide significant psychological benefits, significantly reducing depression over the long term. Notably, even those with more advanced dementia benefit, challenging the prevailing notion that non-pharmacological interventions are primarily effective in the early stages of the disease (Jevtic et al. 2025).

7.4.5 Maintenance and Monitoring

For proper use and enjoyment of the therapeutic garden, it is essential to maintain a comfortable, welcoming, clean, and orderly environment that supports the healthy growth of plant species, preferably using treatments based on natural, non-toxic products. A well-maintained garden will help patients feel pampered and welcome. This will not only promote the patient's recovery but also enhance and facilitate communication.

To make Healing Gardens sustainable, it is necessary to use plant species with low water and nutrient requirements. Pesticides, which pollute and harm the small fauna associated with the garden, should be avoided. The inclusion of exotic plants, such as tropical ones, should be minimal or absent. Because they need different soil and climate for proper growth, they require excessive maintenance and can waste precious resources and cause unnecessary contamination.

7.4.6 Challenges and Limitations

When designing a therapeutic garden, it is important to avoid large empty spaces, as these can also cause anxiety (Cannara et al. 2004). It is a good idea to provide easily recognizable rest areas, both for socializing, such as a gazebo, and for individual meditation, with spaced-out benches. Intersections are important because they provide a natural gathering place for patients and therefore serve as the focus of group activities, such as horticultural ponds. The garden should be clearly delineated to facilitate orientation and avoid conveying a sense of insecurity.

Placing the garden in an internal courtyard overlooked by most bedrooms and corridors strongly encourages patients to use it, because if they were unaware of it, they would obviously never use it. Furthermore, staff can observe patients without them noticing (Ulrich 1984).

Another important aspect is the choice of plant species based on their different flowering periods. Creating sensory areas to stimulate the senses is essential for reconnecting patients with the outside world (Pomposini 2004).

7.4.7 Case Studies

The Fontenuovo Healing Garden (Figs. 7.19 and 7.20) was designed to be directly accessible from the activity room, creating a sense of continuity between inside and outside and avoiding disorientation for patients. The main garden extends around a circular path, thus ensuring a safe space for walking. The main entrance (1) features a winding path with no intersections or sudden changes in direction. Patients can use the handrail along the entire path, enhancing the sense of safety. There are no intersections or sharp corners, and the path is paved with a light-coloured, non-slip, anti-freeze and anti-reflective material. Along the main path, there are several rest

Fig. 7.19 Sensory Garden for Alzheimer's patients at the Fontenuovo Nursing Home (Perugia, Italy). (Photo by A. Ranfa)

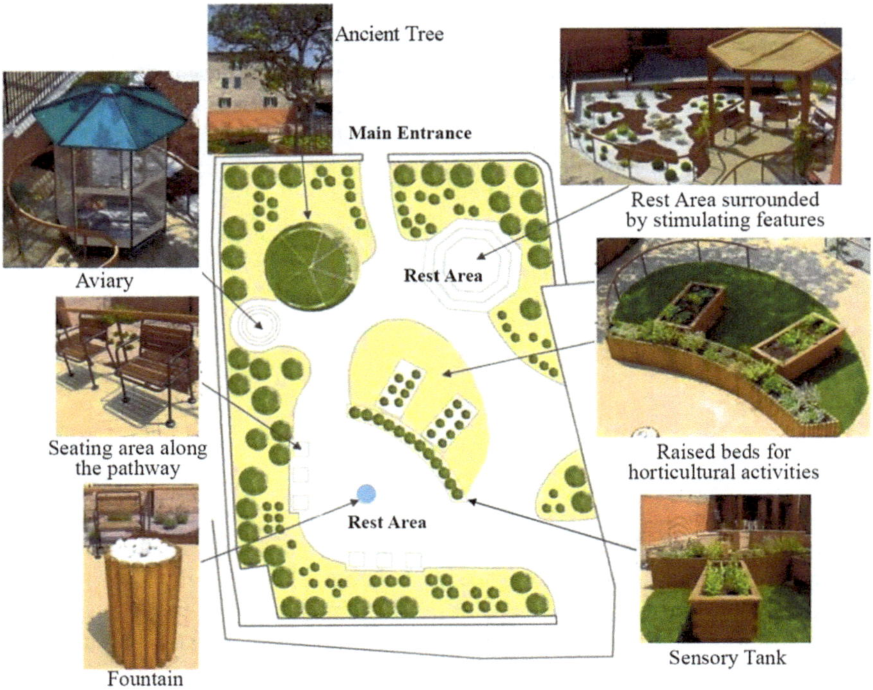

Fig. 7.20 The details of the healing garden of Fontenuovo (Perugia, Italy). (Courtesy by IJAER)

areas (2) equipped with benches surrounded by scented plants to make the break even more relaxing. There are also individual seats that provide a private space, along with a small wooden gazebo to encourage socialisation (2).

The central part (3) is characterised by two sensory pools (4) for growing vegetables, stimulating attention and observation of the growth and transformation of the plants. There is also a fountain for auditory stimulation (5) and an aviary (7).

The lighting has been designed to avoid glare and excessive contrast between light and shadow, and to allow evening access to the garden. Five sensory zones have been created to further stimulate the senses, as colours, scents, and sounds help Alzheimer's patients reconnect with the outside world (Pomposini 2004). Plant species were chosen primarily for color and scent, ease of handling, and seasonal availability, while also considering their non-toxic or allergenic properties.

Aromatic species, such as *Salvia rosmarinus* Spenn., *Mentha* × *rotundifolia* (L.) Huds., and *Myrtus communis* L., were chosen to stimulate the sense of taste. To stimulate the sense of touch, *Stachys lanata* Crantz and *Festuca glauca*, with their rough or silky-woolly foliage, were chosen. To stimulate the sense of smell, species such as *Lavandula dentata* L., *Aloisya citriodora* Paláu, and some mint species (*Mentha* spp.), which release their scent when the leaves are rubbed between the fingers, were planted toward the outer edges of the garden. These species are planted

Fig. 7.21 Oasis of Fragrances and Breath (Ponte San Giovanni—Perugia, Italy). (Photo by A. Ranfa)

next to species whose scent can be detected from a short distance, followed by others such as *Viburnum tinus* L., whose scent spreads further.

The fountain's running water was designed to stimulate the sense of hearing, and an aviary near the garden's only existing tree, a majestic Japanese medlar (*Rhaphiolepis bibas* (Lour.) Galasso & Banfi), whose leaves rustle in the wind.

The alternating colors that change with the seasons stimulate the sense of sight, which is the fundamental aesthetic aspect of a garden. The fountain's flowing water was designed to stimulate the sense of hearing.

Another example of a therapeutic garden is the one created in Ponte San Giovanni (Perugia), the "Oasis of Scents and Breath—Urban Bee Park". The garden, conceived and designed by the Perugia Host Lions Club, is shaped like a lung and features 304 aromatic and medicinal plants from seven different species. Its goal is to help improve breathing, clean the air in areas with high human impact, and introduce children, teens, and adults to the wonderful world of bees and environmental issues (Fig. 7.21).

7.4.8 Conclusion

Throughout history, it has been demonstrated that access to aspects of "nature" is a fundamental human need and that the natural environment is not only a source of food but also of psychological and physical well-being. As Barnes and Marcus stated, the design of an outdoor therapeutic area arises from the combination of two concepts—"healing process" and "space in which the process occurs"—and this only occurs when human needs and scientific research findings go hand in hand. The positive effects these spaces have on patient health and the cost-effectiveness of healthcare infrastructures encourage further study of therapeutic gardens as a therapeutic support for various pathologies. Awareness that the environment, and landscape in particular, can play a fundamental role in improving health and perhaps preventing disease has been slow to emerge. However, there are clear signs that changes are underway.

A current problem is the need for funding, as gardens are often created thanks to private donations or funding that are not always readily available; therefore, other forms of public funding should be encouraged. These gardens could also be established in public open spaces, thereby becoming important centres of social therapy outside the confines of traditional medicine. This would provide access for those who cannot afford private care institutions.

In the following words, Adrian Burton demonstrates how this scientific stance has been recognized: "[...] *if [gardens] can be shown to shorten hospital stays, reduce the need for pain medication or other drugs, hasten (and therefore reduce the cost of) the rehabilitation process, or reduce staff stress and burnout (as initial research suggests), financing bodies might look upon them favorably.*" (Burton 2014, p. 448).

7.5 Community Gardens (F.B., G.C.)

7.5.1 Definition and Classification

Community gardens can be defined as shared spaces where individuals collectively grow plants for food, recreation, education, and cultural purposes (Fig. 7.22).

Initially conceived as emergency relief measures during economic crises, community gardens have evolved into strategic instruments for urban sustainability. Their transformation from temporary interventions to permanent green infrastructure underscores their growing recognition in planning discourse (Lawson 2004). They exhibit diverse organizational models and purposes, reflecting their

Fig. 7.22 Examples of community gardens in Rome (Italy), with garden plot, plant support, cultivated soil with diverse crops and fruit trees. (Photos by F. Bartoli)

multifaceted role in urban landscapes and delivering ecological, social, and economic benefits. Scholars emphasize that classification is essential for planning and management because it determines governance structures, resource allocation, and social outcomes (Caneva et al. 2020; Raneng et al. 2023).

Community gardens can be established on public land, such as municipal parks or vacant lots, on private property, often through agreements with landowners, or on institutional grounds, including schools, hospitals, and housing complexes. Ownership influences tenure security—a critical factor for long-term sustainability (Lawson 2004). The aims of community gardens extend beyond food production. They encompass ornamental horticulture, educational activities, therapeutic functions, and the preservation of cultural heritage. Recent research highlights their role in promoting food security, mitigating urban heat, and fostering social inclusion (Raneng et al. 2023). Community gardens generally adopt two primary organizational structures, each reflecting distinct social, cultural, and ecological aims.

Allotment gardens allocate individual plots to members for autonomous cultivation. This model promotes personal responsibility, fosters crop diversity, and enables gardeners to tailor practices to their preferences. It is widely associated with food security, self-sufficiency, and the preservation of agrobiodiversity, particularly in urban contexts where private green space is limited (Guitart et al. 2012). Studies also highlight that allotment gardens can serve as micro-scale strategies for climate adaptation and urban resilience by reducing food miles and supporting sustainable diets (Byrne et al. 2017).

Collective gardens, conversely, emphasize shared planting and harvesting areas. These spaces function as hubs for social interaction, collaborative learning, and community engagement, often serving educational and therapeutic purposes. They reinforce social cohesion and inclusivity, providing opportunities for environmental education and cultural exchange (Caneva et al. 2020; Kingsley et al. 2020). Collective models are particularly effective in fostering community empowerment and addressing urban inequalities through participatory governance (Lawson 2004).

Importantly, these typologies are not mutually exclusive. Hybrid configurations frequently emerge, integrating individual allotments with communal zones to balance autonomy and collective participation. Such mixed models are increasingly recognized as effective strategies for maximizing ecological benefits while strengthening social networks and cultural identity (Raneng et al. 2023; Anderson et al. 2019).

7.5.2 Specific Benefits

Community gardens deliver a wide range of benefits that extend beyond food production, contributing to environmental sustainability, social cohesion, public health, education, and local economies. These benefits are increasingly recognized in urban planning and sustainability frameworks (Guitart et al. 2012; Raneng et al. 2023). Community gardens function as vital components of urban green infrastructure, delivering multiple ecological benefits. They enhance air quality by filtering airborne pollutants and increasing oxygen production, and help mitigate the urban heat island effect, lowering ambient temperatures and improving microclimatic conditions (Holland 2004; Voicu and Been 2008; Ebissa et al. 2024).

Additionally, these spaces contribute to sustainable water management by promoting stormwater infiltration and reducing surface runoff, thereby alleviating urban flooding risks. By providing diverse floral resources and nesting habitats, they sustain bees and other beneficial insects, which are essential for pollination, biodiversity conservation, and the maintenance of ecosystem services in densely built environments (Felderhoff et al. 2023; Baldock et al. 2015). They foster interaction among culturally and socioeconomically diverse groups, strengthen social capital, and promote trust through collaborative decision-making, thereby reducing isolation and enhancing neighbourhood resilience (Kingsley et al. 2020). As Ferris et al. (2002) argue, they embody the social dimension of sustainable development by linking food security, education, and health with urban green space, while promoting environmental justice and equity.

Community gardens contribute significantly to public health by improving access to fresh, nutrient-dense produce, supporting healthier diets, and reducing food insecurity (Huq and Deacon 2025; Alaimo et al. 2008). Their benefits include restorative environments that support mental well-being, therapeutic programs for vulnerable populations, and encouragement of physical activity. Ecological restoration initiatives, such as creek rehabilitation and habitat renewal, often intersect with community gardening, transforming degraded urban spaces into biodiverse ecosystems. Ghose and Pettygrove (2014) emphasize that community gardens often become spaces where people can shape their surroundings, build relationships, and express shared identities, offering opportunities for residents—especially those who are often excluded from formal planning—to participate in decisions and reclaim neglected urban areas.

Recent systematic reviews confirm that community gardening fosters social cohesion and life satisfaction, making it an effective, low-cost strategy for promoting holistic health in urban settings (Berg et al. 2023). Exposure to natural environments during gardening has been linked to improved mental well-being, reduced stress, and enhanced resilience, offering therapeutic benefits for individuals facing psychological challenges (Koay and Dillon 2020; Okvat and Zautra 2011). Furthermore, gardening activities promote physical exercise through tasks such as digging, planting, and weeding, which are comparable to moderate-intensity physical activity and help reduce the risk of chronic disease (Gregis et al. 2021).

Gardens function as dynamic living laboratories, providing experiential learning opportunities in horticulture, ecology, and nutrition. They foster environmental literacy by enabling learners to observe ecological processes, plant growth cycles, and biodiversity interactions in real time (Guitart et al. 2012). In educational and community contexts, gardens are widely implemented to promote environmental stewardship, sustainable land-use practices, and cultural heritage linked to food systems (Irvine et al. 1999; Holland 2004).

By lowering household food expenditures and promoting localized food production, community gardens can help mitigate economic stress among urban families. In addition to reducing reliance on external food systems, surplus harvests can be marketed through farmers' markets or integrated into Community-Supported Agriculture (CSA) schemes, which involve direct partnerships between consumers and local producers, enabling pre-paid seasonal shares of farm produce, creating supplementary income opportunities and reinforcing local economic networks (Voicu and Been 2008; Raneng et al. 2023).

7.5.3 Methodological Approach for Plan Guidelines and Taxa Selection

Planning community gardens requires a structured, evidence-based approach that balances ecological, social, and cultural objectives. Caneva et al. (2020) underlined six primary aims—savings and productivity, food security, environmental restoration, environmental education, sociability, and cultural value—supported by twelve indicators grouped into three categories: environmental parameters, pollution risk factors, and social accessibility (Fig. 7.22).

Environmental indicators such as plot size, slope, soil quality, sunlight, water availability, and biodiversity value determine the feasibility and productivity of gardens, while pollution-related factors (distance from traffic and previous land use) ensure food safety and ecological integrity. Social context indicators, such as accessibility, proximity to schools, community centers, and green spaces, enhance inclusivity and educational potential. Optimal conditions include flat terrain, fertile soils, a reliable water supply, low risk of contamination, and good connectivity to urban green networks. Integrating these parameters into planning not only maximizes ecosystem services—such as biodiversity conservation and microclimate regulation—but also fosters cultural continuity and social cohesion, positioning community gardens as multifunctional assets in sustainable urban development (Fig. 7.23).

According to the current concept, sustainable forest resource management (both urban and non-urban) aims primarily to manage them to maintain and improve their biodiversity, Selecting plant species for community gardens requires a multifactorial and evidence-based approach that integrates ecological, social, and cultural dimensions (Fig. 7.24). At their core, climate and soil analyses are indispensable for ensuring species compatibility with local conditions, thereby reducing resource inputs and enhancing resilience (Caneva et al. 2020). For example, in the Milan Urban Gardening Project, soil pH and texture assessments guided the introduction of drought-tolerant legumes, reducing irrigation needs by 25% during summer heatwaves (Gregis et al. 2021).

Seasonality planning is critical for maintaining year-round productivity, particularly in temperate and Mediterranean climates where crop rotation and succession planting optimize yields. In Barcelona, seasonal crop rotation combined with

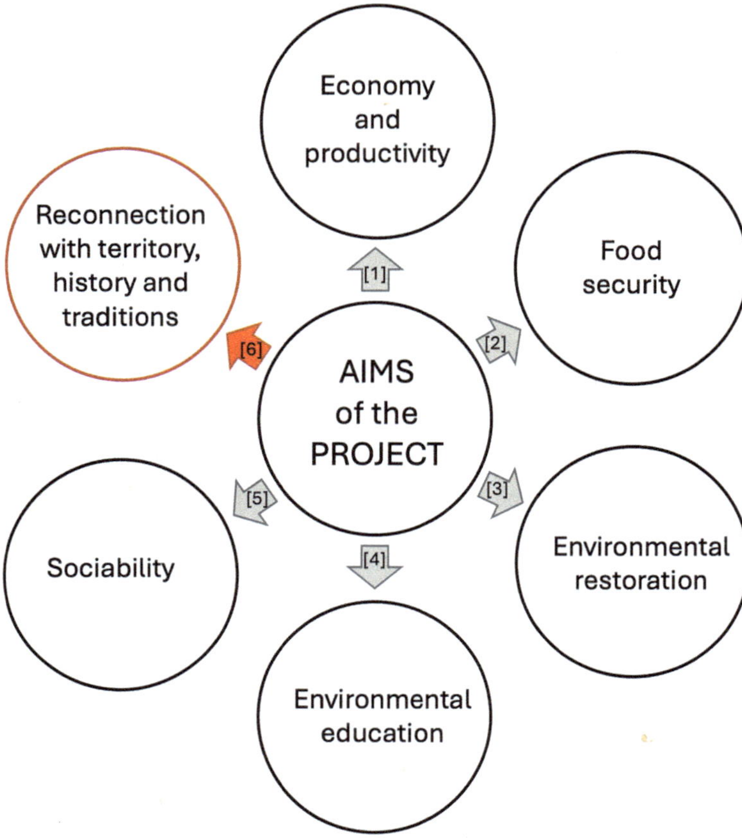

Fig. 7.23 Conceptual framework illustrating the main aims and functions of community urban gardening projects (modified by Caneva et al. 2020)

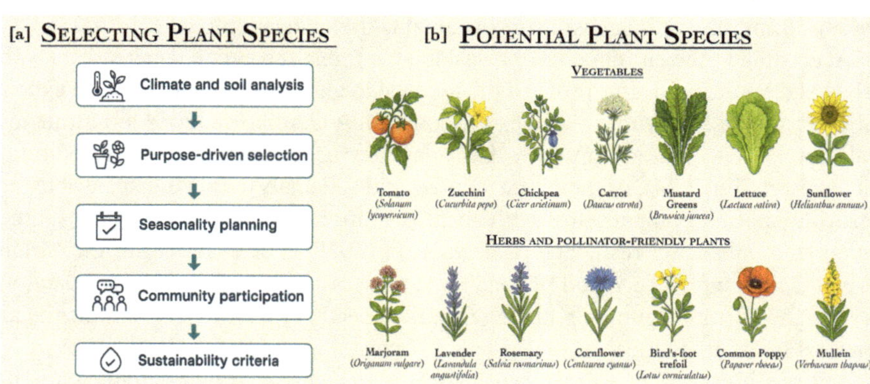

Fig. 7.24 Framework for plant species selection and examples of potential species for community gardens. (**a**) Key steps in selecting plants: (**b**) Examples of recommended plant species grouped into vegetables and herbs/pollinator-friendly plants. (Elaboration by authors)

intercropping improved soil nitrogen levels and reduced pest outbreaks by 30%, demonstrating the ecological benefits of temporal diversity (Guitart et al. 2012). This temporal strategy not only sustains the food supply but also stabilizes pollinator resources across seasons and gardens with higher plant species richness and structural complexity, supporting greater pollinator diversity and functional traits, including nesting behaviours and pollen specialization (Felderhoff et al. 2023; Baldock et al. 2015). Incorporating structural diversity and pollinator-friendly features into planting designs maximizes biodiversity and functional ecosystem services (Normandin et al. 2017).

Community participation plays a pivotal role in ensuring that crop selection reflects cultural preferences and traditional knowledge, thereby fostering social inclusion and reinforcing local identity. Integrating culturally significant species enhances the educational and cultural dimensions of gardens while promoting intergenerational knowledge exchange (Guitart et al. 2012). This participatory approach aligns with sustainability principles, as evidenced by initiatives that combine ecological restoration with community engagement (Irvine et al. 1999). For example, the Melbourne Community Garden Network incorporated Mediterranean herbs and Asian greens to reflect the cultural diversity of its members, thereby increasing participation and facilitating cooking workshops that linked gardening to culinary heritage (Kingsley et al. 2020). Similarly, Toronto's Alex Wilson Community Garden introduced heritage tomato cultivars and indigenous medicinal plants as part of a cultural restoration project, fostering collaboration between local communities and First Nations groups while preserving traditional ecological knowledge (Irvine et al. 1999).

Finally, sustainability criteria should guide species choice, favouring plants with low water demand, pest resistance, and adaptability to organic practices. However, few studies examine seed sourcing, crop diversity, and the role of traditional knowledge in shaping planting strategies, which are critical aspects for enhancing resilience and food sovereignty (Raneng et al. 2023; Guitart et al. 2012).

7.5.4 Maintenance and Monitoring

Effective maintenance and monitoring are essential for the long-term success of community gardens, ensuring plant health, productivity, and sustainability. Routine tasks such as watering, weeding, composting, and integrated pest management form the foundation of garden upkeep (Raneng et al. 2023). Collectively, these practices form an integrated approach that safeguards ecosystem functionality and optimizes resource efficiency within urban green spaces. These practices not only sustain crop productivity but also contribute to ecological stability by promoting nutrient cycling and reducing reliance on synthetic inputs. By sustaining crop productivity and promoting ecological stability, they enhance nutrient cycling and reduce reliance on synthetic inputs (Caneva et al. 2020; Raneng et al. 2023). For example, the Alex Wilson Community Garden in Toronto implemented composting and crop rotation, which increased soil organic matter by 25% over three years, reducing the need for chemical fertilizers (Irvine et al. 1999). Similarly, a study of community gardens in Melbourne found that integrated pest management, combined with organic mulching, reduced pesticide use by 40% while maintaining biodiversity and pollinator presence (Kingsley et al. 2020).

Regular implementation of these activities is critical for mitigating biotic stressors and maintaining the functional integrity of urban green spaces. These responsibilities are often organized through shared systems, including rotating schedules or volunteer teams, which foster community engagement and equitable workload distribution (Kingsley et al. 2020). Technological innovations, including soil moisture sensors and mobile applications for growth tracking, are increasingly integrated into community garden management to optimize irrigation and monitor plant development, thereby reducing resource waste and improving crop yields (Felderhoff et al. 2023; Huq and Deacon 2025). For instance, a case study in Melbourne demonstrated that the use of IoT-based soil moisture sensors reduced water consumption by 30% while maintaining optimal soil hydration, significantly enhancing crop resilience during drought periods (Kingsley et al. 2020). Similarly, community gardens in New York City adopted mobile apps to track real-time growth and send pest alerts, enabling volunteers to coordinate interventions efficiently and achieve a 15% increase in seasonal yield (Anderson et al. 2019).

Seasonal reviews play a critical role in adaptive management by guiding adjustments to planting schedules, crop selection, and soil amendments based on observed performance and climatic variability (Gregis et al. 2021). For example, a longitudinal study of community gardens in Barcelona demonstrated that seasonal evaluations, combined with crop-rotation strategies, improved soil nutrient balance and reduced pest incidence by 35% over two years (Guitart et al. 2012). Similarly, in Milan, adaptive planning informed by seasonal reviews enabled gardeners to shift toward drought-tolerant species during heatwaves, resulting in a 20% increase in survival rates and water-use efficiency (Gregis et al. 2021).

7.5.5 Challenges and Limitations

Community gardens encounter a range of structural and operational challenges that constrain their long-term viability. Land access and tenure insecurity remain critical barriers, as temporary agreements discourage investment and continuity (Lawson 2004). For example, in Cape Town, South Africa, 34 urban gardens operated under informal or short-term arrangements, leaving them vulnerable to eviction and limiting infrastructure investment (Kanosvamhira and Tevera 2023). Similarly, in Berlin and Munich, the absence of policy protection led to the loss of multiple gardens (Egerer et al. 2024). Financial limitations, including insufficient funding for tools, seeds, and infrastructure, further restrict scalability and equitable participation (Raneng et al. 2023). The XYZ Community Garden in the U.S.A. illustrates this challenge, as inadequate irrigation resources led to crop failure and reduced participation (Drake and Lawson 2015). Social dynamics introduce governance complexities, with conflict management often required to address disputes over responsibilities and the allocation of harvests (Kingsley et al. 2020). In Austin, Texas, bottom-up governance models achieved higher social satisfaction and ecological performance, whereas Shenzhen's state-led "We Garden" experiment generated tensions between planning authorities and community autonomy (Ponstingel 2022; Zhang et al. 2022).

Environmental risks such as soil contamination and climate-related stressors pose significant threats to crop safety and productivity. In Philadelphia and Pittsburgh, 36–60% of gardens exceeded Canadian lead safety guidelines, highlighting persistent contamination risks despite raised beds (Bassetti et al. 2023), while E.A.T. South's teaching farm in Alabama faced costly remediation due to prior industrial use, ultimately leading to closure (Roling and Edwards 2025). Sustaining engagement over time is equally challenging, as volunteer fatigue and resource depletion frequently lead to project discontinuity (Voicu and Been 2008). Collectively, these limitations underscore the need for integrated planning frameworks that balance social equity, environmental safety, and economic feasibility to ensure resilience and multifunctionality in urban green spaces (Raneng et al. 2023).

References

Alaimo K, Packnett E, Miles RA, Kruger DJ (2008) Fruit and vegetable intake among urban community gardeners. J Nutr Educ Behav 40(2):94–101. https://doi.org/10.1016/j.jneb.2006.12.003

Anderson EC, Egerer MH, Fouch N et al (2019) Comparing community garden typologies of Baltimore, Chicago, and New York City (USA) to understand potential implications for socio-ecological services. Urban Ecosyst 22:671–681. https://doi.org/10.1007/s11252-019-00855-9

André E (1879) L'Art des jardins. Traité général de la composition des parcs et jardins, Paris, Masson

Antonelli R, Casavecchia C (2005) Il tempo ritrovato. Primo giardino sensoriale, Osimo, Ancona

Báez JC, Torreblanca D (2024) Archaeological sites and historical monuments as refuges for threatened species: a call for integrated conservation management. Front Conserv Sci 5:1495157. https://doi.org/10.3389/fcosc.2024.1495157

Baldock KC, Goddard MA, Hicks DM, Kunin WE et al (2015) Where is the UK's pollinator biodiversity? The importance of urban areas for flower-visiting insects. Proc R Soc B Biol Sci 282(1803):20142849. https://doi.org/10.1098/rspb.2014.2849

Bassetti OG, McDonough RA, Shakya KM (2023) Soil contamination in community gardens of Philadelphia and Pittsburgh, Pennsylvania. Environ Monitor Assess 195(6):782. https://doi.org/10.1007/s10661-023-11329-z

Berg AC, Padilla HM, Sanders CE, Garner CT, Southall HG, Holmes G et al (2023) Community gardens: a catalyst for community change. Health Promot Pract 24(1_suppl):92S–107S. https://doi.org/10.1177/15248399221120808

Blasi C, Capotorti G, Del Vico E, Bonacquisti S, Zavattero L (2025) Native Flora and Potential natural vegetation references for effective Forest restoration in Italian Urban Systems. Plants 14(15):2396. https://doi.org/10.3390/plants14152396

Brannon L, Feist J (1997) Health psychology, 3rd edn. Brooks/Cole, Pacific Grove, CA

Burton A (2014) Gardens that take care of us. Lancet Neurol 13(5):447–448. https://doi.org/10.1016/S1474-4422(14)70002-X

Byrne J (2017) Urban parks, gardens and greenspace. In the routledge handbook of environmental justice. Routledge pp 437–448
Caneva G (1999) A botanical approach to the planning of archaeological parks in Italy. Conserv Manag Archaeol Sites 3(3):127–134
Caneva G, Cutini M, Pacini A, Vinci M (2002) Analysis of the Colosseum's floristic changes during the last four centuries. Plant Biosyst 136(3):291–311. https://doi.org/10.1080/11263500212331351199
Caneva G, Ceschin S, Bartoli F, Cianci MG, Montuori L (2016) A botanical network for the sustainable development of the Roma Tre University campus area. In VI international conference on landscape and urban horticulture 1189 pp 297–302. https://doi.org/10.17660/ActaHortic.2017.1189.58
Caneva G, Cicinelli E, Scolastri A, Bartoli F (2020) Guidelines for urban community gardening: proposal of preliminary indicators for several ecosystem services (Rome, Italy). Urban For Urban Green 56:126866. https://doi.org/10.1016/j.ufug.2020.126866
Caneva G, Hosseini Z, Bartoli F (2023) Risk, hazard, and vulnerability for stone biodeterioration due to higher plants: the methodological proposal of a multifactorial index (RHV). J Cult Herit 62:217–229. https://doi.org/10.1016/j.culher.2023.05.032
Caneva G, Hosseini Z, Bartoli F, Capotorti G, Attorre F, Blasi C (2025) Revaluating CUNA places (CUltural place of high relevance for NAture): Rome as a multifaced example of outstanding values and potentials. Land 14(2):1–25. https://doi.org/10.3390/land14020226
Cannara A, Brizioli C, Brizioli E (2004) Progettare l'ambiente per l'Alzheimer. Specifiche progettuali per l'ambiente terapeutico, Franco Angeli Editore
Cano E, Piñar Fuentes JC, Cano-Ortiz A, Quinto Canas R, Rodríguez Meireles C, Raposo M et al (2024a) Research and management of thermophilic cork forests in the central-south of the Iberian Peninsula. Plant Biosyst 158(5):942–962. https://doi.org/10.1080/11263504.2024.2379366
Cano E, Piñar Fuentes JC, Cano-Ortiz A, Quinto Canas R, Rodríguez Meireles C, Raposo M et al (2024b) Contribution to the Iberian thermomediterranean oak woods (Spain, Portugal): the importance of their teaching for the training of experts in environmental management. Plant Biosyst 158(6):1285–1305. https://doi.org/10.1080/11263504.2024.2404431
Cano-Ortiz A, Musarella CM, Cano E (2025a) Socio-economic development and eco-education for urban planning committed to sustainability. Sustainability 17(6):2442. https://doi.org/10.3390/su17062442
Cano-Ortiz A, Piñar-Fuentes JC, Musarella CM, Peña-Martínez J (2025b) Education for environmental sustainability component: innovative strategies for experiential learning in natural contexts. Educ Sci 15(6):697. https://doi.org/10.3390/educsci15060697
Cicinelli E, Salerno G, Caneva G (2018) An assessment methodology to combine the preservation of biodiversity and cultural heritage: the San Vincenzo al Volturno historical site (Molise, Italy). Biodivers Conserv 27(5):1073–1093. https://doi.org/10.1007/s10531-017-1480-z
Cooper Marcus C, Barnes M (1995) Gardens in health care facilities: uses, therapeutic benefits, and design considerations. The Center of Health Design, Martinez, CA
Cooper Marcus C, Barnes M (1999) Healing gardens: therapeutic benefits and design recommendations, New York. John Wiley & Sons
D'Andrea SJ, Batavia M, Sasson N (2008) Effect of horticultural therapy on preventing the decline of mental abilities of patients with Alzheimer's type dementia. J Ther Hortic 18:8–17
Dafis S (2001) Urban forestry. Publications ARTOFTEXT, Thessaloniki, p 198
Dickinson JL, Shirk J, Bonter D, Bonney R, Crain et al (2012) The current state of citizen science as a tool for ecological research and public engagement. Front Ecol Environ 10(6):291–297. https://doi.org/10.1890/110236
Diener A, Mudu P (2021) How can vegetation protect us from air pollution? A critical review on green spaces' mitigation abilities for air-borne particles from a public health perspective—with implications for urban planning. Sci Total Environ 796:148605. https://doi.org/10.1016/j.scitotenv.2021.148605

Drake L, Lawson L (2015) Best practices in community garden management to address participation, water access, and outreach. J Exten 53(6):7. https://doi.org/10.34068/joe.53.06.07

Dreija K (2012) Historic gardens and parks: challenges of development in the context of relevant regulations, definitions and terminology. Moksl- Liet Ateitis/Sci —Future Lith 4:167–175

Ebissa G, Yeshitela K, Desta H, Fetene A (2024) Urban agriculture and environmental sustainability. Environ Dev Sustain 26(6):14583–14599. https://doi.org/10.1007/s10668-023-03208-x

EC (2021) New EU Forest Strategy for 2030. https://knowledge4policy.ec.europa.eu/publication/communication-com2021572-new-eu-forest-strategy-2030_en

Edwards CA (2013) An evaluation of a therapeutic garden's influence on the quality of life of aged care residents with dementia. Dementia 12:494–510. https://doi.org/10.1177/1471301211435188

Egerer M, Karlebowski S, Conitz F, Neumann AE, Schmack JM, Sturm U (2024) In defence of urban community gardens. People and Nature 6(2):367–376. https://doi.org/10.1002/pan3.10612

EU (2024) Regulation (EU) 2024/1991 of the European Parliament and of the Council of 24 June 2024 on nature restoration and amending Regulation (EU) 2022/869. Off J Eur Union L 309

Felderhoff J, Gathof AK, Buchholz S, Egerer M (2023) Vegetation complexity and nesting resource availability predict bee diversity and functional traits in community gardens. Ecol Appl 33(2):e2759. https://doi.org/10.1002/eap.2759

Forest Europe (2008). Pan-European guidelines for afforestation and reforestation with a special focus on the provisions of the UNFCCC

Funsten C, Borsellino V, Schimmenti E (2020) A systematic literature review of historic garden management and its economic aspects. Sustainability 12(24):10679. https://doi.org/10.3390/su122410679

Géhu JM (2006) Dictionnaire de sociologie et synécologie végétales. In: Cramer J, Berlin/Stuttgart, Germany. ISBN 978-3-443-50030-6. https://www.schweizerbart.de/publications/detail/isbn/9783443500306/Dictionnaire-de-Sociologie-et-Syncologie-Vgtales

Ghose R, Pettygrove M (2014) Actors and networks in urban community garden development. Geoforum 53:93–103. https://doi.org/10.1016/j.geoforum.2014.02.009

Giusti MA (2004) Restauro dei Giardini. Teorie e storia, Firenze, Alinea

Gregis A, Ghisalberti C, Sciascia S, Sottile F, Peano C (2021) Community garden initiatives addressing health and well-being outcomes: a systematic review of infodemiology aspects, outcomes, and target populations. Int J Environ Res Public Health 18(4):1943. https://doi.org/10.3390/ijerph18041943

Guitart D, Pickering C, Byrne J (2012) Past results and future directions in urban community gardens research. Urban For Urban Green 11(4):364–373. https://doi.org/10.1016/j.ufug.2012.06.007

Gullino P, Pomatto E, Gaino W, Devecchi M, Larcher F (2020) New challenges for historic gardens' restoration: a holistic approach for the Royal Park of Moncalieri Castle (Turin metropolitan area, Italy). Sustainability 12:10067. https://doi.org/10.3390/su122310067

Hobhouse P (2004) Plants in garden history: an illustrated history of plants and their influences on garden style. Pavilion, London

Holland L (2004) Diversity and connections in community gardens: a contribution to local sustainability. Local Environ 9(3):285–305. https://doi.org/10.1080/1354983042000219388

Hosseini Z, Caneva G (2022) Lost gardens: from knowledge to revitalization and cultural valorization of natural elements. Sustainability 14(5):2956. https://doi.org/10.3390/su14052956

Hosseini Z, Bartoli F, Pontrandolfi MA, Caneva G (2024) Identity of a historic garden: the view of the international literature and committees on the role of the botanical elements. Urban For Urban Green 101:128501. https://doi.org/10.1016/j.ufug.2024.128501

Hudson J (1907) A Japanese garden in England. J R Hortic Soc 32:1–10

Huq FF, Deacon L (2025) A systematic review of community gardens and their role in urban food security and resilience. Discov Sustain 6(1):696. https://doi.org/10.1007/s43621-025-01628-5

ICOMOS (1982) The Florence charter on historic gardens. Art:1–4. https://icomosjapan.org/static/homepage/charter/charter1982.pdf
IR (2018) UN Climate Change Annual Report 2018. URL: https://unfccc.int/sites/default/files/resource/UN-Climate-Change-Annual-Report-2018.pdf
IR (2019) Decree-Law No. 111/2019 converted, with amendments, by law 141/2019 on air quality. https://climate-laws.org/document/decree-law-no-111-2019_634a
Irvine S, Johnson L, Peters K (1999) Community gardens and sustainable land use planning: a case-study of the Alex Wilson community garden. Local Environ 4(1):33–46. https://doi.org/10.1080/13549839908725579
ISPRA (2015) Linee guida di forestazione urbana sostenibile di Roma Capitale: Manuali e linee guida. ISPRA
Jevtic S, Wittlinger M, Teimann S, Wiltfang J, Scherbaum N, Benninghoff J (2025) Impact of dementia-landscaped therapy garden on psychological well-being—a pilot study. J Neural Transm 132(6):877–885. https://link.springer.com/article/10.1007/s00702-025-02917-z
Kanellou E, Papafotiou M (2025) The historical evolution of the role of vegetation in the enhancement and conservation of archaeological sites. Plants 14(15):2302. https://doi.org/10.3390/plants14152302
Kanosvamhira TP, Tevera D (2023) Urban community gardens in Cape Town, South Africa: navigating land access and land tenure security. GeoJournal 88(3):3105–3120. https://doi.org/10.1007/s10708-022-10793-3
Kim JJ, Atique U, An KG (2019) Long-term ecological health assessment of a restored urban stream based on chemical water quality, physical habitat conditions and biological integrity. Water 11(1):114. https://doi.org/10.3390/w11010114
Kingsley J, Foenander E, Bailey A (2020) "It's about community": exploring social capital in community gardens across Melbourne, Australia. Urban Forest Urban Green 49:126640. https://doi.org/10.1016/j.ufug.2020.126640
Kip J (1720) Britannia illustrata or views of several of the royal palaces…, London, Mortier
Koay WI, Dillon D (2020) Community gardening: stress, well-being, and resilience potentials. Int J Environ Res Public Health 17(18):6740. https://doi.org/10.3390/ijerph17186740
Koniak-Griffin D (1994) Aerobic exercise, psychological well-being, and physical discomforts during adolescent pregnancy. Res Nurs Health 17(4):253–263. https://doi.org/10.1002/nur.4770170404
Latz P (2001) Landschaftspark Duisburg-Nord: a guide to the park. Latz
Lawson L (2004) The planner in the garden: a historical view into the relationship between planning and community gardens. J Plan Hist 3(2):151–176. https://doi.org/10.1177/1538513204264752
Lian J, Nijhuis S, Bracken G, Wu X, Wu X, Chen D (2024) Conservation and development of the historic garden in a landscape context: a systematic literature review. Landsc Urban Plan 246:105027. https://doi.org/10.1016/j.landurbplan.2024.105027
Lynch K (1960) The image of the City. The MIT Press, Cambridge
Maovaz M, Legé AS (2024) Il mondo in un giardino: La Villa della Selva di Hugo Cahen d'Anvers ad Allerona. Perugia, Quattroemme
Masson G (1961) Italian Gardens. Harry N. Abrams, New York
MATTM—Committee for the Development of Public Green Areas (2018) National urban green strategy: resilient and heterogeneous urban forests for the health and well-being of citizens. Rome. https://www.mase.gov.it/portale/documents/d/guest/strategia_verde_urbano-pdf
Montford A (2004) Health, sickness, medicine and the friars in the thirteenth and fourteenth centuries. Ashgate, Aldershot, UK. (pp. 318)
Murroni V, Cavalli R, Basso A, Borella E, Meneghetti C, Melendugno A, Pazzaglia F (2021) Effectiveness of therapeutic gardens for people with dementia: a systematic review. Int J Environ Res Public Health 18(18):9595. https://doi.org/10.3390/ijerph18189595
Nassauer JI (1995) Culture and changing landscape structure. Landsc Ecol 10(4):229–237. https://doi.org/10.1007/BF00129257
Norman DA (1988) The Design of Everyday Things. Doubleday, New York

Normandin É, Vereecken NJ, Buddle CM, Fournier V (2017) Taxonomic and functional trait diversity of wild bees in different urban settings. PeerJ 5:e3051. https://doi.org/10.7717/peerj.3051

Nowak DJ, Dwyer JF (2007) Understanding the benefits and costs of urban forest ecosystems. In: Kuser J (ed) Urban and community forestry in the northeast. Springer, Dordrecht

Nowak DJ, Crane DE, Stevens JC (2006) Air pollution removal by urban trees and shrubs in the United States. Urban For Urban Green 4(3–4):115–123. https://doi.org/10.1016/j.ufug.2006.01.007

Okvat HA, Zautra AJ (2011) Community gardening: a parsimonious path to individual, community, and environmental resilience. Am J Community Psychol 47(3):374–387. https://doi.org/10.1007/s10464-010-9404-z

Olivadese M, Dindo ML (2022) Historic and contemporary gardens: a humanistic approach to evaluate their role in enhancing cultural, natural and social heritage. Land 11(12):2214. https://doi.org/10.3390/land11122214

Panitsa M, Tsakiri M, Kampiti D, Skotadi M (2024) Archaeological areas as Habitat Islands: plant diversity of Epidaurus UNESCO world heritage site (Greece). Diversity 16(7):403. https://doi.org/10.3390/d16070403

Pomposini R (2004) Il giardino sensoriale terapia anti Alzheimer. Assistenza anziani 146:19

Ponstingel D (2022) The governance of community gardens as commons and its role in the socio-ecological outcomes of gardening in Austin, Texas, USA. Socio-Ecol Prac Res 4(4):355–376. https://doi.org/10.1007/s42532-022-00133-7

Raneng J, Howes M, Pickering CM (2023) Current and future directions in research on community gardens. Urban For Urban Green 79:127814. https://doi.org/10.1016/j.ufug.2022.127814

Ranfa A, Rubeca MG, Bodesmo M (2014) Il Giardino sensoriale. Fontenuovo residenza di ospitalità per anziani, Francesco Tozzuolo Editore, Perugia

Ranfa A, Rubeca MG, Cesarini S, Bodesmo M (2017) A healing garden for people with Alzheimer's disease: a case of green therapy in Umbria region (Central Italy). Int J Agric Environ Res 3(2):2672–2686

Ren X, Zhang G, Chen Z, Zhu J (2023) The influence of wind-induced response in urban trees on the surrounding flow field. Atmos 14(6):1010. https://doi.org/10.3390/atmos14061010

Rivasseau Jonveaux T, Batt M, Fescharek R, Benetos A, Trognon et al (2013) Healing gardens and cognitive behavioral units in the management of Alzheimer's disease patients: the Nancy experience. J Alzheimer's Dis 34(1):325–338. https://doi.org/10.3233/JAD-121657

Roling C, Edwards A (2025) E.A.T. South: a contamination story. Thriving earth. Exchange. Retrieved from https://thrivingearthexchange.org

Romanato MG (2024) Contemporary urban green space and its historical roots. In: Contemporary heritage lexicon. Springer, pp 409–433. https://doi.org/10.1007/978-3-031-61245-9_19

Salbitano F, Borelli S, Conigliaro M, Chen Y (2016) Guidelines on urban and peri-urban forestry. FAO, Rome

Samara T, Tsitsoni T (2007) Road traffic noise reduction by vegetation in the ring road of a big city. In: Proceedings of the International Conference on Environmental Management, Engineering, Planning and Economics. CEMEPE, Skiathos, pp 2591–2596

Samara T, Tsitsoni T (2011) The effects of vegetation on reducing traffic noise from a city ring road. Noise Contr Eng J 59(1):68–74. https://doi.org/10.3397/1.3528970

Samara T, Tsitsoni T (2014) Selection of forest species for use in urban environment in relation to their potential capture to heavy metals. Global NEST J 16(5):966–974. https://doi.org/10.30955/gnj.001523

Singh P (2020) Restoring and nurturing the 'nature–human' bond through conservation of historic gardens. In Heritage Conservation in Postcolonial India Routledge. https://doi.org/10.4324/9781003109426-21

Söderström M (2000) The possibility garden. A Garden for All People, The Swedish Institute for Handicapped and The Swedish Broadcasting AB

Spampinato G, Malerba A, Calabrò F, Bernardo C, Musarella C (2021) Cork oak Forest spatial valuation toward post Carbon City by CO_2 sequestration. In: Bevilacqua C, Calabrò F, Della

Spina L (eds) New metropolitan perspectives. NMP 2020. Smart innovation, systems and technologies, vol 178. Springer. https://doi.org/10.1007/978-3-030-48279-4_123
Stalter R, Kincaid D (2021) The flora on the High Line, New York City, New York: a 17-year comparison. J Torrey Bot Soc 148(3). https://doi.org/10.3159/TORREY-D-21-00007.1
Stavrinou V, Tsitsoni T (2018) The development of green networks in order to link urban areas with the Peri-urban forests. Curr Environ Eng 5(IKEEART-2020-3003). https://doi.org/10.2174/2212717805666181106155206
Tan PY, Wang J, Sia A (2013) Perspectives on five decades of the urban greening of Singapore. Cities 32:24–32. https://doi.org/10.1016/j.cities.2013.02.001
The High Line: Biodiversity and Design. (2020). https://www.thehighline.org
Tsitsoni T (2010) Urban forestry. University Notes, Thessaloniki, p 150
Tsitsoni T, Tsakaldimi M, Simeliadou E, Fouska M (2009) Silvicultural research of *Pinus brutia* mixed stands with reforestation: 12 years after fire. Proc. of the 14th Panhellenic Forestry Conf. of the Hellenic Forestry Society
Tsitsoni T, Gounaris N, Kontogianni A, Xanthopoulou-Tsitsoni V (2015) Creation of an integrated system model for governance in urban MTEs and for adapting cities to climate change—preliminary results. Ecologia mediterranea: revue d'écologie terrestre et limnique 41(2):33–44
Ulrich R (1984) View through a window may influence recovery from surgery. Science 224:420–421. https://doi.org/10.1126/science.6143402
UNESCO-SCBD (2010) Riguarda il joint programme on the links between biological and cultural diversity (JP-BiCuD). https://www.unesco.org/en/articles/linking-biological-andcultural-diversity-europe
UNESCO (2021) Reimagining our futures together: a new social contract for education. UNESCO Publishing, Paris
Valente R, Cooper Marcus C (2015) Giardini che guariscono: processi progettuali e realizzazioni di ambienti benefici. Press, Firenze Univ. https://doi.org/10.13128/Techne-16120
Valla P (2002) Alzheimer: architetture e giardini come strumento terapeutico. Guerrini e Associati, Milano
Vigorelli M (2005) Il lavoro della cura nelle istituzioni. Progetti, gruppi e contesti nell'intervento psicologico. Edizioni Franco Angeli, Milano
Voicu I, Been V (2008) The effect of community gardens on neighboring property values. Real Estate Econ 36(2):241–283. https://doi.org/10.1111/j.1540-6229.2008.00213.x
Wagenhofer G, Sahl R, Novotny K (2025) Cooling our heritage: enhancing biodiversity in historic gardens and parks to reduce urban heat in European heritage sites. In: Eu, Conf. Software process improvement. Springer, pp 52–63. https://doi.org/10.1007/978-3-032-04291-0_4
Waters A (2017) Coming to my senses: the making of a counterculture cook. Clarkson Potter
Wolch JR, Byrne J, Newell JP (2014) Urban green space, public health, and environmental justice: the challenge of making cities 'just green enough'. Landsc Urban Plan 125:234–244. https://doi.org/10.1016/j.landurbplan.2014.01.017
Yuan C, Norford L, Ng E (2017) A semi-empirical model for the effect of trees on the urban wind environment. Landsc Urban Plan 168:84–93. https://doi.org/10.1016/j.landurbplan.2017.09.029
Zangari G, Bartoli F, Lucchese F, Caneva G (2023) Plant diversity in archaeological sites and its bioindication values for nature conservation: assessments in the UNESCO site Etruscan Necropolis of Tarquinia (Italy). Sustainability 15(23):16469. https://doi.org/10.3390/su152316469
Zeisel J (2001) Health outcomes. Improvements from Alzheimer's care design. International academy for design and health, design and health: the therapeutic benefit of design. The Swedish Building Center Publishing, Stockholm
Zeisel J, Welch P (1981) Housing designed for families: a summary of research. Joint Center for Urban Studies MIT & Harvard Univ, Cambridge
Zhang Y, Li H, Chen X (2022) State-led governance and community tensions in Shenzhen's we garden project. Cities 120:103–118

Chapter 8
Green Ways

Flavia Bartoli, Livia Bonciarelli, Giulia Caneva, Simona Ceschin, Luca D'Amato, Enrica Roccotiello, and Thekla K. Tsitsoni

Abstract The "Green Ways" are essential components of urban green infrastructure (GI). Three specific nature-based solutions—street tree lines, rain gardens, and green roundabouts—and their contributions to climate resilience, ecological connectivity, and urban well-being are discussed. The chapter highlights their role in promoting multifunctional and biodiversity-richness in urban environments, positioning them within the broader context of European policy frameworks.

Street tree lines are presented as critical regulators of air quality, microclimate, and ecological continuity. Through the outlines of historical development, methodological approaches for species selection are provided, prioritising native, climate-resilient, and low-maintenance trees. It further discusses ecosystem services such as carbon sequestration, noise mitigation, and habitat provision, alongside ecosystem disservices and the importance of monitoring, phytosanitary control, and integrated management. Case studies illustrate how species traits and urban constraints shape tree performance, infrastructure interactions, and long-term sustainability.

Rain gardens (RGs) are explored as small-scale bioretention systems addressing stormwater quantity and quality issues linked to soil sealing and extreme rainfall. The chapter details their structural components, plant selection principles, pollutant removal mechanisms, and climate-adaptive design. Challenges include spatial

---

F. Bartoli (✉)
Institute of Heritage Science (ISPC), National Research Council (CNR), Montelibretti, Rome, Italy
e-mail: flavia.bartoli@cnr.it

L. Bonciarelli
Department of Civil and Environmental Engineering, University of Perugia, Perugia, Italy

G. Caneva · S. Ceschin · L. D'Amato
Department of Sciences, University Roma Tre, Rome, Italy

E. Roccotiello
Department of Earth, Environment and Life Sciences, University of Genoa, Genoa, Italy

T. K. Tsitsoni
Department of Forestry and Natural Environment, Aristotle University of Thessaloniki, Thessaloniki, Greece

G. Caneva et al. (eds.), *Urban Botany*, Future City 22,
https://doi.org/10.1007/978-3-032-28986-5_8

boundaries, maintenance needs, and limited long-term performance data. Case studies demonstrate how RGs enhance infiltration, pollutant filtration, and biodiversity while revitalizing public spaces.

Green roundabouts are reframed as ecological stepping stones enhancing biodiversity, pollinator networks, microclimate regulation, stormwater management, and traffic-related emission reductions. Planting strategies emphasise drought-tolerant natives, structural and seasonal diversity, and safety-driven visibility constraints. Case studies demonstrate how strategic vegetation design transforms traffic nodes into functional ecological assets within larger urban green networks.

Together, these three GI solutions form a coherent framework for designing adaptive, connected, and multifunctional urban landscapes.

Keywords Urban green infrastructure · Street tree lines · Rain gardens · Green roundabouts · Ecological connectivity · Sustainable mobility landscapes

8.1 Introduction (F.B.)

Urban green infrastructure (GI) has become a cornerstone of sustainable city planning, addressing the growing challenges posed by urbanization, climate change, and biodiversity loss. Among the most effective nature-based solutions, elements such as street tree lines, rain gardens, and green roundabouts play a pivotal role in enhancing ecological connectivity, mitigating environmental stressors, and improving the quality of life in urban areas. These interventions not only provide essential ecosystem services—such as air purification, stormwater management, and microclimate regulation—but also contribute to cultural and aesthetic values, fostering healthier and more resilient communities.

At the European level, green infrastructure is formally defined by the European Commission as "*a strategically planned network of natural and semi-natural areas...designed and managed to deliver a wide range of ecosystem services while enhancing biodiversity*".

This concept underpins major EU strategies such as the EU Green Infrastructure Strategy (2013), the EU Biodiversity Strategy for 2030, and the recently adopted Nature Restoration Law (Regulation 2024/1991), which mandates restoring 20% of EU land and sea areas by 2030 and improving urban green spaces and tree canopy cover. Additionally, the *Green City Accord*, launched under the *European Green Deal*, commits cities to measurable targets for air and water quality, biodiversity, circular economy, and noise reduction by 2030. These frameworks emphasize nature-based solutions as cost-effective alternatives to grey infrastructure, promoting climate resilience, ecological connectivity, and improved urban liability.

This chapter explores the multifunctional benefits and design principles of street tree lines, rain gardens, and green roundabouts, and offers methodological approaches for plant selection, maintenance strategies, and case studies illustrating their practical implementation. By aligning local interventions with EU regulations and funding mechanisms, the chapter provides a comprehensive guide for transforming urban spaces into adaptive, biodiversity-rich environments capable of meeting contemporary sustainability challenges.

8.2 Street Trees Lines (L.D.A, F.B, T.T., G.C.)

8.2.1 Definition and Classification of Street Trees Lines

Street trees have become vital elements of urban green infrastructure (GI) in recent decades. The importance of street trees is growing due to urban expansion and the consequent loss of nature within cities, as well as the increasing warming caused by the urban heat island effect and climate change (Roy et al. 2012; Capotorti et al. 2020). Since street trees represent a distinctive component of urban ecological systems (Roy et al. 2012), the ecosystem services and disservices they provide are equally distinctive (Salmond et al. 2016). Being integral elements of urban ecosystems, street trees provide both environmental and social benefits, improving environmental conditions and promoting community well-being (Mullaney et al. 2015; Salmond et al. 2016) (Fig. 8.1).

Urban trees play a relevant role in climate mitigation, which has led to policies aimed at improving urban forests and realizing nature-based solutions (Capotorti

Fig. 8.1 Street trees in the city of Münster (Noth Germany) and their role in addressing the cycle paths (**a**) and reducing the speed of the traffic in residential areas. (Photos by G. Caneva)

et al. 2020). The development of urban green spaces, aimed at improving ecological conditions, contributes to the creation of more human-friendly urban centers, while their extent, operation, and efficiency constitute important criteria for evaluating residents' quality of life (Tsitsoni and Samara 2002; Tsitsoni et al. 2015). In ancient times, the arboreal component of cities was mostly limited to sacred woodlands, private gardens, and public spaces, with little evidence of systematic street tree planting due to spatial constraints in densely built urban centers (Forrest and Konijnendijk 2005). A criterion of practical use seems to have inspired the selection of the first street trees in Rome, when Pope Sisto V, at the end of the sixteenth century, decided to plant elms (*Ulmus minor*) along the larger roads connecting the main churches within the city to give shelter to pilgrims during the Jubilee. Street trees became more common and widespread during the nineteenth century, following major urban transformations across Europe, establishing tree-lined streets as an important component of the modern urban landscape (Attorre et al. 2000; Caneva et al. 2020).

8.2.2 Methodological Approach for Plant Selection

Street trees exhibit varying degrees of tolerance to environmental stressors such as drought, heat, air pollution, and root confinement. These factors are essential considerations for promoting root development, tree growth, canopy expansion, and enhancing overall tree stability in urban settings (Mullaney et al. 2015; Ow et al. 2018). Biodiversity is another critical factor in the selection of urban tree species. Planting diverse native tree species helps prevent the homogenization of urban flora, while vegetation complexity and connectivity support higher levels of urban biodiversity (Mullaney et al. 2015; Bartoli et al. 2022). Regarding methodological approaches for selecting appropriate tree species in urban environments, a recent study provides valuable guidance for promoting native species in the Italian context, with principles that can be extended to other territorial settings (D'Amato et al. 2025) (Fig. 8.2). This study presents a methodological framework designed to identify native tree species potentially suitable for urban streets. Given the challenges of selecting street trees in contemporary urban contexts, this approach aims to enhance urban livability and sustainability by introducing more natural vegetation and increasing urban biodiversity.

8.2.3 Specific Benefits of Street Trees Lines

To assess the benefits of street trees, the literature refers to ecosystem services. In recent decades, ecosystem services (ES) have gained increasing prominence in ecology, environmental science, and sustainability research due to their diverse benefits derived from natural environments, ranging from tangible resources such as food and water to intangible contributions such as climate regulation and disease

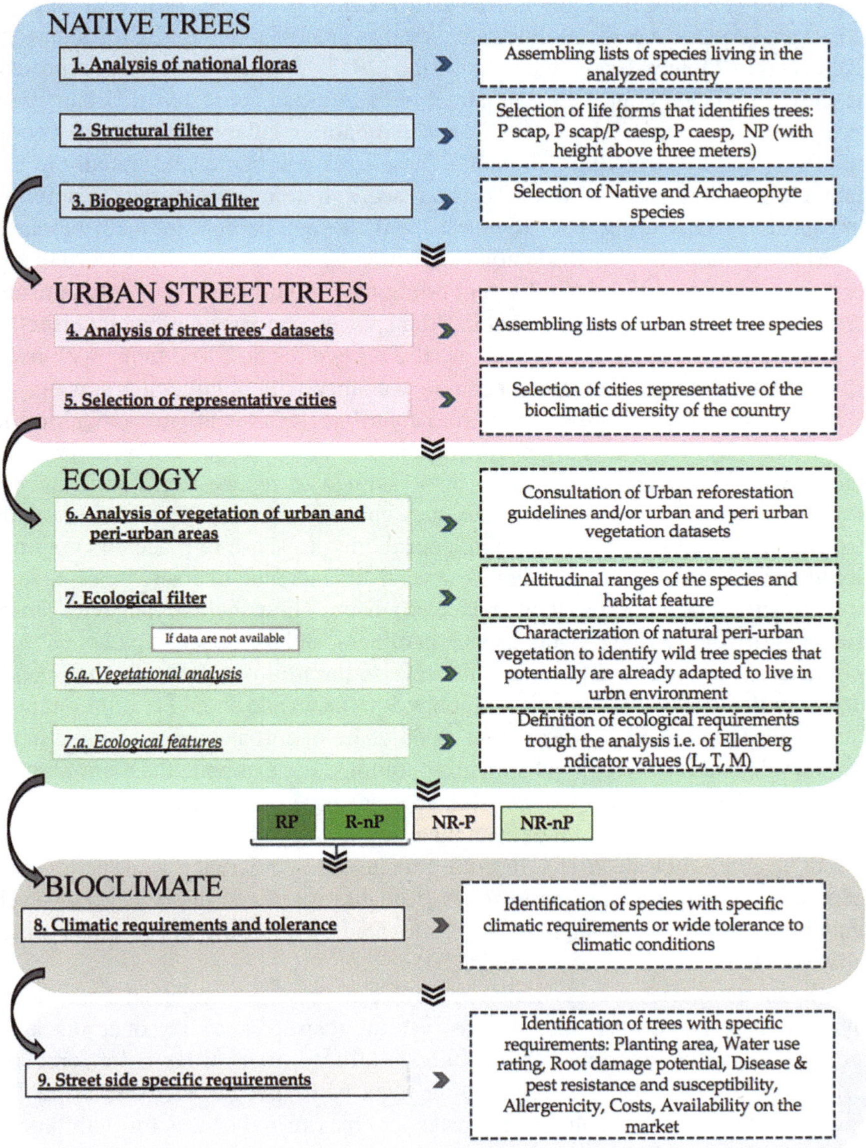

Fig. 8.2 Structure of the selection keys and filters for choosing street tree species. Keys 1–3 generate the first species pool, while keys 4–5 generate a second one. Both pools are then compared against any available lists of species suitable for urban environments (from guidelines and/or research) or further filtered using the ecological filters (either 6/7 or 6a/7a). The resulting datasets are cross-referenced to produce four street tree species groups: [RP]: Recommended and already Present. [R-nP]: Recommended but not Present. [NR-P]: Not Recommended but Present. [NR-nP]: Not Recommended and not Present. L = Light; T = Temperature; M = Moisture. (Workflow from D'Amato et al. 2025)

control (Gómez-Baggethun and Barton 2013). Urban street trees provide a range of ecosystem services across multiple categories, as identified in the TEEB framework (2011) and by Gómez-Baggethun and Barton (2013). Regulating services represent the most significant contribution of street trees to urban ecosystems. Urban street trees play a crucial role in local climate and air quality regulation through their cooling effects and by removing pollutants. These trees contribute substantially to climate change mitigation through carbon sequestration and storage, capturing atmospheric CO_2 during photosynthesis and storing carbon in their biomass. Additionally, street trees provide noise reduction by acting as acoustic barriers and contribute to biological control by supporting natural pest regulation mechanisms. Street trees also provide essential habitats for species, supporting urban biodiversity by offering nesting sites, food resources, and *refugia* for flora and fauna within the urban matrix, thereby maintaining ecological connectivity in built environments.

From a cultural perspective, street trees contribute significantly to recreation and mental and physical health by providing opportunities for outdoor activities and contact with nature, which has been demonstrated to reduce stress and improve psychological well-being. Furthermore, they enhance urban aesthetics and provide inspiration for culture, art, and design, contributing to sense of place and community identity. Urban street trees require several fundamental characteristics beyond those directly linked to ecosystem service provision. These include longevity, structural integrity of branches, visual appeal, hardiness, and resilience to pathogens and pollutants. Additionally, they should not release harmful or allergenic compounds such as pollen or Volatile Organic Compounds (VOCs) and should require minimal maintenance interventions (Sæbø et al. 2003). The importance of selection criteria may vary depending on regional climate conditions. For example, the thermal regulation provided by trees offers considerable advantages in warm climate zones, while it may be less beneficial or even undesirable in cooler regions or during cold periods (Salmond et al. 2016). In cold climate areas, tolerance to de-icing salts becomes essential, whereas in arid zones, drought resistance is a priority. However, many climate-related challenges can be mitigated by choosing indigenous species native to the urban area (Sæbø et al. 2003).

Indigenous species are generally preferable as they have evolved under local environmental conditions and possess natural resistance to regional diseases. Furthermore, native trees are more frequently utilized by local fauna for foraging, nesting, and shelter. Therefore, selecting trees with specific attributes from the regional flora represents a strategic approach to maximize benefits for both human populations and wildlife.

8.2.4 Maintenance and Monitoring

Urban street tree management requires a multifaceted approach that integrates scientific expertise, operational capacity, and systematic monitoring protocols. Beyond the fundamental considerations of project design and species selection appropriate to various urban contexts (Savo et al. 2025), several critical components must be coordinated to ensure successful long-term outcomes. A cornerstone of effective street tree management is the establishment of close collaboration between botanical experts and municipal "*Garden services*" responsible for the entire tree supply chain. This partnership should encompass all stages from procurement to maintenance, including direct engagement with local nursery operations to ensure appropriate sourcing of plant material and quality control throughout the cultivation process.

The implementation of comprehensive municipal guidelines for urban green space monitoring represents an essential framework for standardized management. These guidelines should establish clear protocols for both planting and ongoing maintenance tailored to specific street typologies and environmental conditions. Such documents provide the regulatory foundation necessary for consistent decision-making across different urban zones and temporal scales.

Adequate training of horticultural staff in botanical identification and species-specific requirements is fundamental to the proper implementation of management protocols. Personnel must be equipped with sufficient knowledge to recognize individual species characteristics, growth patterns, and specific care requirements to ensure appropriate interventions. Attention must be devoted to irrigation management during the establishment phase following transplantation. The initial years after planting are critical for root development and overall tree health, requiring carefully calibrated watering regimes adapted to species requirements and local climatic conditions. Ongoing collaboration between municipal authorities and research institutions is essential for effective phytopathological surveillance and response. This partnership should facilitate the rapid identification of disease outbreaks, pest infestations, and other health concerns affecting urban tree populations. The development and implementation of computerized monitoring systems represent a modern approach to urban forest management.

These digital platforms can integrate real-time data on tree health, growth patterns, and pathogen distribution, enabling more informed decision-making and timely interventions. Such systems facilitate simultaneous information sharing among stakeholders and provide valuable datasets for long-term urban forestry planning.

8.2.5 Case Studies

Urban areas are major contributors to climate change, and street trees play a key role in mitigating these effects (Mullaney et al. 2015). As distinctive components of urban ecosystems (Roy et al. 2012), street trees provide important ecosystem services—including climate regulation, air purification, and biodiversity enhancement—while also generating ecosystem disservices (Salmond et al. 2016; Masini et al. 2023). Their performance, however, is constrained by multiple stressors, including degraded soils, restricted rooting space, pollution, thermal stress,

Fig. 8.3 Ecosystem services provided by different species. On the top left, a visualization of ecosystem services provided in general by trees (in different colors); on the top right, a visualization of ecosystem services specifically provided by *Quercus ilex* L. (a Mediterranean species); on the bottom left, *Platanus* × *hispanica* Mill. ex Münchh.; on the bottom right *Robinia pseudoacacia* L. (in white are indicated the ecosystem services that the species do not provide) (from Savo et al. 2025)

irregular water availability, and anthropogenic disturbances. Thus, effective street tree planning requires careful assessment of factors influencing tree survival and functionality (Czaja et al. 2020) (Fig. 8.3).

To illustrate the range of ecosystem services provided by urban trees, three species characteristics of Italian streetscapes were examined: the native *Quercus ilex*, the invasive *Robinia pseudoacacia*, and the widely planted *Platanus* × *hispanica*. *Quercus ilex* contributes to freshwater retention and biodiversity due to its canopy shape, bark texture, evergreen foliage, and indigenous status, though hydrophobic leaves may somewhat limit water retention. Its noise mitigation capacity increases with proper height management (Savo et al. 2025). *Platanus* × *hispanica* enhances stormwater runoff efficiency because of its smooth bark, while both *Platanus* × *hispanica* and *Robinia pseudoacacia* offer limited benefits for biological control and noise reduction.

Despite their contributions, street trees also generate ecosystem disservices (von Döhren and Haase 2019), notably maintenance demands and infrastructure damage (Salmond et al. 2016). An evaluation across the five provinces of Latium assessed tree-induced impacts on pavements, urban artifacts, and road surfaces using a three-point damage scale adapted from Masini et al. (2023). *Platanus* × *hispanica* demonstrated the highest potential for damage due to its vigorous surface root system (Dervishi et al. 2023). *Quercus ilex* caused minimal structural impairment, consistent with its deep root system and slow growth, while *Robinia pseudoacacia* produced moderate damage, primarily from sidewalk uplift. The observation that a non-native species such as *Robinia pseudoacacia* can sometimes cause less structural damage than native alternatives indicate the need for integrated maintenance in addition to upstream species selection. This underscores the importance of balancing ecosystem services and disservices to ensure long-term infrastructure compatibility, citizen safety, and biodiversity conservation (Savo et al. 2025).

Beyond species performance, street trees face multiple constraints that affect their establishment and survival, making systematic monitoring, maintenance, and adaptive management essential. Key issues include mechanical damage from vehicles and human activities, root system restriction, and airspace limitations.

These can be mitigated through trunk protection, the correct placement of underground utilities (Fig. 8.4a, b), appropriate species selection, and proper pruning. Additional challenges include vehicle parking in planting areas, interference from

Fig. 8.4 Examples of Thessaloniki city urban street tree conditions and constraints: (**a**) Newly planted trees along a sidewalk with protective guards; (**b**) Tree pit with metal fencing in a narrow pedestrian area; (**c**) Tree trunk positioned very close to a parked car, limiting root space; (**d**) Street with heavily pruned trees and dense vehicular parking; (**e**) Mature street trees providing canopy cover along a busy urban road with parked cars. (Photos by T. Tsitsoni)

overhead or adjacent cables, construction works, and traffic accidents (Fig. 8.4c, d). Pruning practices must preserve natural form and function, remaining visually unobtrusive to maintain both aesthetics and ecosystem functionality (Fig. 8.4d). New street tree installations must conform to technical and traffic-related requirements, such as maintaining a clear height of approximately 4.5 m (Fig. 8.4e) and ensuring visibility near intersections and signage (Fig. 8.5a–d). Root growth areas must be designed with protective coverings, vegetated surfaces, and stabilized substrates (Fig. 8.5e). Adequate root space is essential and varies according to tree size: 2.50–3.00 m planting strips for small or narrow-crowned trees, and 3.50–4.50 m for tall or wide-crowned species.

Parking-lot plantings require additional spacing, particularly where trees share space with vertical or elongated rows of parking stalls. Tree spacing in street alignments depends on mature species characteristics—including crown diameter, height, root spread, and growth rate—and typically ranges from 6–12 (15) m.

Continuous monitoring and maintaining an urban tree register are crucial for timely, evidence-based interventions. Inventory data should document health conditions, structural defects, the presence of pests or diseases, soil compaction, and conflicts with visibility or circulation, supporting informed management decisions (Tsitsoni et al. 2015). Because species vary in their tolerance to pollution, compaction, drought, and human impacts, such monitoring is essential for assessing adaptation. Additionally, documenting public preferences contributes to more effective planning and management, reinforcing that municipal responsibility extends beyond initial planting (Tsitsoni and Samara 2002).

Proper street tree care includes irrigation, fertilization, soil and crown management, pruning, structural stabilization, and pest and disease control. High-quality management begins with a thorough whole-tree diagnosis rather than an evaluation

Fig. 8.5 Examples of different tree pit designs and surface treatments in urban environments: in Thessaloniki (**a**) Interlocking pavement blocks with minimal open soil area; (**b**) Circular gravel border integrated into a paved surface; (**c**) Decorative metal grates for pedestrian safety and soil protection; and in Paris (**d**) Paved areas featuring circular grates (**e**) Raised beds, with additional green space. (Photos by T. Tsitsoni)

limited to visibly damaged parts. To support this process, the *GreenTree* application (Tsitsoni et al. 2015) compiles tree inventories, records silvicultural characteristics, provides citizens with access to tree data (including dynamic CO_2 sequestration values), and integrates with municipal mapping systems, accompanied by staff training.

8.3 Rain Gardens (S.C., L.B.)

8.3.1 Definition and Classification

In urban contexts, the increasing level of overbuilding has negative effects not only on infrastructure but also on the environment. Indeed, soil impermeabilization leads to difficulties in replenishing groundwater and a progressive increase in the amount of stormwater that, unable to be absorbed into the ground, accumulates and flows on the surface. In addition, the existing drainage systems, both artificial (sewers) and natural (green areas, fields, etc.), that remain active within the urban matrix, are increasingly inadequate for fully channeling all surface runoff water (ISPRA 2013). Climate change is worsening these phenomena by impacting precipitation patterns. Over the past sixty years, it has contributed to a noticeable rise in the frequency and intensity of heavy rainfall events across the globe, affecting both arid and humid regions alike (Donat et al. 2016). This results not only in a higher frequency of flooding but also in the direct or indirect release into the environment of large quantities of pollutants originating from vehicular traffic and industrial activities (ISPRA 2013). In fact, rainwater, as it flows over urban asphalt roads, picks up a wide variety of pollutants, resulting from both dry and wet atmospheric deposition. These pollutants include both inorganic toxic elements, such as heavy metals (lead, zinc, copper, cadmium), nitrogen oxides, sulfates, chlorides, phosphates, and organic compounds, mainly hydrocarbons, such as Polycyclic Aromatic Hydrocarbons (PAHs), deriving mainly from oil, gasoline, deicing substances, and vehicle emissions (Davis and McCuen 2005).

Nature-based solutions to stormwater management began to emerge worldwide in the final two decades of the twentieth century, tackling urban drainage systems through innovative approaches, regulating water quantity and quality more sustainably (Fletcher et al. 2015). These practices, with differing scopes and technologies, can be found under different names, such as Low-Impact Development (LID) or Green Infrastructures (GI) in the USA, Water Sensitive Urban Design (WSUD) in Australia and Sustainable Urban Drainage Systems (SuDS) in the UK (Fletcher et al. 2015). The range of solutions spans from small-scale interventions near runoff-generation points to extensive landscapes intentionally designed to accommodate excess water during heavy rainfall events (Ferreira et al. 2020). Examples of small-scale solutions include rainwater bioretention systems (rain gardens), tree trenches, green roofs, permeable pavements, filter strips, vegetated swales, detention and retention basins, and infiltration systems (Woods Ballard and Wilson 2015). All of these "water accepting landscapes" are based on a set of structures and methodologies designed to recreate the natural hydrological cycle in human-altered environments, mitigate the effects of increased stormwater runoff, and provide technical, economic, ecological, recreational, and aesthetic additional benefits (Davis and McCuen 2005).

Rain gardens (RGs) are LID systems mainly found in Anglo-Saxon countries (UK, the United States, Australia), and represent a sustainable, semi-natural method for managing stormwater flow, especially in urban areas (Ahiablame et al. 2012). Rain gardens can be a simplified version, or a synonym, of the broader category of Bioretention Systems (BRs), defined as "*shallow landscaped depressions that can reduce run-off rates and volumes, and treat pollution through the use of engineered soils and vegetation*" (Woods Ballard and Wilson 2015). Specifically, RGs are small ground depressions designed to collect and retain rainwater runoff from paved and cemented surfaces. Typically located along roadsides and in car parks, RGs capture water, promote its infiltration into the soil, and slow down runoff speed. When also addressing water quality enhancement, thanks to the substrate (often specifically engineered) and the use of specially selected filtering and purifying plants, they help reduce the pollutant load entering the sewage system (Ahiablame et al. 2012).

"Full" BRs are depressions with a ponding depth of 15–30 cm, composed of selected plants (which bioremove pollutants, prevent surface erosion, maintain the permeability of the filter medium), a filter medium (normally sand-based, of at least 40 cm of depth until 100 cm, which enhances water filtration), a transition layer or geotextile, and a drainage layer around the perforated underdrain (Woods Ballard and Wilson 2015). There are different variations of these configurations, including "typical" rain gardens, which can be less engineered of the full BR components. More in detail, a rain garden is structurally composed of several distinct units (Cavinato 2014):

1. *Buffer area*: located at the edges of the depression, it is typically made up of grass-like species and serves to slow down runoff velocity and physically trap sediments, particulate matter or oily substances.
2. *Protective layer* (such as a mulch or gravel cover): positioned on the surface of the accumulation area, it is essential for supporting bacterial activity and maintaining soil moisture. It acts as a filter for pollutants, protects the soil from erosion, and provides an optimal environment for microorganisms involved in the biodegradation and assimilation of substances.
3. *Plant component*: vascular plants that tolerate periodic flooding and moderately high levels of pollution. These plants can adsorb dissolved substances and bioaccumulate various pollutants. They play both functional ecological and ornamental roles due to their aesthetic value, which is considered in their selection.
4. *Edaphic matrix*: substrate that supports root development and meets the nutritional needs of the selected plant species. It usually replaces the filter and drainage layer of BRs and is composed of a thin (20–50 cm) layer of sand/compost-amended native soil or specific engineered soil mixes, intended to

enhance water filtration and/or regulate the balance between water infiltration and retention.

5. *Depression or runoff water accumulation area*: this is the surface with the highest infiltration capacity. Bounded by a soil layer or other rocky material, it retains runoff from rainwater within the depression, thereby increasing the contact time between the water and the selected plant species.

8.3.2 Methodological Approach for Plant Selection

Plants are the most important component of RGs in enhancing the system's efficiency. Indeed, the plants in RGs determine both quantitative (enhancing media porosity over time and therefore the infiltration) and qualitative (enhancing and providing water filtration) stormwater control. The plant species selection, therefore, must consider a broad range of criteria, meeting specific biological, ecological, and functional requirements (Greksa et al. 2024; Andruczyk and Fox 2009).

Biological, physiological, and ecological criteria

- Usually, perennial herbaceous species are used, but also suitable shrubs and trees can be inserted to enhance the overall environmental benefits (Chaves et al. 2024).
- All the species should be native for their strong adaptation to local climatic and edaphic conditions and to the specific environmental stresses. The selection of suitable aliens is possible, but only after careful consideration of their potential invasiveness (Corduan and Kühn 2024).
- Species should be able to tolerate both occasional flooding and periods of drought, which can be more or less pronounced depending on climatic conditions, water infiltration, and retention capacity of the substrate, and on the positioning of the plants in the RG.
- Species should be resistant to the typical contaminants of the run-off from the specific surrounding surfaces.

Functional criteria

- For water quantity regulation, plants characterized by deep and thick (also fleshy) roots are preferred to promote infiltration, while higher plants and broad-leaved species to maximize water "extraction" through evapotranspiration (Bonciarelli et al. 2025).
- The species should be capable of phytoremediating run-off pollutants, particularly bioaccumulating heavy metals, such as lead, copper, zinc, and cadmium, which are commonly found in vehicle runoff.
- The plants selected should also be aesthetically pleasing, so that rain gardens serve also as ornamental features that enhance the visual quality of the surrounding area where they are installed. For example, combining species with different flowering periods, flower types, and colors can create a more aesthetically pleasing and longer-lasting visual effect (Andruczyk and Fox 2009).

The selected plant species must be arranged within the RGs according to a specific zonation pattern that defines the system's structural layout. Plants that do not require high soil moisture should be placed along the edges of the rain garden, while progressively more hygrophilous species, those tolerant of possible temporary submersion in water, should be positioned toward the central, deeper parts of the garden (Andruczyk and Fox 2009). To optimize pollutant removal, nitrophilous species (those that thrive in nitrogen-rich conditions) should be placed along the outer edges of the system, as these outer zones typically contain higher concentrations of organic matter and mineral salts (Di Stasio et al. 2025). In addition to these criteria, it is important to consider the plant species' growth habit and size. Taller plants should be placed in the central zones and not along the edges of the structure, as this can create shading over the central plants, reducing their photosynthetic efficiency, vitality, and vegetative growth. Taking into consideration the structural and functional characteristics described above, an example of a rain garden that can be achieved in urban areas of the Mediterranean region is reported (Fig. 8.6).

In the most external zone, a buffer strip can be created with a border of grass-like plants of the genera *Carex* and *Cyperus*. In the intermediate wet zone, where nitrogenous substances usually accumulate in the soil, populations of *Typha latifolia*, *Iris pseudacorus* or *Sparganium erectum* can be located. In the innermost and deepest part, hygrophilous species with phytoremediation properties, such as *Lythrum salicaria*, *Lysimachia vulgaris* and *Veronica anagallis-aquatica*, can be placed (Fig. 8.7).

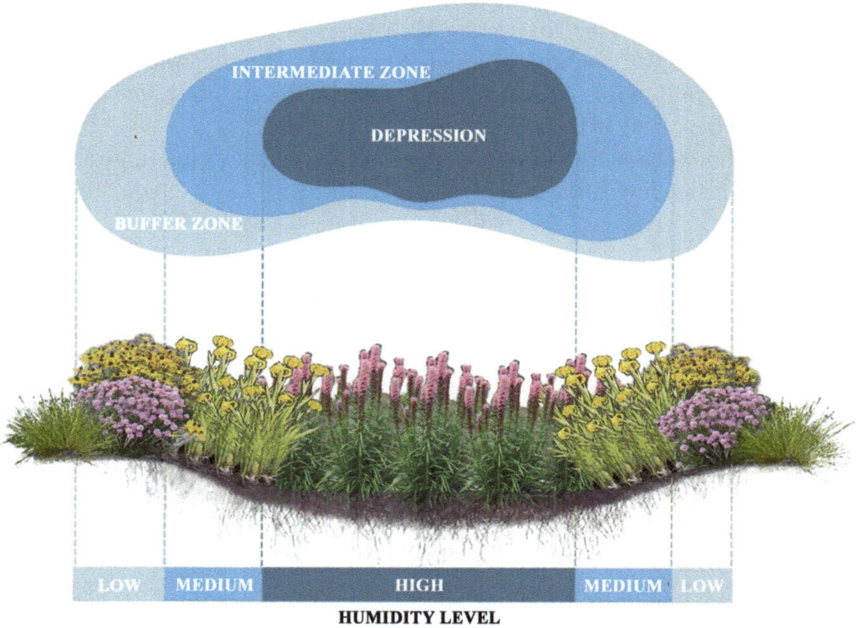

Fig. 8.6 Example of RGs with species selection consistent with moisture and nitrogen amount gradient. (Figure created by F. Bartoli)

Fig. 8.7 Plant species suitable for the setup of a RG. From above, *Sparganium erectum*, *Iris pseudacorus*, *Typha latifolia*, *Lysimachia vulgaris*, *Veronica anagallis-aquatica*, *Lythrum salicaria*. (Photos courtesy by Acta Plantarum)

The selection of these species ensures continuous blooming from April to October, with a variety of aesthetically pleasing flower colors ranging from purple to yellow.

8.3.3 Maintenance and Monitoring

For sustained effectiveness, RGs require routine maintenance and adaptive monitoring protocols. Regular clearing of accumulated debris and sediment preserves flow paths and filtration efficiency, while periodic pruning and replanting of plants (to maintain optimal density and species composition) are required for plant management (EPA 2021). Monitoring actions help control the potential occurrence of pests and invasive species. Also, plant health assessment is relevant, as it involves observing plant vigor, disease symptoms, and survival rates. Furthermore, water-level management helps maintain pollutant-removal efficiency. Monitoring protocols include water quality testing through regular sampling and hydrological monitoring by tracking inflow, outflow, and retention times (Cording et al. 2017).

8.3.4 Challenges and Limitations

The most challenging aspect is the design phase of efficient RGs, which should consider all the biotic and abiotic factors, their interactions, and the system's evolution over time. The choice of a well-adapted, resilient and functional native plant species is often not supported by scientific literature, especially for Europe (Bonciarelli et al. 2025). The species' remediation potential is another significant gap, especially in field studies, and the selection criteria and ecological functioning are often not properly addressed (Chaves et al. 2024). Literature also shows that infiltration rates can worsen over time due to clogging, and the media's filtration capacity can be exhausted, necessitating costly replacement. In this sense, the lack of a consistent literature on long-term performance reflects a relevant degree of uncertainty, both operational and economic (Bonciarelli et al. 2025). Also, spatial limitations of the RGs can create significant constraints to their implementation. For example, if infiltration is the target, a higher RG surface-to-catchment area ratio must be used. This also limits performance during heavier storms, worsened by climate change, which are often poorly managed (Davis and McCuen 2005).

8.3.5 Case Studies

1. *Case study: Potential realization of Rain gardens in university outdoor spaces*

Within outdoor areas managed by Roma Tre University (Rome), RGs have been hypotesized in car parking areas near the buildings of various Department (Fig. 8.8). In these areas, the asphalted surface is extensive and receives rainwater both directly and from runoff flowing from the gutters of nearby buildings. As it flows, the rainwater washes over the roof surfaces and asphalt of the parking areas, carrying with

Fig. 8.8 Potential areas for the installation of RGs within car parking areas of Roma Tre University. In red and yellow, the parking areas of the Department of Engineering (**a**, **b**) and Mathematics and Physics (**c**), respectively. (Figure by the authors)

Fig. 8.9 In red, six rectangular areas (2 × 15 m) identified in the parking Area, suitable for the installation of RGs. On the right, an example of one of these six areas where a buffer (grey) and a depression zone (green) are shown. (Elaboration by the authors)

it both atmospheric pollutants deposited on the ground and contaminants from local vehicular traffic.

As an example, the proposal for parking area A is shown. Here, there are six rectangular green spaces, approximately 2 × 15 m (Fig. 8.9), that separate the rows of parking spaces and contain specimens of *Albizia julibrissin*. In each of these selected areas, a rain garden of approximately 1 × 8 m could be installed (Fig. 8.8), while preserving *A. julibrissin* specimens at each end of the plot and leaving buffer space to prevent vehicles from hitting the rain garden structure during parking maneuvers. Each rain garden installation could include a perimeter buffer area planted with *Carex* and a stepped depression area containing *Iris pseudacorus*, *Sparganium erectum* and *Lythrum salicaria*, which provide flowering from April to October, with a range of colors from the bright yellow of the *Iris*, to the white of the *Sparganium* and the pink-violet of the *Lythrum* (Fig. 8.5). Given their different heights, it is preferable to place the smaller *S. erectum* in the more peripheral and elevated parts of the depression and the taller *I. pseudacorus* and *L. salicaria* in the central zones.

2. *Case study: Sheffield's Grey to Green*

The Grey to Green project is a pioneering example of urban regeneration and sustainable drainage (SuDS) in Sheffield (UK). Through this project, the former highway space was transformed into multifunctional green corridors, tackling pressing urban challenges, such as flood risk management, urban heat island mitigation, biodiversity enhancement, and the revitalization of public spaces in post-industrial neighborhoods (CIRIA Susdrain 2021).

Implemented in stages, phase 1 (2014–2016) converted redundant carriageways occurring in the Riverside Business District into a continuous sequence of rain gardens, bioswales, and wildflower meadows, integrated with improved walking and cycling routes. In phase 2 (2019–2020), green infrastructure coverage was increased, with a focus on surrounding streets, diversifying the planting palette, and setting more ambitious targets for ecological performance and biodiversity gains.

The planting beds (bioswales) installed in phase 1 act as RGs (Fig. 8.10).

They comprise a series of sunken, interconnected sections that capture surface water and pollutants. Each section is fitted with an outlet, allowing excess water to move along the chain of beds as levels rise. Water is gradually absorbed into the soil and filtered through the layered vegetation. During periods of heavy rainfall, these linear biosystems efficiently manage and mitigate water runoff by temporarily holding water and allowing suspended solids to settle and pollutants to be filtered by the vegetation and soil. They are engineered to drain, either through infiltration into the underlying soil or via subsurface drainage systems (Sheffield City Council 2022).

The GRs of Grey to Green capture contaminants, such as oils, heavy metals, and microplastics, before they enter the River Don, thereby safeguarding aquatic biodiversity. The engineered soil used is composed of low-fertility, free-draining, locally sourced recycled materials, including recycled compost and crushed glass blended with sandstone and a small proportion of loam. This provides significantly higher infiltration rates, while its filtration capacity helps trap sediments and remove pollutants from runoff.

The project's multi-layered planting and specially engineered soils are also tailored to support a rich diversity of flora and fauna. The project establishes safe green corridors for wildlife, which allow birds, bats, and insects to move more freely across the city, while flower-rich beds provide nectar and pollen essential to pollinators. Biodiversity monitoring in phase 2 has indicated a more than 500% increase in biodiversity value compared to pre-project conditions (Sheffield City Council 2022).

The innovative perennial planting design is coherent with the "different" conception of RGs, centered on fast drainage, drought tolerance, and alternating wet and dry conditions. Drought-resistant, pollen-rich species for pollinators are paired with these fast-draining engineered soils to manage water. A multi-layered perennial planting maximizes biodiversity, seasonal interest, and resilience, and also provides pollution abatement. As in the UK horticultural tradition, native and alien (non-invasive) ornamental plants are combined to deliver ecological and aesthetic benefits.

A general plant adaptation to fast drainage is required, since the soil matrix is highly permeable. The most drought-resistant plant species are positioned on the upper sides of the RG; hygrophilous plants are located where an intermediate humidity level occurs; the most waterlogging-tolerant species are positioned in the lower, central part of the infrastructure. Both native and alien (non-invasive) trees and shrubs, and herbaceous plants, are used, even if the woody species are more sporadic (Sheffield City Council 2022).

Among trees, *Quercus palustris*, *Gleditsia triacanthos*, and *Betula pendula* can tolerate waterlogging conditions, while *Cercis siliquastrum* is highly drought-tolerant. *Pinus mugo*, *Euonymus alatus* and *Sarcococca hookeriana* var. *digyna*. are some of the shrubs forming the middle height layer, tolerating moist but not

Fig. 8.10 Example of RGs realized according to the Grey to Green in Sheffield city (UK). (Photos by L. Bonciarelli)

waterlogged soils. Various herbaceous perennial species, such as *Narcissus poeticus*, *Gladiolus byzantinus*, *Allium sphaerocephalon*, *Iris sibirica*, are combined to offer ornamental and long-lasting flowers for pollinators. However, the visual structure of these RGs is also largely based on perennial graminoid plants, such as *Carex secta*, a wetland species suitable for the moist to wet areas, *Deschampsia caespitosa* and *Molinia caerulea*, tolerant of waterlogging, while *Calamagrostis acutiflora* and *Stipa gigantea* are respectively typical of moist and moist to dry soils (Yuan and Dunnett 2018).

Helictotrichon sempervirens is mainly used in free-draining drier areas. Among forbs, spring-flowering species of humid environments, such as *Pulsatilla vulgaris*, *Primula veris*, *Rudbeckia fulgida*, and *Echinacea pallida*, are also used. In the driest areas, ornamental plants such as *Echinops ritro*, *Clinopodium nepeta*, *Achillea filipendulina*, and species of the genera *Saponaria* and *Euphorbia* are found. For the waterlogged zone, *Eupatorium cannabinum* and *Veronicastrum virginicum* are usually used.

8.4 Green Roundabouts: Integrating Ecology and Functionality (F.B., L.D.D., E.R.)

8.4.1 Definition and Classification

Modern roundabouts have evolved beyond their original purpose as traffic management devices to become multifunctional components of urban green infrastructure. Their design inherently supports mobility efficiency by reducing delays, fuel consumption, and emissions compared to signalized intersections, while simultaneously offering opportunities for ecological enhancement (Ariniello and Przybyl 2010; Ciampa et al. 2020). When carefully designed from a landscape perspective, roundabouts can become small ecosystems that connect different green areas within the city. This allows them to support urban biodiversity, improve rainwater management, and reduce the "heat island" effect typical of urban environments (Dall'Ara et al. 2019).

Within contemporary planning discourse, green roundabouts can be explicitly framed as Nature-Based Solutions (NbS), as defined by the European Commission and the IUCN, delivering simultaneous benefits for biodiversity conservation, climate change mitigation and adaptation, and human well-being (European Commission 2015; IUCN 2020). This framing positions roundabouts not merely as ornamental green spaces, but as functional components of climate-resilient urban systems operating at the neighborhood scale.

Their significance lies precisely in this dual role: on one hand, they ensure efficient traffic flow, while on the other hand, they enhance the environmental quality of their surroundings. This is why green roundabouts are becoming increasingly central to climate-conscious urban planning.

From a sustainability perspective, roundabouts help reduce greenhouse gas emissions by reducing the time vehicles spend idling and accelerating. Research shows that busy intersections converted to roundabouts can save up to 400 metric tons of CO_2 annually (Ariniello and Przybyl 2010). Newer designs, such as turbo roundabouts and flower roundabouts, go further, improving both safety and traffic capacity while reducing pollutant emissions by as much as 20% through smoother traffic flow with fewer stops and starts (Ciampa et al. 2020). What makes roundabouts particularly interesting is how landscape design can transform what would otherwise be leftover space into something valuable both ecologically and culturally (Fig. 8.11). By incorporating native plants, rain gardens, and small urban groves, these spaces offer real ecosystem benefits alongside aesthetic value (Dall'Ara et al. 2019). This shows that transport infrastructure doesn't have to be purely functional; it can actively contribute to making cities more resilient and livable. From a landscape ecology perspective, green roundabouts can function as ecological "stepping-stones" within highly fragmented urban matrices, facilitating species movement and genetic exchange across sealed surfaces and infrastructure barriers (Beninde et al. 2015). When strategically distributed, they contribute to urban ecological networks by improving functional connectivity between parks, street trees, private gardens, and peri-urban green areas, a factor increasingly recognized as critical for pollinator persistence and urban biodiversity resilience (Helden and Leather 2004).

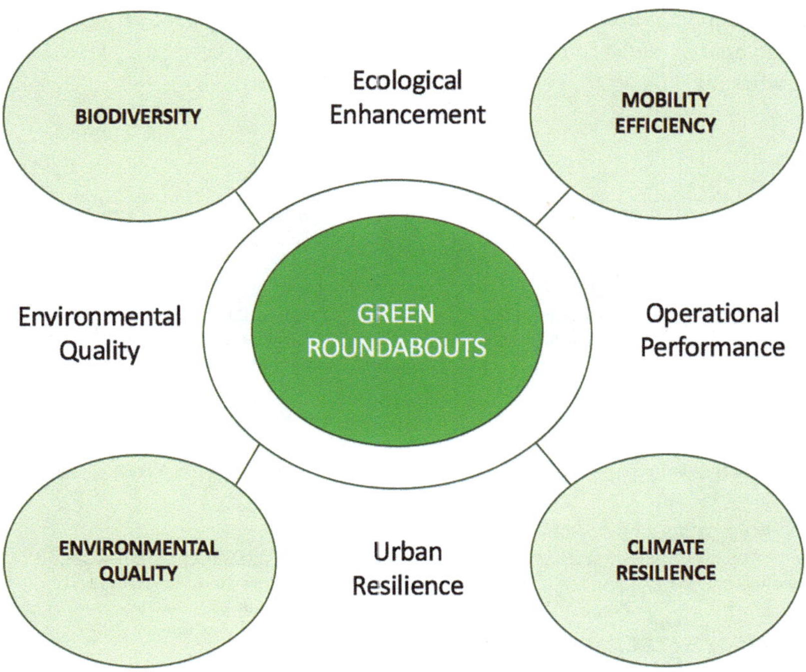

Fig. 8.11 Multifunctional benefits of green roundabouts: biodiversity, environmental quality, climate resilience, urban resilience, mobility efficiency, and operational performance. (Figure created by the authors)

8.4.2 Greening Solutions for Roundabouts

Integrating vegetation into a roundabout design transforms these traffic nodes into multifunctional green infrastructure elements (Fig. 8.12). One effective approach is to create native plant meadows with local species that can withstand urban stress and Mediterranean climates.

Recent climate projections indicate an increasing frequency of heatwaves, prolonged droughts, and extreme rainfall events in Mediterranean and temperate regions; therefore, plant palettes should be evaluated not only for average climatic conditions but also for tolerance to climatic extremes, episodic flooding, and thermal stress (IPCC 2022). Designing for climate variability rather than climatic means enhances long-term resilience and reduces the risk of vegetation failure under future scenarios. Plants like *Achillea millefolium* L. (yarrow), *Onobrychis viciifolia* (sainfoin), *Salvia pratensis* (meadow sage), and *Leucanthemum vulgare* (oxeye daisy) work well because they need less irrigation and maintenance. Beyond the practical benefits, these species strengthen ecological resilience while maintaining the character of the local landscape (Martínez et al. 2014).

- **Pollinator-Friendly Plantings:** Incorporate nectar-rich species, such as *Lavandula angustifolia* (lavender), *Salvia officinalis* (sage), *Echinacea purpurea* (coneflower), *Verbena bonariensis* (verbena), to support bees, butterflies, and other beneficial insects. These plantings strengthen ecosystem services, improve biodiversity, and contribute to urban pollinator corridors (Helden and Leather 2004).

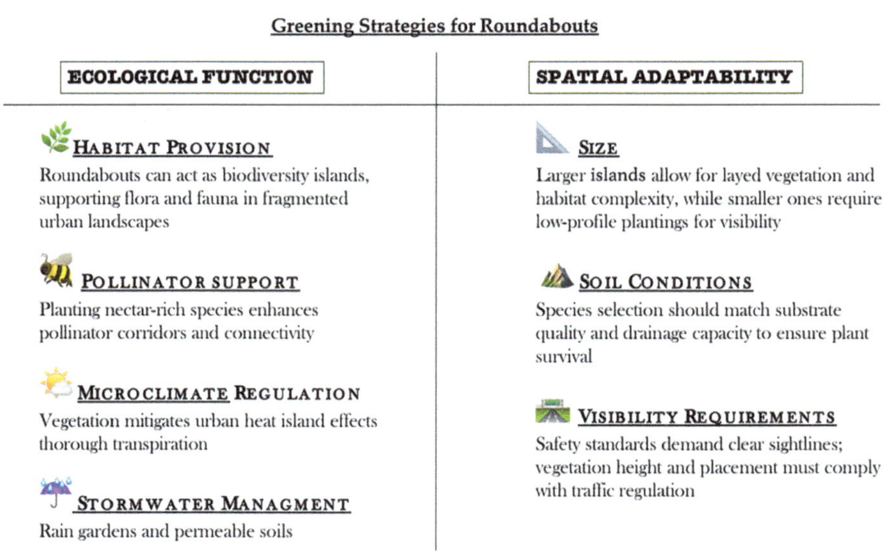

Fig. 8.12 Greening strategies for roundabouts categorized by ecological function and spatial adaptability. (Elaboration by F. Bartoli)

- **Structural Vegetation Diversity:** Combine groundcovers, i.e., *Thymus vulgaris* (thyme), *Santolina chamaecyparissus* (cotton lavender), shrubs, i.e., *Viburnum tinus* (laurustinus), *Buxus sempervirens* (boxwood), and small trees. i.e., *Acer campestre* (field maple), *Quercus ilex* (holm oak), to mimic natural habitats and increase ecological complexity. Layered vegetation improves microclimate regulation, provides shelter for fauna, and enhances visual quality (Dall'Ara et al. 2019).
- **Low-Maintenance Greenery:** Use drought-tolerant perennials and xerophytic species, such as *Santolina, Lavandula, Rosmarinus officinalis* (rosemary), or ornamental grasses such as *Stipa tenuissima* and *Pennisetum alopecuroides,* to ensure resilience under water scarcity and minimize upkeep. These species offer aesthetic appeal while reducing resource consumption (Martínez et al. 2014).

In addition, non-conventional roundabout configurations—such as turbo-roundabouts and flower-roundabouts—introduce advanced geometric layouts that optimize lane discipline, reduce conflict points, and enhance capacity. These designs not only improve traffic flow and safety but also contribute to sustainability by lowering vehicle delays, fuel consumption, and associated emissions (Ciampa et al. **2020**). By merging operational efficiency with environmental performance, they exemplify how innovative intersection design can serve as a cornerstone of climate-responsive and resilient urban infrastructure.

8.4.3 Methodological Approach for Plant Selection

Designing resilient and ecologically functional roundabouts begins with a comprehensive site analysis, which considers soil quality, drainage capacity, pollution levels, sunlight exposure, and proximity to traffic. These factors define stress conditions and microclimatic characteristics, guiding species selection and maintenance strategies (Stipanović and Andonovski 2024). Once the environmental baseline is established, the next step is ensuring ecological fit by prioritizing native or non-invasive species adapted to Mediterranean climates and urban stressors. Such species not only reduce irrigation needs but also enhance biodiversity and maintain ecological integrity (Martínez et al. 2014).

Beyond ecological compatibility, plant choice should reflect functional traits aligned with site conditions. Applying Grime's CSR theory—Competitor, Stress-tolerator, Ruderal—helps match species strategies to environmental constraints. Stress-tolerators thrive in nutrient-poor soils and polluted areas, while competitors suit fertile, stable sites, and ruderals adapt to disturbed zones (Grime 1977). To strengthen ecosystem services, designers should also pursue **biodiversity goals**, combining structural diversity—groundcovers, shrubs, and small trees—with temporal diversity through staggered flowering periods. This approach supports pollinators, creates habitat complexity, and improves visual appeal (Helden and Leather 2004).

Finally, successful greening strategies require community input to align plant palettes with cultural identity and aesthetic preferences. Integrating local heritage and public expectations fosters acceptance and reinforces the sense of place (Dall'Ara et al. 2019). Together, these criteria form a holistic framework for plant selection that balances ecological resilience, functional performance, and social value (Fig. 8.13).

Flowering meadows and xeriscaping offer a sustainable solution, reducing irrigation needs by up to 2000 L/m^2/year while attracting pollinators and enhancing both ecological and visual quality (Martínez et al. 2014). Incorporating drought-tolerant perennials and ornamental grasses further minimizes maintenance costs and improves resilience under prolonged dry periods, creating landscapes that combine

Fig. 8.13 Methodological approach for plant selection in greening roundabouts, including site analysis, ecological fit, functional traits, biodiversity goals, and community input. (Figure created by F. Bartoli)

functionality, biodiversity, and aesthetic appeal. Mediterranean contexts demand alternatives to traditional lawns, which are water-intensive and biodiversity-poor, making them unsuitable for climate-resilient design.

Designing green roundabouts requires integrating **evidence-based typologies** that balance ecological benefits with road safety standards. Approaches such as **wildflower meadows** enhance biodiversity while reducing maintenance costs, using species like *Papaver rhoeas* (common poppy), *Achillea millefolium* (yarrow), and *Onobrychis viciifolia* (sainfoin) (Blackmore and Goulson 2014; Hitchmough 2017). **Urban groves** of native trees—such as *Quercus robur* (English oak), *Fraxinus ornus* (manna ash), and *Acer campestre* (field maple)—contribute to carbon sequestration and provide shade (Nowak et al. 2006). **Rain gardens** improve soil permeability and manage stormwater runoff with plants like *Iris pseudacorus* (yellow iris), *Carex pendula* (pendulous sedge), and *Salix purpurea* (purple willow) (Davis et al. 2006). For **dry gardens (xeriscapes),** drought-tolerant species such as *Lavandula angustifolia* (lavender), *Helichrysum italicum* (curry plant), and *Rosmarinus officinalis* (rosemary) support climate adaptation (IPCC 2022). To ensure safety, vegetation in central islands should not exceed 80 cm in height, with taller species positioned in peripheral zones to maintain visibility (Pratelli and Souleyrette 2009). Planting strategies should include mixed seed blends for nitrogen enrichment and prioritize seasonal color succession—spring (*Papaver rhoeas*), summer (*Lavandula, Achillea*), autumn (*Helichrysum*) (Hitchmough 2017). Designers must avoid invasive species, combine native plants with drought-tolerant varieties for resilience, and use mulching and permeable soils to retain moisture (Dunnett and Hitchmough 2004). These measures not only enhance aesthetics and pollinator support but also ensure functionality and safety within urban infrastructure.

8.4.4 Specific Benefits

Roundabout greening is increasingly recognized as a **nature-based solution** that integrates ecological, climatic, and functional benefits within urban infrastructure. By transforming traffic islands into green spaces, cities can address multiple sustainability challenges simultaneously. Larger roundabouts and those located farther from major traffic corridors have been shown to support **higher species richness and more stable insect communities**, as structural complexity and native plant diversity enhance habitat quality for arthropods and small vertebrates (Helden and Leather 2004; Beninde et al. 2015). Moreover, **pollinator support** is a critical function of these spaces: flower-rich planting schemes attract bees, butterflies, and hoverflies, contributing to pollination services in fragmented urban landscapes (Martínez et al. 2014; Blackmore and Goulson 2014). Seasonal succession of blooms—spring poppies (*Papaver rhoeas*), summer lavender (*Lavandula angustifolia*) and yarrow (*Achillea millefolium*), autumn curry plant (*Helichrysum italicum*)—ensures continuous nectar and pollen availability, mitigating pollinator decline.

Beyond biodiversity, roundabout vegetation plays a vital role in **microclimate regulation**, reducing urban heat island intensity through shading and evapotranspiration. Tree canopies and groundcover plants also improve air quality by intercepting particulate matter and absorbing gaseous pollutants (Nowak et al. 2006). Hydrological benefits are equally significant: rain gardens and permeable soils integrated into roundabouts reduce surface runoff, promote infiltration, and recharge groundwater while filtering pollutants (Davis et al. 2006). From a functional perspective, design strategies such as bypass lanes and optimized geometry can reduce congestion and fuel consumption by up to 5%, lowering emissions of CO_2, NO_n, and particulate matter in high-traffic contexts (Guerrieri et al. 2015). These combined benefits position green roundabouts as a cornerstone of **urban resilience**, merging ecological performance with traffic efficiency.

Finally, green roundabouts offer **educational and aesthetic value**, serving as living laboratories that foster ecological awareness and enhance urban landscapes (Hitchmough 2017). To maximize these benefits, design must adhere to evidence-based principles: prioritize native and drought-tolerant species for resilience, avoid invasive plants, maintain visibility and safety standards (vegetation height $\leq$ 80 cm in central islands), incorporate soil amendments and mulching for water retention, and monitor ecological performance through biodiversity surveys and hydrological assessments. When implemented strategically, roundabout greening becomes more than an ornamental intervention—it is a multifunctional solution that advances sustainability goals while enriching the urban experience.

8.4.5 Maintenance and Monitoring

Long-term success of green roundabouts depends on adaptive maintenance supported by measurable ecological and functional indicators (Key Performance Indicators, KPIs) rather than on aesthetic criteria alone, and on systematic monitoring to sustain ecological functionality and safety. One effective strategy is implementing reduced mowing regimes, which allow natural succession and flowering cycles, thereby supporting pollinator communities and reducing maintenance costs (Helden and Leather 2004; Hitchmough 2017). Complementary practices such as seasonal pruning and weeding are essential to maintain plant health, prevent the spread of invasive species, and ensure compliance with visibility standards for road safety (Pratelli and Souleyrette 2009). Soil health monitoring represents a critical yet often overlooked component; recent advances in metagenomic tools enable assessment of microbial diversity and nutrient cycling, providing insights into soil resilience and ecosystem services (Fierer 2017). Engaging local communities through citizen science initiatives and mobile applications can enhance biodiversity monitoring, foster environmental stewardship, and generate valuable datasets for urban ecology research (Bonney et al. 2014).

Typical KPIs for green roundabouts include pollinator abundance and species richness, vegetation cover stability, soil organic carbon content, reductions in surface temperature, air pollutant interception, and stormwater infiltration rates (Nowak et al. 2006; Fierer 2017). Increasingly, these parameters can be monitored using low-cost environmental sensors, remote sensing techniques, and citizen science platforms, improving data availability for urban ecological assessment (Bonney et al. 2014). Finally, adaptive management frameworks—where maintenance practices are adjusted in response to ecological feedback—are vital for responding to dynamic urban conditions and climate variability (Holling 1978). Together, these strategies ensure that green roundabouts evolve as resilient, multifunctional elements of sustainable urban infrastructure.

8.4.6 Challenges and Trade-Offs

Despite their ecological and functional potential, green roundabouts face several constraints and trade-offs that influence design and long-term performance. Space limitations pose a primary challenge, as small central islands reduce species richness and structural complexity, thereby limiting opportunities for layered vegetation and habitat diversity (Helden and Leather 2004). Furthermore, pollution stress is a significant concern: proximity to high-traffic roads exposes plants and soils to heavy metals, hydrocarbons, and elevated noise levels, which can impair plant health and reduce faunal diversity (Olajire and Ayodele 1997). Maintenance funding poses another barrier, as limited municipal budgets often hinder consistent irrigation, pruning, and invasive species control, compromising ecological integrity over time (Dunnett and Hitchmough 2004). Safety considerations introduce additional complexity: unmanaged vegetation height may obstruct driver visibility, increasing accident risk and necessitating strict adherence to design standards such as maintaining vegetation below 80 cm in central islands (Pratelli and Souleyrette 2009). Finally, data gaps persist due to the lack of standardized biodiversity monitoring protocols for traffic islands, making it difficult to assess ecological performance and inform adaptive management strategies (Beninde et al. 2015). Addressing these challenges requires integrated planning that balances ecological ambition with operational feasibility, ensuring that green roundabouts remain both functional and resilient within urban transport networks. At last, green roundabouts represent a synergistic solution for urban sustainability, merging ecological, aesthetic, and functional goals.

When explicitly designed as Nature-Based Solutions and embedded within broader urban green networks, green roundabouts shift from marginal landscape elements to strategic infrastructure components that simultaneously support biodiversity conservation, climate adaptation, and sustainable mobility. By adopting native plant palettes, low-maintenance typologies, and integrated design strategies, cities can transform traffic nodes into biodiversity hotspots and climate-resilient infrastructure. Combining ecological greening with functional innovations—such as bypass lanes for traffic efficiency—maximizes environmental and operational benefits, paving the way for multifunctional urban landscapes.

8.4.7 Case studies

Across contexts and scales, the evidence from Kansas, Imola, and Skopje shows how roundabouts can shift from mere traffic devices to compact, high-impact green infrastructure (Fig. 8.14).

In Kansas (Mandavilli et al. 2008), converting stop-controlled intersections into single-lane modern roundabouts proved highly effective in improving both traffic efficiency and environmental performance. By reducing queuing and idling, these interventions achieved substantial emission reductions—up to 59% less CO_2, 48% less NO_n, and 65% less hydrocarbons—while also cutting delays and enhancing air

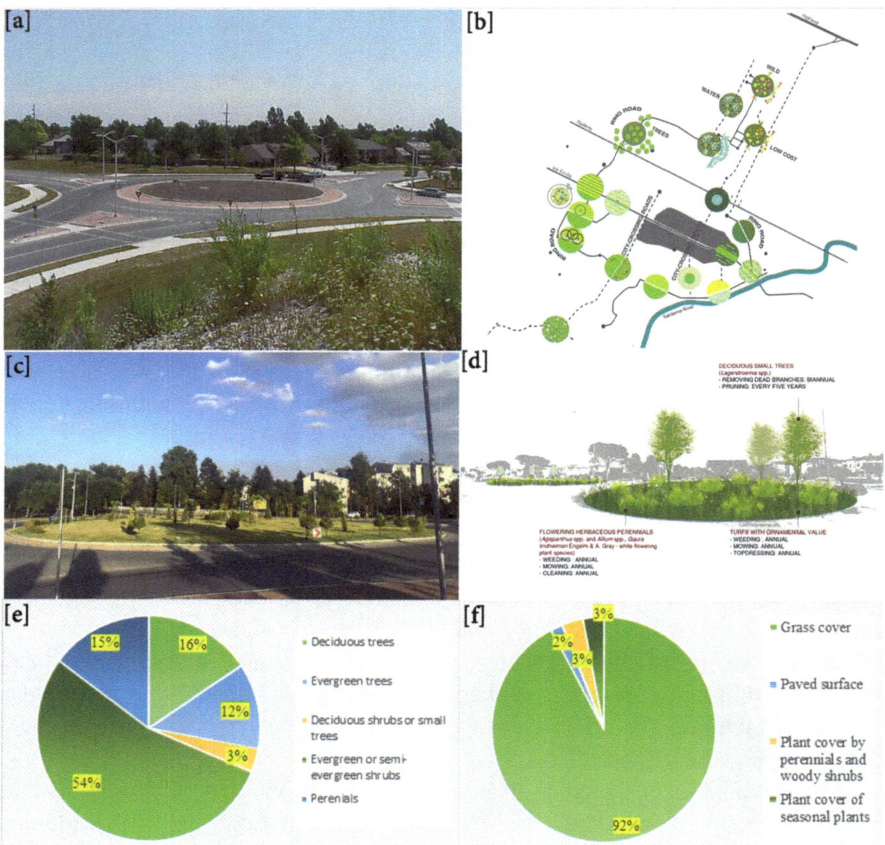

Fig. 8.14 Case studies linked image (**a**) Modern roundabout context (operational focus and low-input planting); (**b**) Imola's landscape-potential network linking roundabout islands; (**c**) large central island in Skopje; (**d**) Imola design scenario cross-section with layered vegetation and safety-compliant heights; (**e**) Skopje plant-type composition (share of evergreen shrubs, trees, perennials); (**f**) Skopje ground-cover allocation (grass, paved areas, perennials/woody shrubs, seasonal beds). (Images from Mandavilli et al. 2008; Dall'Ara et al. 2019; Stipanović and Andonovski 2024)

quality (Fig. 8.14a). The success of this approach rests on clear design principles: adopting simple single-lane geometry for safety and operational fluidity, and complementing traffic improvements with low-maintenance landscaping. To ensure long-term sustainability, these conversions should be paired with native vegetation to strengthen ecological benefits, incorporate drought-tolerant species to minimize irrigation needs, and include monitoring systems to track emissions and traffic performance. Collectively, the Kansas experience demonstrates that modern roundabouts can deliver measurable climate benefits while maintaining functional and aesthetic integrity, making them a cornerstone of resilient urban mobility planning.

In Imola (Dall'Ara et al. 2019), a city-wide masterplan reimagined 48 roundabouts as interconnected ecological nodes, embedding them within a broader green infrastructure network to deliver environmental and aesthetic benefits. The approach combined typology-guided planting strategies—wildflower meadows with species such as *Achillea millefolium*, *Leucanthemum vulgare*, and *Papaver rhoeas* to support pollinators and reduce maintenance; urban groves featuring *Quercus robur*, *Fraxinus ornus*, and *Acer campestre* to provide shade and mitigate heat islands; rain gardens with *Iris pseudacorus*, *Carex pendula*, and *Lythrum salicaria* for stormwater filtration and infiltration; and dry or xeric gardens using Mediterranean perennials like *Santolina*, *Helichrysum*, and *Rosmarinus* to minimize irrigation needs (Fig. 8.14b, c). These interventions were guided by essential design principles: integrating roundabouts into a city-wide ecological network, layering vegetation for structural diversity, and maintaining central-island plant heights below 80 cm to ensure visibility and traffic safety. Sustainability recommendations emphasized prioritizing native and non-invasive species for resilience, replacing water-intensive lawns with xeriscaping, and adopting maintenance regimes that reduce mowing and encourage pollinator habitats. Collectively, Imola's strategy demonstrates how thoughtful planting design and long-term care can transform traffic nodes into multifunctional green spaces that enhance biodiversity, regulate microclimates, and strengthen urban resilience.

In Skopje (Stipanović and Andonovski 2024), the introduction of six eco-roundabouts added approximately 7000 m^2 of vegetated space, creating green nodes within the city's traffic network (Fig. 8.14d).

Current ground cover is dominated by grass (~92%), with only 2% paved surfaces and small shares of perennials and woody shrubs (~3%) alongside seasonal beds (~3%) (Fig. 8.14e). Plant composition reflects a strong prevalence of evergreen and semi-evergreen shrubs (~54%), complemented by deciduous trees (16%), evergreen trees (12%), perennials (15%), and minor deciduous shrubs or small trees (3%) (Fig. 8.14f).

While this expansion improves visual quality and offers potential for pollutant interception, field assessments reveal critical gaps: the persistence of high-turnover seasonal plantings and the vulnerability of certain deciduous species to urban stress and exposure to exhaust. To enhance resilience and ecological function, recommendations include replacing seasonal displays with pollution-tolerant natives and durable evergreens such as *Betula pendula*, *Carpinus betulus*, and *Photinia × fraseri*, introducing rain gardens and xeriscape pockets to manage stormwater and drought, and enforcing strict height limits (≤80 cm) to maintain visibility and traffic safety. Species such as *Cupressus sempervirens, Platycladus orientalis*, and perennials such as *Santolina chamaecyparissus* and *Jacobaea maritima* can provide structure and year-round greenery, while shrubs such as *Buxus sempervirens* and *Nandina domestica* add texture and biodiversity. Ultimately, Skopje's experience underscores that maximizing green coverage must be paired with thoughtful species selection, hydrological design, and structured maintenance to transform roundabouts into resilient, multifunctional components of urban green infrastructure.

References

Ahiablame LM, Engel BA, Chaubey I (2012) Effectiveness of low impact development practices: literature review and suggestions for future research. Water Air Soil Pollut 223(7):4253–4273. https://doi.org/10.1007/s11270-012-1189-2

Andruczyk M, Fox L (2009) Rain garden plants. College of Agriculture and Life Sciences, Virginia Polytechnic Institute and State University, publ, pp 420–443

Ariniello A, Przybyl B (2010) Roundabouts and sustainable design. In: Green streets and highways 2010: an interactive conference on the state of the art and how to achieve sustainable outcomes, pp 82–93

Attorre F, Bruno M, Francesconi F, Valenti R, Bruno F (2000) Land- scape changes of Rome through tree-lined roads. Landscape Urban Plan 49(3–4):115–128. https://doi.org/10.1016/S0169-2046(00)00069-4

Bartoli F, Savo V, Caneva G (2022) Biodiversity of urban street trees in Italian cities: a comparative analysis. Plant Biosyst 156(3):649–662. https://doi.org/10.1080/11263504.2021.1906347

Beninde J, Veith M, Hochkirch A (2015) Biodiversity in cities needs space: a meta-analysis of factors determining intra-urban biodiversity variation. Ecol Lett 18(6):581–592. https://doi.org/10.1111/ele.12427

Blackmore LM, Goulson D (2014) Evaluating the effectiveness of wildflower seed mixes for pollinator conservation. J Insect Conserv 18(4):509–518. https://doi.org/10.1111/icad.12071

Bonciarelli L, Orlandi F, Muscas D, Fornaciari M (2025) Sustainable stormwater management and bioretention: an overview of reviews of the last 10 years. Land 14(4):736. https://doi.org/10.3390/land14040736

Bonney R, Shirk JL, Phillips TB, Wiggins A, Ballard HL, Miller-Rushing AJ, Parrish JK (2014) Next steps for citizen science. Science 343(6178):1436–1437. https://doi.org/10.1126/science.1251554

Caneva G, Bartoli F, Zappitelli I, Savo V (2020) Street trees in italian cities: story, biodiversity and integration within the urban environment. Rendiconti Lincei Scienze Fisiche e Naturali 31(2):411–417. https://doi.org/10.1007/s12210-020-00907-9

Capotorti G, Bonacquisti S, Abis L, Aloisi I, Attorre F, Bacaro G et al (2020) More nature in the city. Plant Biosyst 154(6):1003–1006. https://doi.org/10.1080/11263504.2020.1837285

Cavinato F (2014) Hydrological and hydraulic measurements at the experimental rain garden of Agripolis (Legnaro, PD) [PhD dissertation, University of Padua]

Chaves MTR, Farias TRL, Eloi WM (2024) Comparative analysis of bioretention design strategies for urban runoff infiltration: a critical overview. Ecol Eng 207:107352. https://doi.org/10.1016/j.ecoleng.2024.107352

Ciampa D, Diomedi M, Giglio F, Olita S, Petruccelli U (2020) Effectiveness of unconventional roundabouts in the design of suburban intersections. Eur Transport 80:1–16

CIRIA Susdrain (2021) Grey to green phase 1, Sheffield. https://www.susdrain.org/case-studies/pdfs/suds_awards/006_18_03_28_susdrain_suds_awards_grey_to_green_phase_1_sheffield.pdf
Cording A, Hurley S, Whitney D (2017) Monitoring methods and designs for evaluating bioretention performance. J Environ Eng 143. https://doi.org/10.1061/(asce)ee.1943-7870.0001276
Corduan D, Kühn N (2024) Planting for the urban rain—vegetation in urban bioretention systems for stormwater management under temperate climate conditions—a systematic review. Sustainability 16:8861. https://doi.org/10.3390/su16208861
Czaja M, Kołton A, Muras P (2020) The complex issue of urban trees—stress factor accumulation and ecological service possibilities. Forests 11(9):932. https://doi.org/10.3390/f11090932
Dall'Ara E, Maino E, Gatta G, Torreggiani D, Tassinari P (2019) Green Mobility Infrastructures: a landscape approach for roundabouts' gardens applied to an Italian case study. Urban For Urban Green 37:109–125. https://doi.org/10.1016/j.ufug.2018.03.011
D'Amato L, Bartoli F, Savo V, Caneva G (2025) Promoting native biodiversity: an evaluation of multifactorial and bioclimatic selection criteria for street trees in Italian cities. Urban For Urban Green 107:128784. https://doi.org/10.1016/j.ufug.2025.128784
Davis AP, McCuen RH (2005) Stormwater management for smart growth. Springer US, Boston
Davis AP, Shokouhian M, Sharma H, Minami C (2006) Water quality improvement through bioretention media: nitrogen and phosphorus removal. Water Environ Res 78(3):284–293. https://doi.org/10.2175/106143005X94376
Dervishi V, Fleckenstein C, Rahman MA, Pauleit S, Ludwig F, Pretzsch H, Rötzer T (2023) Trees in planters–growth, structure and ecosystem services of Platanus x hispanica and Tilia cordata and their reaction to soil drought. Urban For Urban Green 86:128024. https://doi.org/10.1016/j.ufug.2023.128024
Di Stasio L, Gentile A, Tangredi DN, Piccolo P, Oliva G, Vigliotta G et al (2025) Urban phytoremediation: a nature-based solution for environmental reclamation and sustainability. Plants 14(13):2057. https://doi.org/10.3390/plants14132057
Donat MG, Lowry AL, Alexander LV, O'Gorman PA, Maher N (2016) More extreme precipitation in the world's dry and wet regions. Nat Clim Chang 6(5):508–513. https://doi.org/10.1038/nclimate2941
Dunnett N, Hitchmough J (2004) The dynamic landscape: design, ecology and management of naturalistic urban planting. Spon Press, Taylor & Francis, London and New York
Environmental Protection Agency (EPA) (2021) NPDES: stormwater best management practice, Bioretention (Rain Gardens). https://www.epa.gov/system/files/documents/2021-11/bmp-bioretention-rain-gardens.pdf
European Commission (2015) Towards an EU research and innovation policy agenda for nature-based solutions & re-naturing cities. In: Final report of the horizon 2020 expert group. Publications Office of the European Union, Luxembourg. https://doi.org/10.2777/765301
Ferreira CS, Mourato S, Kasanin-Grubin M, Ferreira AJ, Destouni G, Kalantari Z (2020) Effectiveness of nature-based solutions in mitigating flood hazard in a Mediterranean peri-urban catchment. Water 12(10):2893. https://doi.org/10.3390/w12102893
Fierer N (2017) Embracing the unknown: disentangling the complexities of the soil microbiome. Nat Rev Microbiol 15(10):579–590. https://doi.org/10.1038/nrmicro.2017.87
Fletcher TD, Shuster W, Hunt WF, Ashley R, Butler D, Arthur S et al (2015) SUDS, LID, BMPs, WSUD and more–the evolution and application of terminology surrounding urban drainage. Urban Water J 12(7):525–542. https://doi.org/10.1080/1573062X.2014.916314
Forrest M, Konijnendijk C (2005) A history of urban forests and trees in Europe. In: Konijnendijk CC, Nilsson K, Randrup TB, Schipperijn J (eds) Urban forests and trees: a reference book. Springer, Berlin, pp 23–48
Gómez-Baggethun E, Barton DN (2013) Classifying and valuing ecosystem services for urban planning. Ecol Econ 86:235–245. https://doi.org/10.1016/j.ecolecon.2012.08.019
Greksa A, Ljubojević M, Blagojević B (2024) The value of vegetation in nature-based solutions: roles, challenges, and utilization in managing different environmental and climate-related problems. Sustainability 16(8):3273. https://doi.org/10.3390/su16083273

Grime JP (1977) Evidence for the existence of three primary strategies in plants and their relevance to ecological and evolutionary theory. Am Nat 111(982):1169–1194. https://doi.org/10.1086/283244

Guerrieri M, Corriere F, Rizzo G, Lo Casto B, Scaccianoce G (2015) Improving the sustainability of transportation: environmental and functional benefits of right turn by-pass lanes at roundabouts. Sustainability 7(5):5838–5856. https://doi.org/10.3390/su7055838

Helden AJ, Leather SR (2004) Biodiversity on urban roundabouts—Hemiptera, management and the species–area relationship. Basic Appl Ecol 5(4):367–377. https://doi.org/10.1016/j.baae.2004.06.004

Hitchmough J (2017) Sowing beauty: designing flowering meadows from seed. Timber Press

Holling CS (1978) Adaptive environmental assessment and management. John Wiley and Sons, Chichester, UK

IPCC (2022) Climate change 2022: impacts, adaptation and vulnerability. Cambridge University Press

ISPRA (2013) Quality of the urban environment. Focus on water and the urban environment. In: Ninth annual report. State of the Environment 46/13. ISPRA Publishing

IUCN (2020) Global Standard for Nature-based Solutions: a user-friendly framework for the verification, design and scaling up of NbS. International Union for Conservation of Nature, Gland. https://doi.org/10.2305/IUCN.CH.2020.08.en

Mandavilli S, Rys MJ, Russell ER (2008) Environmental impact of modern roundabouts. Int J Ind Ergon 38(2):135–142. https://doi.org/10.1016/j.ergon.2006.11.003

Martínez MDP, Castro MC, Pinto-Gomes C (2014) Flowering meadows: a biodiverse alternative to lawns in Mediterranean urban spaces. The Garden As A Lab, p 130

Masini E, Tomao A, Corona P, Fattorini L, Giuliarelli D, Portoghesi L, Agrimi M (2023) The ecosystem disservices of trees on sidewalks: a study based on a municipality urban tree inventory in Central Italy. Urban For Urban Green 86:128007. https://doi.org/10.1016/j.ufug.2023.128007

Mullaney J, Lucke T, Trueman SJ (2015) A review of benefits and challenges in growing street trees in paved urban environments. Landsc Urban Plan 134:157–166. https://doi.org/10.1016/j.landurbplan.2014.10.013

Nowak DJ, Crane DE, Stevens JC (2006) Air pollution removal by urban trees and shrubs in the United States. Urban For Urban Green 4(3–4):115–123. https://doi.org/10.1016/j.ufug.2006.01.007

Olajire AA, Ayodele ET (1997) Contamination of roadside soil and grass with heavy metals. Environ Int 23(1):91–101. https://doi.org/10.1016/S0160-4120(96)00080-3

Ow LF, Ghosh S, Mohamed Lokman Mohd Y (2018) Effects of varying establishment approaches on the growth of urban street trees. Arboricult J 40(4):201–209. https://doi.org/10.1080/03071375.2018.1528092

Pratelli A, Souleyrette RR (2009) Visibility, perception, and roundabout safety. WIT Trans Built Environ 107:577–588. https://doi.org/10.2495/UT090511

Roy S, Byrne J, Pickering C (2012) A systematic quantitative review of urban tree benefits, costs, and assessment methods across cities in different climatic zones. Urban For Urban Green 11:351–363. https://doi.org/10.1016/j.ufug.2012.06.006

Sæbø A, Benedikz T, Randrup TB (2003) Selection of trees for urban forestry in the Nordic countries. Urban For Urban Green 2(2):101–114. https://doi.org/10.1078/1618-8667-00027

Salmond JA, Tadaki M, Vardoulakis S, Arbuthnott K, Coutts A, Demuzere M et al (2016) Health and climate related ecosystem services provided by street trees in the urban environment. Environ Health 15(Suppl 1):S36. https://doi.org/10.1186/s12940-016-0103-6

Savo V, D'Amato L, Bartoli F, Zappitelli I, Caneva G (2025) Evaluation of main regulating, provisioning, and supporting ecosystem services of urban street trees: a literature review. Ecosyst Serv 71:101690. https://doi.org/10.1016/j.ecoser.2024.101690

Sheffield City Council (2022) Grey to Green: Mitigating flood risk and regenerating Sheffield's city centre. https://www.uk100.org/pdf_export/302

Stipanović VB, Andonovski V (2024) Eco-roundabouts: assessing and amplifying environmental benefits in Skopje's urban green infrastructure. For Sci

TEEB (2011) The economics of ecosystems and biodiversity in national and international policy making. Edited by P ten Brink. Earthscan, London and Washington

Tsitsoni T, Samara T (2002) The existing situation and management of urban forests and trees in western Macedonia. In: Proceedings of 10th Panhellenic Forest Science Conference, pp 136–147

Tsitsoni T, Gounaris N, Kontogianni AB, Xanthopoulou-Tsitsoni V (2015) Creation of an integrated system model for governance in urban MTEs and for adapting cities to climate change: preliminary results. Ecologia Mediterranea 41(2):33–43

von Döhren P, Haase D (2019) Risk assessment concerning urban ecosystem disservices: the example of street trees in Berlin, Germany. Ecosyst Serv 40:101031. https://doi.org/10.1016/j.ecoser.2019.101031

Woods Ballard B, Wilson S (2015) Udale-Clarke H, Illman S, Scott T, Ashley R, et al. The SuDS Manual

Yuan J, Dunnett N (2018) Plant selection for rain gardens: response to simulated cyclical flooding of 15 perennial species. Urban For Urban Green 35:57–65. https://doi.org/10.1016/j.ufug.2018.08.005

Chapter 9
Green Buildings

Enrica Roccotiello, Flavia Bartoli, Teresa M. Batista, Amii Bellini, Francesca Bretzel, Giulia Caneva, Carmelo Maria Musarella, Marta Pianta, and Giovanni Spampinato

Abstract This chapter presents vegetated building systems as Nature-based Solutions (NbS), emphasizing the role of plants as active components of the built environment. A common foundation across green roofs, green façades and walls, and bioarchitecture lies in the integration of vegetation as living infrastructure capable of delivering multiple ecosystem services. These systems are not treated merely as design features, but as dynamic ecological assemblages whose performance depends on plant traits, species interactions, and environmental conditions.

A key unifying aspect is the importance of plant functional traits in determining system effectiveness. Characteristics such as water-use efficiency, rooting depth, phenology, and tolerance to thermal and hydric stress directly influence the capacity of vegetated systems to regulate microclimate, mitigate the Urban Heat Island effect, retain stormwater, and capture airborne pollutants. This trait-based approach highlights the need for ecologically informed design, where plant selection is central to both performance and resilience under changing climatic conditions.

E. Roccotiello (✉) · M. Pianta
Departmen of Earth, Environmental and Life Sciences, University of Genoa, Genoa, Italy
e-mail: enrica.roccotiello@unige.it

F. Bartoli
Institute Heritage Science (ISPC), National Research Council (CNR), Montelibretti, Rome, Italy

T. M. Batista
Mediterranean Institute for Agriculture, Environment and Development & Change – Global Change and Sustainability Institute, Institute Advanced Studies and Research, University of Évora, Évora, Portugal

A. Bellini · G. Caneva
Department of Sciences, University Roma Tre, Rome, Italy

F. Bretzel
Institute Research Terrestrial Ecosystem (IRET), National Research Council (CNR), Pisa, Italy

C. M. Musarella · G. Spampinato
Department of AGRARIA, Mediterranean University of Reggio Calabria, Reggio Calabria, Italy

G. Caneva et al. (eds.), *Urban Botany*, Future City 22,
https://doi.org/10.1007/978-3-032-28986-5_9

Across all applications, vegetation supports urban biodiversity by creating habitats and enhancing ecological connectivity within fragmented urban landscapes. At the same time, these systems contribute to human well-being by improving environmental quality and enabling continuous interaction with vegetation in everyday spaces.

Within this shared framework, green roofs function as horizontal ecosystems where substrate–plant interactions regulate hydrological and thermal processes. Green façades and living walls extend these functions vertically, requiring careful alignment between plant traits and façade-specific microclimates. Bioarchitecture, including street furniture and indoor greening, represents a more distributed application, embedding vegetation into both outdoor and indoor environments at multiple spatial scales.

Overall, the chapter underscores that the success of green building systems relies on botanical knowledge, ecological principles, and context-sensitive plant selection, positioning vegetation as a primary driver of resilient and adaptive urban environments.

Keywords Bioarchitecture · Green roofs · Plant functional traits · Urban biodiversity · Urban Nature-based Solutions · Vertical greening systems

9.1 Introduction (E.R.)

Green buildings, including vegetated roofs, living façades, and bioarchitectural elements, are recognised as Nature-based Solutions (NbS) that harness ecological processes to address urban environmental challenges (Capotorti et al. 2020).

By integrating plants into architecture, these systems contribute to mitigating the Urban Heat Island effect (UHI), improving air quality, moderating stormwater runoff, and enhancing urban biodiversity, while simultaneously supporting human psycho-physical well-being and climate adaptation goals (EEA 2021; UNEP 2022).

Vegetated building surfaces act as biofilters, capturing airborne particulate matter and reducing air pollutant concentrations, while substrate-root systems intercept and retain rainfall, delaying runoff and reducing pressure on urban drainage networks.

The effectiveness of these processes strongly depends on a functional trait-based approach to plant selection, including life cycle, leaf area index, substrate-rooting depth, flowering phenology, and drought tolerance, which together determine both ecological performance and climatic regulation.

Cities increasingly exposed to water scarcity and thermal extremes due to climate change, such as those in Mediterranean regions, can be examined to find plant-based adaptation measures that can respond to these changes. In this chapter, several case studies are presented as examples of approaches from other climatic regions (IPCC 2023; EEA 2021; De Cristo et al. 2025). These represent vegetated buildings and their components that have demonstrated a significant capacity to reduce roof surface temperatures and preserve indoor heat, thereby lowering energy demand. They also provide habitat for urban fauna (De Cristo et al. 2025). The integration of plant-centred strategies to transform buildings into green, multifunctional infrastructure can enhance environmental quality, increase biodiversity, and improve human well-being in a resilient, climate-adaptive manner (EEA 2021; UNEP 2022).

9.2 Green Roofs (F.B., A.B., F.B., G.C., E.R.)

9.2.1 Definition and Classification

Green Roofs (GRs) are vegetated layers installed on building rooftops and can be classified along a gradient of structural and botanical complexity into extensive (EGRs) and intensive systems (IGRs), with semi-intensive solutions occupying an intermediate position that combine features of both approaches (Getter and Rowe 2006; Oberndorfer et al. 2007). These systems differ primarily in substrate depth, structural load, plant composition and maintenance requirements. The first modern applications of GRs were developed in Europe during the mid-twentieth century and subsequently spread to America and Asia, with a rapid increase in implementation over recent decades. Their diffusion was initially driven by engineering and architectural objectives, especially thermal mitigation and runoff reduction, while botanical aspects and plant selection criteria have historically received less attention (Bellini et al. 2024a).

From a regulatory perspective, GRs are addressed by national and international guidelines that define design, construction, and maintenance requirements, including load-bearing capacity, waterproofing, and drainage. Standards such as UNI 11235:2015, the Green Roof Organization Code of Practice (GRO 2021), and the Landscape Development and Landscaping Research Society e.V. (FLL 2018) guidelines provide reference frameworks for system classification, functional performance, and long-term durability.

EGRs (Fig. 9.1) are lightweight planted systems (generally <150 kg/m^2) characterized by shallow substrates, typically 8–15 cm, designed for low maintenance. They are usually inaccessible to the public and particularly suitable for retrofitting existing buildings without major structural reinforcement. Plant composition on EGRs is predominantly drought-tolerant, including succulents, forbs, mosses, and hardy grasses, selected for their ability to thrive with limited substrate depth and minimal irrigation. EGRs offer a cost-effective ecological benefit, including stormwater retention, thermal insulation, and UHI mitigation, although limited substrate depth restricts plant diversity and aesthetic complexity (Getter and Rowe 2006; Oberndorfer et al. 2007).

IGRs (Fig. 9.1) are designed as accessible, multifunctional spaces comparable to ground-level gardens and parks. They require deep substrates (generally ≥25 cm) and high load-bearing capacity, often exceeding 300–500 kg/m^2, allowing the

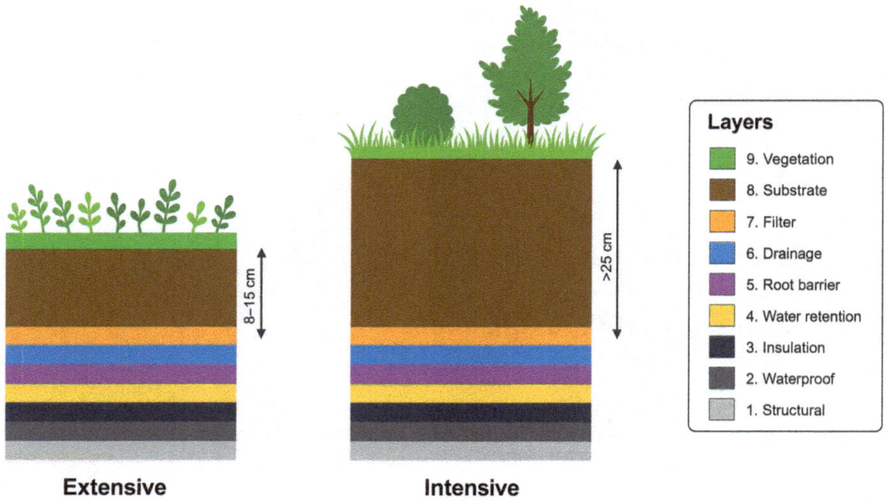

Fig. 9.1 Layers often used for the main green roof type: extensive and intensive. (Image AI-generated from a picture with Microsoft 365 Copilot)

establishment of diverse plant assemblages including shrubs and small trees. Although these systems require greater investment in installation and maintenance (e.g., irrigation, fertilization, and pruning), they provide more significant ecological and social benefits, such as biodiversity support, microclimate regulation, recreational use, and, in some cases, urban agriculture. Due to their structural requirements, which allow the incorporation of additional natural elements, such as woody materials or soil and sand mounds, intensive systems are more frequently applied in new developments (Reyes et al. 2016; Giacomello 2021).

Semi-intensive green roofs represent a hybrid solution, combining characteristics of both extensive and intensive systems, offering a balance between technical feasibility and ecological performance. Functionally, these systems provide stormwater retention, thermal regulation, and biodiversity support, while limiting public access and recreational use. As such, they are suitable for commercial and institutional buildings seeking multifunctional green infrastructure without the high costs and structural demands associated with intensive systems (Giacomello 2021).

Recent developments include Biosolar Roofs (BSRs), which integrate photovoltaic or thermal panels above the plant layer. This configuration can improve panel efficiency through evaporative cooling, while panels create heterogeneous shading conditions that may enhance plant performance and biodiversity (Nash et al. 2016).

Blue Green Roofs (BGRs) further expand GRs' functionality by incorporating water-retention layers that temporarily store rainfall and by using Internet of Things technologies for optimized remote water management. This configuration reduces peak runoff and supports climate adaptation in highly sealed urban areas, while contributing to UHI mitigation and reducing volatile pollutants.

9.2.2 *Methodological Approach for Plant* Taxa *Selection*

Selecting an appropriate plant *taxa* for GRs require consideration of eco-physiological traits, local climate, expected ecosystem services and maintenance (Dunnett and Kingsbury 2008), as urban rooftop environments are characterised by harsher conditions than ground-level sites (Table 9.1).

The methodology aligns with FLL standards (2018) and integrates key bioclimatic and structural parameters as follows:

1. Microclimate assessment. Roof orientation, solar exposure, substrate composition, and wind regimes determine thermal and hydric constraints affecting plant survival and performance. Roofs with high solar exposure and shallow substrates require extreme-drought-tolerant strategies, including CAM (Crassulacean Acid

Table 9.1 Recommended plant traits for GRs synthesized from literature

Green roof type	Substrate depth and load	Main stressors	Key plant functional traits	Main functional outcomes
Extensive (EGRs)	8–15 cm; <150 kg m^{-2}	High solar radiation, wind, heat, and limited water	High Water-Use Efficiency (WUE); drought-tolerant physiology (e.g. CAM); shallow rooting; slow growth	High survival under low inputs, low maintenance, basic stormwater retention, and thermal buffering
Semi-intensive	15–25 cm; moderate load	Moderate exposure; occasional irrigation	Moderate–high WUE; medium-depth roots; functional diversity (perennials, low shrubs); canopy cover reducing evaporation	Balanced ecosystem services; increased biodiversity vs EGRs; lower structural demand than IGRs
Intensive (IGRs)	≥25 cm; 300–500 kg m^{-2}	Variable microclimates; public access	Wide range of life forms; deep rooting; high biomass and recovery capacity; tolerance to regular maintenance	High multifunctionality (recreation, habitat, education); enhanced biodiversity; strong thermal and stormwater performance
Biosolar (BSRs)	Variable -often EGR or semi-intensive	Patchy shade; variable moisture	Shade tolerance combined with sun tolerance, morphological plasticity, and compact growth near panels	Niche diversification; cooler panel temperatures; potential gains in energy efficiency
Blue-green (BGRs)	Variable, with retention layers	Wet–dry cycles; high runoff input	Roots tolerant to moisture gradients; persistent vegetative cover	Advanced stormwater management; thermal regulation; improved resilience to extreme rainfall

Metabolism) *taxa*, rock-dwelling *taxa*, or annual life forms (Van Mechelen et al. 2014).
2. Plant functional traits. Species selected for EGRs should exhibit high water-use efficiency for drought resilience; shallow roots, compatible with limited substrate depth, CAM, for nocturnal CO_2 uptake, and reduced water loss (e.g., the succulents *Sedum* sp.pl., *Sempervivum tectorum* L., the drought-tolerant grass *Brachypodium retusum* (Pers.) P.Beauv.
3. Native biodiversity integration. Prioritizing indigenous *taxa* to strengthen ecological connectivity and system resilience. Habitat template approach, which aligns plant selection with natural community structures (Lundholm and Walker 2018), and ecomimicry principles (i.e., guidelines that translate key characteristics of local ecosystems into design decisions to create functional, resilient, and biodiversity-supporting built environments) that replicate local ecological processes (Nash et al. 2019) can enhance pollinator support and reduce long-term management needs.
4. Climate-resilient design. In line with IPCC (2023) projections, plant assemblages should be able to withstand heatwaves, prolonged droughts, and intense rainfall events. Thoughtful plant *taxa* selection thus represents a key link between horticultural practice and climate adaptation policy, as combining drought-tolerant succulents with stress-resilient grasses and forbs increases functional diversity.
5. Maintenance. Plant *taxa* should be selected based on the level of maintenance owners can realistically provide over time. From an eco-sustainable perspective, plant *taxa* should have low water requirements and slow growth rates.

Synecological, structural, and auto-ecological characteristics should be integrated to identify *taxa* that are potentially suitable, based on their syntaxonomical classification and ecological behaviour (Caneva et al. 2015).

9.2.3 Specific Benefits of Green Roofs

Plant-based GRs provide multifunctional benefits that directly support urban climate adaptation. Vegetation evapotranspiration reduces the temperature of irrigated roof surfaces, mitigating UHI (EEA 2021). In addition to its thermal-isolating function, the root–substrate layer delays stormwater runoff, reducing sewer overflow risk during heavy rainfall events (IPCC 2023). In terms of biodiversity, GRs create habitat for insects, reptiles, and birds, reinforcing pollination networks disrupted by urbanization, increasing plant richness and diversity. GR insulation effect enhances indoor thermal comfort, lowering energy consumption for cooling and heating. Moreover, photosynthetic and microbial processes within the substrate contribute to carbon sequestration and to the capture of pollutants, improving air quality. By integrating vegetated roofs into the urban fabric, cities implement cost-effective NbS that strengthen resilience to climate extremes while improving quality of life and biophilic design (EC 2021).

9.2.4 Maintenance and Monitoring

Maintenance strategies for GRs vary by roof type. EGR systems require limited interventions such as seasonal inspections, weed removal, and, when irrigated, monitoring of substrate moisture, whereas IGRs demand more frequent management, including irrigation, pruning, weed, and nutrient control. Regular biodiversity assessments support the understanding of vegetation dynamics and enable early detection of stress related to climatic anomalies. Maintenance schedules should therefore be adapted to projected increases in temperature and rainfall variability to sustain system performance under climate change scenarios (IPCC 2023).

To optimize management, GR monitoring integrates complementary techniques to evaluate vegetation health, hydrological performance, and microclimatic behaviour. Continuous measurements are obtained from sensor networks (e.g., soil moisture, temperature, and environmental probes) to track water retention, evapotranspiration, and stormwater-mitigation efficiency (Gaffin et al. 2009). Non-destructive assessment of plant vitality and early stress detection are supported by hyperspectral and vegetation-index approaches, including the use of indices such as the Normalized Difference Vegetation Index (NDVI) and Structure Insensitive Pigment Index (SIPI) (Piro et al. 2017), as well as drone-based and multispectral imaging to map canopy condition, biomass distribution, and spatial vegetation dynamics (Moroni et al. 2019).

Recent advances in low-cost Internet of Things (IoT) technologies enable the integration of wireless soil-moisture sensors, temperature probes, and RGB-depth imaging to automate real-time monitoring of environmental and biological parameters, thereby enhancing the sustainability and productivity of vegetated roofs (Akhie and Joksimovic 2023). These digital tools are complemented by physical and micrometeorological techniques, which provide detailed measurements of evapotranspiration and heat fluxes and improve understanding of interactions among vegetation, substrate, and the atmosphere (Raimondi and Becciu 2021). Together, integrated maintenance and monitoring strategies support the long-term resilience, efficiency, and ecological functionality of GRs in urban contexts.

9.2.5 Challenges and Limitations

The success of GRs is constrained by structural load capacity, maintenance costs, and plant adaptability to climatic extremes. Persistent droughts and heat waves may exceed species tolerance, requiring adaptive planting or supplemental irrigation. A 30-year simulation study revealed 70–120 days of annual drought plant stress in semi-arid regions, despite optimized substrate depth and runoff zones (Soni et al. 2025). Economic limitations concern installation costs and access to specialized materials. In several European cities, economic incentives and supportive regulatory measures have been introduced to encourage the adoption of green roofs and vegetated building solutions, helping to offset installation costs and reduce administrative barriers. Biological challenges include root competition, nutrient depletion, and potential colonization by invasive species in semi-intensive GRs (Todeschini and Fett-Neto 2025). From a regulatory perspective, heterogeneity across EU Member States complicates standardization, though initiatives under the Nature Restoration Law (EU 2024) and EU Adaptation Mission (EC 2021) promote unified guidelines for NbS implementation (EEA 2021). Uncertainty in regional climate projections (IPCC 2023) underscores the need for flexible design and continuous monitoring, as demonstrated by Nature's 2024 multi-city modelling of UHI mitigation and EEA-reported temperature reductions (~1.7 °C) in Rethymno and Madrid (Adilkhanova et al. 2024; EEA 2024). These examples confirm that botanical components, when paired with structural adaptation, policy incentives, the selection of greenhouse-tolerant species, and long-term monitoring, can significantly enhance urban resilience over time.

9.2.6 Case Studies

9.2.6.1 Native Species to Enhance ES, Keplero High School (Rome, Italy)

To provide a comprehensive perspective on species-specific adaptations and the delivery of ecosystem services under challenging urban conditions, this synthesis integrates findings from studies conducted on an EGR in Rome, Italy (Bellini et al. 2024b, 2025a, b). Experiments were carried out on the Keplero High School green roof (Fig. 9.2), characterized by a Mediterranean climate with hot, dry Summers and mild, wet Winters. The substrate complied with UNI EN11235:2015 standards and consisted of mineral aggregates (lapillus, pumice, zeolite) and organic matter (peat), with a depth of 12–15 cm and water retention capacity of ≥85 L/m^2. Three complementary experimental approaches were adopted to investigate the GR ecological and functional performance (Table 9.2): (1) species resilience, analysing flowering dynamics, mortality rates, and runoff mitigation; (2) vegetation dynamics, assessing spontaneous colonization and biodiversity trajectories in response to abiotic factors; (3) comparative performance of native species versus *Sedum album* L.

Across all investigations, native Mediterranean species showed higher resilience than *Sedum album* under Mediterranean climatic stress. *Convolvulus cneorum* L. and *Helichrysum italicum* maintained stable health and dense canopies during prolonged drought, while *Thymus serpyllum* (Roth) G.Don, and *Thymus vulgaris* L. had <10% mortality under the most restrictive irrigation regime (10% ET_0). In contrast, *Cerastium tomentosum* L. and *S. album* showed mortality rates exceeding 80% under severe water stress. Flowering periods shifted under drought conditions, with several species blooming earlier or for longer periods. *T. serpyllum* and *Saponaria ocymoides* L. flowered continuously across irrigation regimes, while *H. italicum* and *Anthemis maritima* L. provided extended floral resources,

Fig. 9.2 The display of the experimental roof on the Keplero school in Rome, Italy (**a**). Examples of the containers for experimentation at the planting stage (**b**, **c**). (Photos by F. Bartoli)

Table 9.2 Experimental setup, maintenance, and monitoring at the Keplero school roof (ET = Evapotranspiration), Rome, Italy

Trial time	Species	Maintenance	Monitoring
2017–2019	*Thymus serpyllum* L., *Teucrium chamaedrys* L., *Saponaria ocymoides* L., *Lavandula stoechas* L., *Cerastium tomentosum* L., *Allium schoenoprasum* L.	10%, 20%, and 40% ET_0 irrigation	Flowering, mortality, water retention
2020–2022	Spontaneous colonization	No maintenance	Vegetation, substrate depth, solar exposure, seasonality
2023–2024	*Sedum album* L., *Helichrysum italicum* (Roth) G.Don, *Convolvulus cneorum* L., *Anthemis maritima* L., *Thymus vulgaris* L.	10%, 20%, and 40% ET_0 irrigation. Frequency of 5 min *per* night during the hottest months (from June to September)	Thermal regulation, plant health, reproduction, coverage

enhancing pollinator support. In contrast, *S. album* exhibited limited flowering and reduced biodiversity value. Mediterranean shrubs such as *C. cneorum* and *H. italicum* developed dense, persistent canopies that improved shading and reduced evapotranspiration. Heatmap analyses indicated stable cover for these species, whereas *Sedum* plots showed sparse and declining patterns.

This structural differentiation directly affected hydrological performance: GR retained 77–96% of rainfall during spring and summer, with retention decreasing as plant mortality increased. Intermediate irrigation (20% ET_0) provided the most effective balance between vegetation health and runoff reduction, confirming the synergistic role of vegetation and substrate in stormwater management. Following system abandonment, spontaneous colonization involved 62 species, mainly therophytes and ruderals, driven by microhabitat heterogeneity related to substrate depth and shading. *T. serpyllum* and *Teucrium chamaedrys* L. survived without irrigation, proving their suitability for low-maintenance GRs. Overall, this case study highlights how plant functional traits influence hydrological performance, thermal regulation, and biodiversity outcomes, providing practical insights for the design of resilient GRs in Mediterranean climates.

9.2.6.2 Habitat Template, SMS Library (Pisa, Italy)

The SMS library is located in the Le Piagge urban park in Pisa and was inaugurated in 2013. The roof-space was originally planted with aromatic Mediterranean species. Due to the absence of irrigation in the subsequent years, the plants died, and the roof was spontaneously colonized by herbaceous species (Table 9.3).

This vegetation survives during Winter and Spring and dies down in Summer. Monitoring revealed that the most common plant functional types are annuals and Gramineae. The roof vegetation can be considered a proxy of the potential colonizers of local EGRs (Fig. 9.3).

In autumn 2024, species harvested from a template habitat (the garigue of Monte Pisano) were seeded, and in autumn 2025, 1000 bulbs (commercial origin) of species from the same environment were planted. This project aimed to increase the GR biodiversity while involving people with knowledge of local landscapes and natural sites.

9.2.6.3 GR Substrate Depth and Management as Driver of Vegetation Establishment (Genoa, Italy)

A total of 60 plant *taxa* were recorded during a study of two MGRs with different substrate types in Genoa, Italy (Fig. 9.4). The recorded species were mainly represented by Asteraceae, Fabaceae and Poaceae (40% of the total *taxa*), and belonged to 13 different (CSR) ecological strategies in response to stress and disturbance (Hodgson et al. 1999). The study demonstrated that the taxonomic and functional assemblages of plant communities across the two MGRs consistently differentiate

Table 9.3 Wild Plant species and functional types colonising the SMS roof

Species	Family	LF functional type	Seed dispersal method
Acer negundo L.	Sapindaceae	F/tree	Anemochory
Andryala integrifolia L.	Asteraceae	T/forb	Anemochory
Anisantha madritensis (L.) Nevski	Poaceae	T/Graminae	Anemochory/Zoochory
Blackstonia acuminata (W.D.J.Koch and Ziz) Domin	Gentianaceae	H/forb	Anemochory /Autochory
Blackstonia perfoliata (L.) Huds.	Gentianaceae	T/forb	Anemochory
Crepis capillaris (L.) Wallr.	Asteraceae	T/forb	Anemochory
Crepis sancta cfr. subsp. *nemausensis*	Asteraceae	T/forb	Anemocory/Zoochory
Euphorbia maculata L.	Euphorbiaceae	T/forb	(Mirmechory)
Festuca danthonii Asch. and Graebn.	Poaceae	Ch/Graminae	Anemochory
Hypericum perforatum L. subsp. *perforatum*	Hypericaceae	H/forb	Anemochory
Hypochaeris glabra L.	Hypericaceae	H/forb	Anemochory
Lysimachia arvensis (L.) U.Manns and Anderb.	Primulaceae	T/forb	Idrochory/Autochory/ Zoochory
Senecio vulgaris L.	Asteraceae	T/forb	Anemochory
Trifolium arvense L.	Fabaceae	T/legumes	Autochory/Zoochory/ Anemochory
Trifolium campestre Schreb.	Fabaceae	T/legumes	Autochory/Zoochory/ Anemochory

Fig. 9.3 The garigue of the Monte Castellare (Italy) habitat template (**a**) for the SMS EGR (**b**). (Photos by S. Benvenuti)

over time in response to substrate properties, an effect that is, in turn, mediated by other biotic and abiotic factors, particularly human management. This differentiation was evident from the first stage of MGR substrate colonisation to the mature condition of plant communities. From a planning perspective, a functional trait-based approach for MGR species selection is highly recommended, considering both site-specific factors and expected human management (Pianta et al. 2025).

Fig. 9.4 Mediterranean green roofs in Genoa. MGR with shallower substrate (**a**) composed of a growing mix and lapillus, with prevalent Fabaceae, Asteraceae, and Poaceae *taxa*; MGR with deeper substrate (**b**) composed of inert clay material, with dominant Poaceae *taxa*. (Photos by M. Pianta)

9.3 Green Façades and Walls (T.M.B., C.M.M., E.R., G.S.)

9.3.1 Definition and Classification

Green façades and living walls are the two main typologies of Vertical Greenery Systems (VGS), a broad category that includes vertical gardens, green walls, bio-shaders, and other vegetated vertical surfaces (Dunnett and Kingsbury 2008; Safikhani et al. 2014). Both typologies are increasingly adopted in architecture and urban design to address environmental challenges and improve the quality of the built environment.

Green façades are systems based on climbing or self-clinging plants that grow vertically on or adjacent to buildings. Plants are typically rooted at ground level or in planter boxes and supported by façades, cables, meshes or trellis structures (Perini et al. 2013). According to their construction principle, green façades can be classified as direct, where plants attach directly to the building surface, or indirect, where a dedicated support structure creates a separation between the vegetation and the façade (Fig. 9.5). These systems rely primarily on natural plant growth rather than complex technological components, resulting in relatively low installation and maintenance requirements.

When properly designed and maintained, green façades are characterized by high durability and longevity, with a lifespan that may exceed 50 years (Perini et al. 2013). Many climbing species exhibit rapid vertical growth, reaching heights of 3–10 m within the first 4 years (Köhler 2008), enabling them to gradually cover large surfaces. From an environmental perspective, green façades provide shading, reduce surface temperatures, and regulate microclimate, while also supporting urban biodiversity by providing habitats for fauna.

Living walls, also referred to as green walls or, in a stricter sense, vertical gardens, differ substantially in their technical configuration. They consist of modular

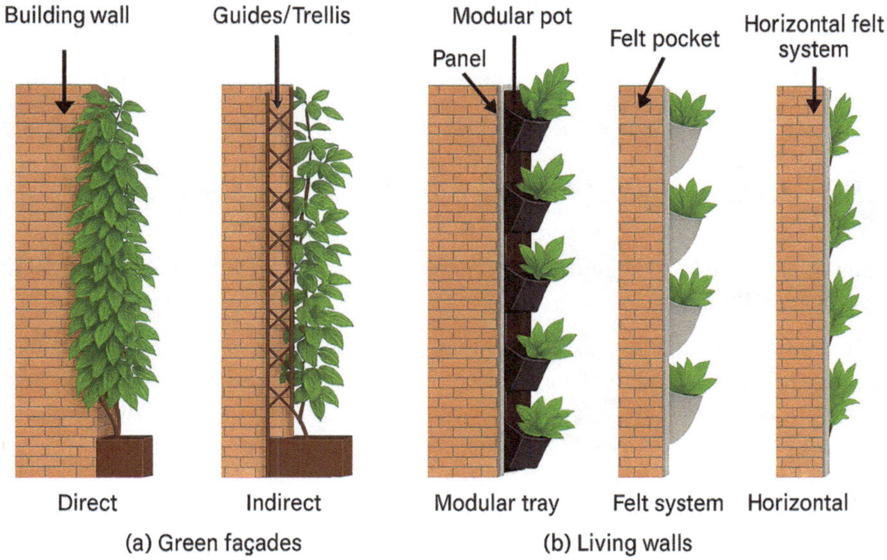

Fig. 9.5 Schematic diagram of examples of different VGS types. (Image AI-generated from a picture with Microsoft 365 Copilot)

systems fixed directly to vertical surfaces, and incorporate planter boxes, trays, felt layers or panels filled with growing media, supplied by mechanised irrigation and, in some cases, fertigation systems. This configuration allows plants to be installed independently of ground-level soil conditions.

A key characteristic of living walls is their high design flexibility. Modular systems can host a wide range of plant species, including herbaceous, perennial, and shrubby *taxa*, and can be arranged in diverse spatial patterns. Both substrate-based and hydroponic planting solutions are commonly adopted, enabling precise control of water and nutrient supply (Perini et al. 2013) and supporting architectural expression and indoor applications.

Despite these characteristics, living walls are technically more complex to install and maintain than green façades. Their lifespan depends on system design and materials; felt-based systems may last around 10 years, whereas modular systems made of high-density polyethylene (HDPE) components can reach lifespans comparable to green façades (Perini et al. 2013).

Both green façades and living walls are increasingly recognised as effective tools for climate change adaptation and sustainable urban development. At the European level, VGS are identified as NbS within key policy frameworks, including the EU Climate Adaptation Mission (EC 2021) and the European Environment Agency framework (EEA 2021). Their benefits include improved air quality, mitigation of the UHI effect, reduced building energy demand through shading and insulation, and enhanced urban resilience to climate change.

Although a fully standardized framework for VGS design and implementation remains lacking, broader sustainability standards, such as ISO 16813:2024, can support the integration of vertical greenery into holistic climate adaptation strategies.

9.3.2 Methodological Approach for Plant Taxa Selection

Plant selection in VGS is guided by eco-physiological, morphological, and climatic parameters (Table 9.4):

1. Façade orientation. Exposure determines the microclimate and the related species' tolerance to thermal and radiative stress: heliophilous and xerophytic *taxa* are suited to sun-exposed walls, whereas sciaphilous and mesophytic *taxa* perform better under shaded conditions.
2. Climbing and rooting traits. Adhesive or adventitious roots or twining stems affect cover density, wall protection and management requirements. In green façades that rely on climbing plants, species with adhesive organs are commonly used (e.g., *Parthenocissus tricuspidata* (Siebold & Zucc.) Planch. and *Hedera helix* L.).
3. Substrate type and irrigation systems. Hydroponic systems require plant *taxa* capable of rapid root turnover and tolerance to fluctuating water availability, while substrate should match species preferences (e.g., calcareous *taxa* on neutral-basic substrates, and acidophilous *taxa* on acidic or sub-acidic substrates).
4. Native and climate-resilient *taxa*. Using local, drought-tolerant species supports long-term adaptability under projected climate scenarios (IPCC 2023) and enhances biodiversity. *Taxa* selection based on functional traits improves VGS resilience, particularly under drought and high solar radiation.

Recent research highlights the importance of matching plant functional traits to ecosystem services. For instance, particulate matter (PM) removal efficiency varies widely across species, as leaf micromorphology, particularly surface waxes and roughness, strongly influences particle deposition (Perini et al. 2017). VGSs mitigate gaseous pollutants and fine dust, thereby improving local air quality and reducing heat-induced emissions (Perini and Roccotiello 2018).

Such species-specific plant and leaf functional traits are also effective in black carbon capture and NO_2 uptake (e.g., *Taxus baccata* L. and *Myrtus communis* L.) (Bosio et al. 2025). An operational model for selecting plant species for green walls under water scarcity and thermal extremes is available, based on a multi-stage screening of native national flora (with specific reference to protected habitats) and ecological criteria, including chorological type, life form, ecological indicator values, altitudinal range, and habitat type (Patti et al. 2025).

Finally, the use of non-native species requires caution. Several studies indicate that many invasive alien species were initially introduced for ornamental purposes

Table 9.4 Recommended plant traits for vertical greening systems synthesized from literature

VGS type	Construction and technology	Rooting system	Key plant functional traits	Main ecosystem services provided
Direct green façade	Plants grow directly on the façade; no intermediate structure; minimal technology	Ground- or planter-rooted at façade base	Self-clinging or adhesive habit; high mechanical resistance; slow–moderate growth; long lifespan; tolerance to façade microclimate	Thermal regulation (shading, surface cooling); long-term durability; low maintenance intensity
Indirect green façade	Cables, meshes or trellises detached from façade	Ground- or planter-rooted	Twining or climbing habit; flexible stems; rapid vertical growth; pruning tolerance	Thermal regulation (shading); façade protection; improved inspection and maintenance access
Indirect green façade (planter-based)	Planter boxes integrated into façades or balconies; structural support required	Confined rooting volume in containers	Tolerance to restricted root space; moderate drought tolerance; compatibility with irrigation; controlled biomass	Design flexibility; suitability for retrofits; medium maintenance intensity
Modular living wall (substrate-based)	Prefabricated panels/modules with engineered substrate	Artificial rooting volume within modules	Compact growth; shallow fibrous roots; tolerance to limited substrate; modular replaceability	Immediate visual impact; thermal regulation; biodiversity support; medium–high maintenance intensity
Living wall—felt-based	Synthetic felt layers; no traditional substrate	Soil-free system (roots embedded in felt)	Rapid root turnover; high phenotypic plasticity; dependence on continuous irrigation; short lifespan	Lightweight vertical greening; indoor/outdoor use; high technological dependence
Living wall—hydroponic	Soil-free systems with automated nutrient supply and recirculation	Soil-free system (exposed roots)	Tolerance to fluctuating water and nutrient availability, high uptake efficiency, and fast physiological response	Precise growth control; air-quality improvement; high cooling potential; high management intensity
Indoor living wall	Systems integrated into interior building environments	Substrate-based or soil-free	Shade tolerance; adaptation to stable temperatures; ornamental foliage; reduced transpiration stress	Indoor environmental quality improvement: biophilic benefits
Architectural living wall	Highly engineered, design-driven systems	System-dependent	Predictable growth form; tolerance to frequent maintenance or replacement	Strong aesthetic and symbolic value; high management intensity

in urban environments, from which they subsequently spread into natural and semi-natural habitats (Caddeo et al. 2020; Spampinato et al. 2022).

9.3.3 Specific Benefits of Green façades and Walls

VGSs can reduce façade surface temperature through combined shading and evapotranspiration effects, with reported reductions up to approximately 10–15 °C compared to conventional shading devices (e.g., Bao et al. 2022) and can contribute to a 10–20% decrease in building cooling demand, under favourable conditions (Mohammed 2022). Cooling effect varies with *taxa* and planting density, with high-transpiration *taxa* showing enhanced effects (Kim et al. 2023; Lin et al. 2024).

VGSs also mitigate the UHI effect and improve outdoor thermal comfort by intercepting solar radiation and enhancing convective airflow, thereby lowering the ambient and mean radiant temperatures at the pedestrian level (Bakhshoodeh et al. 2022).

From an ecological perspective, vertical greenery supports local biodiversity and ecological connectivity by providing habitats and resources for pollinators (Dunnett and Kingsbury 2008; Treder et al. 2024). Using native flowering *taxa* ensures nectar and pollen availability while limiting maintenance requirements.

Beyond environmental functions, visible vegetation is associated with improved psychological well-being and perceived environmental quality, contributing to overall urban liveability (Fonseca et al. 2023).

9.3.4 Maintenance and Monitoring

Maintenance intensity depends on system typology. Green façades are relatively low input, requiring periodic pruning, inspection, and weed removal. Living walls are high-input systems that demand continuous irrigation, nutrient management, and sensor-based monitoring.

Maintenance strategies may integrate soil-moisture sensors, automated irrigation, and remote NDVI monitoring. IoT-based controls can optimize water use, crucial under increasing drought risk, and facilitate adaptive maintenance scheduling. Routine biodiversity audits can monitor pollinator activity and species vitality, ensuring consistent ecological performance under changing climatic conditions.

Modular green wall systems with compact, easily replaceable modules are preferable because they facilitate maintenance: nursery-prepared modules can quickly replace sections of the green wall when gaps or weed growth are observed.

9.3.5 *Challenges and Limitations*

VGS faces both technical and biological constraints. Plant stress from high radiation, wind, and desiccation is expected to intensify with climate change (IPCC 2023). Hydroponic walls risk nutrient imbalance and rapid plant decline in the case of irrigation failure. Economically, high installation and maintenance costs limit widespread adoption, particularly for living wall systems, while limited user acceptance may further constrain implementation.

Botanical challenges include the limited availability of native vertical species and knowledge gaps on long-term pollutant uptake and evapotranspiration performance. Further experiments on species-specific biofiltration and cooling mechanisms are thus required (Bosio et al. 2025; Hoffmann et al. 2021). The Nature Restoration Law (2024) and the Urban Greening Platform (EEA 2021) are expected to enhance standardization and support the scaling up of NbS applications in urban regeneration.

9.3.6 *Case Studies*

9.3.6.1 Habitat Models Living Wall (Mediterranean University of Reggio Calabria, Italy)

The habitat models living wall in Reggio Calabria integrates native and local *taxa* within the PRIN2022-PNRR P202257LNE project "Nature-based solutions for resilient cities using green surfaces in buildings", focusing on climate adaptation and sustainability in Mediterranean urban environments. The use of local plant *taxa* in VGS provides ecological benefits by supporting local biodiversity, reinforcing landscape identity, and reducing risks associated with the introduction of non-native *taxa*. Designing living walls that mimic the ecological conditions of natural habitats is an effective strategy for increasing urban biodiversity and conserving native flora through site-specific, context-sensitive solutions.

For Calabrian cities, plant selection was based on *taxa* characterised by compact growth, high drought tolerance, and rapid vegetative recovery. Selected species include *Lavandula multifida* L., *Jacobaea maritima* (L.) Pelser and Meijden subsp. *gibbosa* (Guss.) Peruzzi, N.G. Passal. and C.E. Jarvis, *Ptilostemon gnaphaloides* (Cirillo) Soják subsp. *gnaphaloides* (Fig. 9.6), and *Helichrysum italicum* (Roth) G. Don subsp. *italicum*. Plants' selection was also linked to the typology of the living wall: cascading species such as *Dianthus rupicola* Biv. are suited to systems with protruding elements, while species such as *Brachypodium retusum* (Pers.) P. Beauv. provide effective ground cover.

Specific attention was given to endangered endemic species with proven vitality and germination capacity, including *Crepis aspromontana* Brullo, Scelsi, and Spamp. and *Pimpinella anisoides*, V. Brig., which can be incorporated into living

Fig. 9.6 Endemic species from Calabria (Italy) at risk of extinction adaptable for a Mediterranean living wall: *Ptilostemon gnaphaloides* (Cirillo) Soják subsp. *gnaphaloides* (Asteraceae) (**a**) *Salvia ceratophylloides* Ard. (Lamiaceae) (**b**). (Photos by C. M. Musarella)

walls as part of an *ex-situ* conservation strategy (Panuccio et al. 2018; Laface et al. 2023a, b, 2024; Patti 2024). Additional endemic *taxa* at risk of extinction, such as *Silene calabra Brullo*, Scelsi and Spamp., Ard., and *Dianthus virgatus* Pasq. were also considered (Fig. 9.6).

Fire safety represents a critical design constraint for VGSs, particularly in Mediterranean climates. Flammable species may pose a risk if maintenance is inadequate; therefore, design solutions emphasise fire-resistant construction materials, controlled plant biomass, and adequate moisture levels, while resinous *taxa* should be avoided. Plant *taxa* with allergenic pollen or invasive behaviour should also be avoided to protect public health, ensure social acceptance, and prevent uncontrolled spread within the system or in the surroundings.

This approach avoids standardised solutions, favouring designs that are ecologically sound, visually distinctive, and applicable to other contexts characterized by water scarcity and thermal extremes. Combining *taxa* with diverse growth forms, colours, and flowering periods increases both ecological and aesthetic value.

Social and perceptual indicators are also considered essential for evaluating impacts on urban quality of life. Participatory design processes support public acceptance, long-term effectiveness, and alignment with local needs and perceptions.

9.3.6.2 Native *Taxa* and Microclimate Regulation CaixaForum (Madrid, Spain)

The CaixaForum Madrid living wall (Fig. 9.7) represents a large-scale example of VGS that contributes to microclimate regulation while supporting local biodiversity. Installed in 2007 and designed by Patrick Blanc, it covers approximately 460 m^2 and hosts over 250 plant *taxa*, comprising about 15,000 plants (World Green Infrastructure Network 2021). Plant *taxa* were selected to thrive in Madrid's continental Mediterranean climate (i.e., hot, dry summers and cold winters), including ferns, succulents, herbaceous perennials, and ornamental taxa that can adapt to soilless vertical growth (World Green Infrastructure Network 2021).

Taxa selected emphasise shallow root systems, tolerance to water variability, and functional diversity, combining species with different growth habits to ensure dense cover and year-round visual interest.

Environmentally, the living wall contributes to façade shading, evapotranspiration, and thermal-extreme buffering. Ecologically, the diverse assemblage of plant *taxa* provides microhabitats for animals.

From a social perspective, the green mosaic of the CaixaForum living wall has become an urban landmark, enhancing visual quality and reinforcing the integration of nature within the urban fabric.

Fig. 9.7 Living wall by Patrick Blanc in CaixaForum, Madrid, Spain. (Photo by Nemo, CC-BY-SA-4.0)

9.3.6.3 myBUILDINGisGREEN - Facade Prototypes Case Study (Portugal and Spain)

The application of vegetated façade systems to increase the climate resilience of educational buildings in Portugal and Spain was investigated within the project LIFE myBUILDINGisGREEN (LIFE17 CCA/ES/000088—LIFE mBiG), which ran from 2018- to 2024. Focusing on educational centres in Southern Europe (Portugal and Spain), the project aimed to improve the climate change resilience of school buildings constructed during the 1970s–1980s (Batista et al. 2022a, b), which are often characterised by limited thermal performance and scarce integration of green infrastructure.

Three schools were selected as pilot sites: Gabriela Mistral Elementary School (Badajoz, Spain); Horta das Figueiras Primary School (Évora, Portugal) and Falcão Elementary School (Porto, Portugal). For each case, architectural and technical retrofit projects were developed, integrating NbS prototype design. The interventions aimed to improve indoor comfort, reduce cooling demand, optimise water use, and enhance air quality and urban biodiversity.

9.3.6.4 Shading of façade, FAVE Prototype, Solana de Los Barros (Spain)

The Vegetal Façade (FAVE) prototype was designed as a substructure fixed to the building façade to support deciduous climbing plant species. Virgin vines (*Parthenocissus tricuspidata* and *P. quinquefolia*) were selected for their seasonal performance, allowing solar radiation to reach the façade during Winter while ensuring effective shading in Summer. The modular and detachable system consists of three-dimensional metal profile structures with standardised dimensions assembled by screwing. This configuration facilitates replication, adaptability, and application to different building types.

9.3.6.5 mBIGToldo Prototype, Évora (Portugal)

Developed by the CARTIF team, the system consists of a low-thickness vegetated vertical surface designed to provide shading and increase ambient humidity. The system (Fig. 9.8) includes an impermeable backing sheet, a nonwoven felt layer, and a semi-woven substrate. Hydroponic drip irrigation is integrated into the supporting structure to optimise water use efficiency. Suitable plant species include *Festuca rubra* L., *Agrostis stolonifera* L., and *Sagina hawaiensis* Pax, selected for their compatibility with shallow substrates and controlled irrigation regimes. The main components comprise: i) a support frame sized to accommodate the prototype; ii) a waterproof support layer resistant to puncture and tearing; iii) a root-fixing layer based on non-woven felt or rock wool, and iv) a substrate-seed mixture applied by hydroseeding.

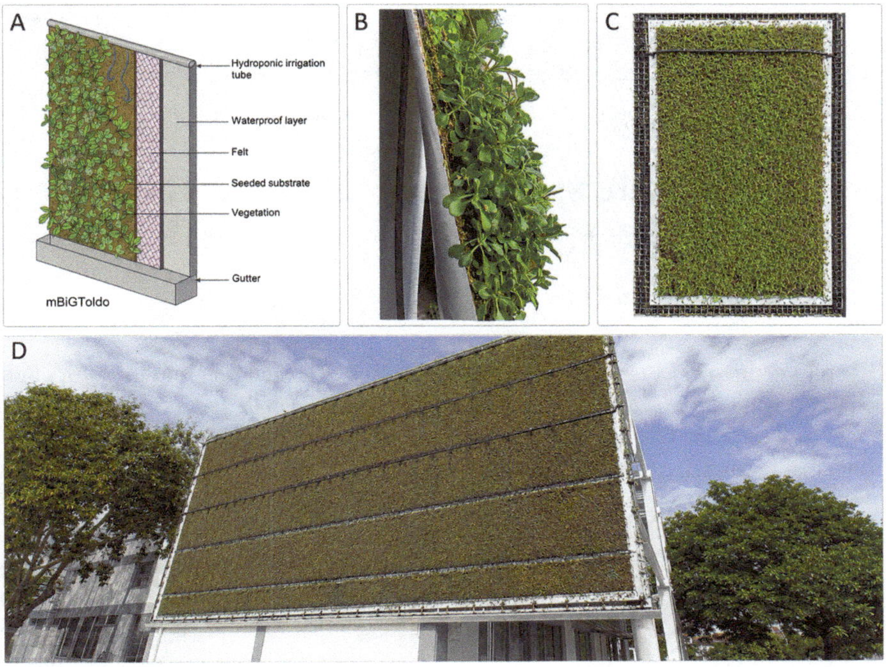

Fig. 9.8 mBiGToldo scheme and implementation (CARTIF and CIMAC) (**a–c**) and complete façade. (Photos by M.T. Batista and J. Sardinha)

9.3.6.6 Green Façade, Wire Rope Structure, Porto (Portugal)

Vegetation was planted at ground level and guided along the building façade by vertical steel cables extending across multiple storeys. An upper-level balcony creates an air gap between the vegetated façade and windows, improving shading efficiency and thermal performance. Deciduous species were selected to provide summer shading, while winter leaf drop allows solar radiation and daylight to reach interior spaces.

Deciduous species were selected to provide summer shading, while winter leaf drop allows solar radiation and daylight to reach interior spaces.

Across all façade-based prototypes, improvements were observed in building energy efficiency, irrigation water optimisation, and heatwave impact mitigation, including reduced indoor air temperatures, as well as an overall improvement in air quality.

Implementation challenges are mainly related to limited experience with VGS, including uneven hydroseeding application and non-uniform vegetation growth, overheating of steel cables during summer, and occasional irrigation malfunctions. These issues can be addressed through controlled monitoring and adaptive management.

Overall, the case studies confirmed that VGSs supported by appropriate botanical diversity contribute effectively to thermal regulation and pollution control, reinforcing their role in climate adaptation and urban wellbeing.

9.4 Bioarchitecture, Street Furniture and Indoor Greening (E.R., M.P.)

9.4.1 Definition and Classification

Bioarchitecture refers to the intentional integration of vegetation into the built environment, where plants function as structural, environmental, and experiential components of buildings and public spaces. This interdisciplinary field rests on two key concepts: biomimicry, which uses nature as a source of design inspiration, and biophilia, which emphasizes the essential role of human–nature connections in supporting health and well-being (Nash et al. 2016; Wilson 1984).

In contemporary cities, characterized by extensive soil sealing and increasing exposure to climate-related stressors, this field extends urban green infrastructure into balconies, terraces, indoor environments, and street furniture elements, embedding ecological processes directly into everyday living and working spaces (Aronson et al. 2017; Wang et al. 2024).

Within this framework, the strategic use of vegetated architectural elements functions as small-scale NbS, contributing to climate adaptation, supporting biodiversity, and enhancing human well-being (Wang et al. 2024; Lima et al. 2025). Their effectiveness, therefore, depends on cumulative implementation across buildings and neighbourhoods (Aronson et al. 2017).

From a conceptual standpoint, bioarchitecture encompasses three complementary perspectives. The first is urban green infrastructure (UGI), which emphasises multifunctionality, connectivity, and adaptive management. From this perspective, balconies, terraces, and vegetated street furniture can function as nodes within a broader ecological network (Aronson et al. 2017; Cabanek et al. 2020; Wang et al. 2024). The second perspective is biophilic design, which emphasises the importance of repeated and meaningful contact with nature for psychological wellbeing, stress reduction, and cognitive restoration (Lima et al. 2025). Outdoor and indoor vegetated architectural elements provide continuous, everyday exposure to natural elements, fulfilling a key requirement of biophilic environments (Wilson 1984; Wijesooriya et al. 2023). The third perspective concerns indoor ecosystem services, recognising that vegetation can contribute to indoor environmental quality and comfort when appropriately integrated into building systems (Zorić et al. 2019).

Bioarchitecture operates from building to urban scale, supporting comfort, biodiversity, and climate adaptation, particularly where ground-level green space is limited (Wang et al. 2024).

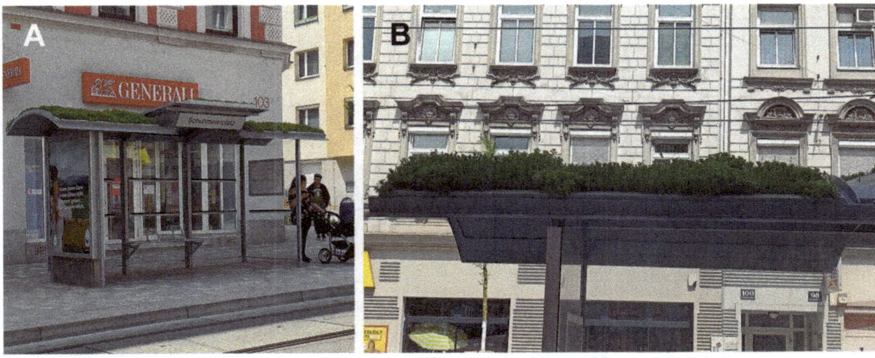

Fig. 9.9 Bus stops at Vienna (Austria) with living roofs for pollinators' support. (Photos by C. Conte)

Street furniture represents a highly visible and socially significant component, and the use of vegetated benches, planters, pergolas, and vegetated bus shelters (e.g., Fig. 9.9) transforms conventional grey infrastructure into multifunctional elements that improve pedestrian comfort, provide micro-habitats for local flora and fauna, and enhance place quality (Cabanek et al. 2020). Such interventions are especially relevant in streetscapes where underground utilities, limited soil volumes, and competing land uses constrain tree planting (Ordóñez Barona and Trammell 2022).

Indoor greening incorporates bioarchitectural principles into enclosed environments, acknowledging that urban populations spend most of their time indoors. Indoor plants, living walls, and hydroponic systems support the biophilic experience and improve indoor environmental quality (Liu et al. 2022a; Priya and Senthil 2025). This aligns with EU adaptation strategies (EEA 2021) and with ISO 16813:2024, which recognises vegetation as a component of healthy indoor environments.

9.4.2 Methodological Approach for Plant Taxa *Selection*

The success of bioarchitecture and vegetated street furniture systems hinges on selecting appropriate plant *taxa*, which can withstand constrained growing conditions, altered microclimates, and long-term, intensive maintenance regimes. Therefore, plant selection requires an ecophysiological and functional trait-based approach rather than a purely aesthetic one (Table 9.5) (Farrell et al. 2022).

Light availability is a primary determinant of plant suitability. South- and west-facing balconies and terraces are exposed to direct solar radiation, high temperatures, and increased evaporative demand, favouring heliophilous and drought-tolerant *taxa*. In contrast, north-facing balconies, shaded courtyards, and indoor environments require sciaphilous *taxa* capable of maintaining physiological function and acceptable visual quality under low irradiance (Liu et al. 2022b). Plants within

Table 9.5 Examples of recommended plant traits for balcony, terrace, and indoor conditions by exposure and substrate type

Space type	Exposure	Substrate depth and type	Functional rationale
Balcony (high-rise)	North-facing	Deep containers, engineered substrate, continuous irrigation	Evergreen *taxa* ensure year-round cover, wind resistance, and visual buffering; suitable for lower solar radiation and stable moisture conditions
Balcony (high-rise)	South- and West-facing	Deep containers, structurally reinforced substrate	Deciduous *taxa* provide seasonal shading, solar regulation, and summer cooling through evapotranspiration
Balcony/ Terrace	Mixed exposure	Medium-depth substrate (20–40 cm), regular but limited irrigation	High tolerance to solar radiation and drought; structural continuity with reduced maintenance
Terrace (extensive)	Full sun	Shallow substrate (≤20 cm), low water availability	Stress-tolerant *taxa* suited to high solar exposure; contribute to thermal buffering and biodiversity with minimal inputs
Terrace (intensive)	Full sun to partial shade	Deep substrate (>40 cm), higher water retention	Increased ecological complexity, shading capacity, and stormwater retention where structural load allows
Indoor (adjacent to terraces)	Indirect light	Containers with controlled substrate and drainage	Visual continuity between indoor and outdoor spaces; enhancement of biophilic experience
Indoor (high daylight)	High solar input	Containers with high drainage capacity	Tolerance to higher light levels and temperature fluctuations near glazed façades

indoor systems must additionally cope with artificial lighting regimes, reduced air movement, and humidity fluctuations driven by heating, ventilation, and air conditioning (HVAC) operation (Liu et al. 2022a).

Wind exposure is another constraint that plants must adapt to. Balconies, particularly at higher elevations, are exposed to strong winds that increase mechanical stress and water loss. Root architecture and growth form are equally critical selection criteria. Shallow or fibrous root systems are essential for containerised balconies and street furniture, where substrate depth is limited, and water availability can fluctuate rapidly. Terrace gardens allow for deeper rooting volumes and therefore support a wider range of growth forms, including shrubs and small trees. According to functional-ecological principles, functional diversity, achieved by combining different life forms and phenological strategies, enhances resilience and stabilises the delivery of ecosystem services under stress (Burlando et al. 2025).

Climate change projections further reinforce the need for resilient plant selection. *Taxa* should be chosen for their tolerance to heatwaves, droughts, and variability in humidity expected under future climate scenarios, rather than relying solely on historical climatic conditions (IPCC 2023; Fineschi and Loreto 2020). Genetic diversity and provenance selection become particularly relevant for long-lived woody plants in terrace systems, where replacement would be costly and disruptive.

Maintenance considerations are integral to the selection process. Low-maintenance *taxa* reduce resource consumption and improve long-term sustainability, particularly in residential and public contexts with limited management capacity (Nazif and Yousefi 2024). Indoor greening requires particular attention to survivability and stability, as evidence indicates that the contribution of certain ornamental *taxa* to indoor air quality trends depends on plant health and adequate maintenance (Zorić et al. 2019).

Structural limitations should also be considered when selecting plants. The weight capacity of balconies and terrace floors may prevent the use of large plants and trees, as these require a large amount of soil, which increases in weight when watered. Similarly, available space must be considered to ensure that the adult plant and its root system fit within it. Vertical gardens or vertical-growing plants could be a solution where horizontal space is limited (Priya and Senthil 2025).

Proper plant species selection is essential for different perspectives. For plant health reasons, plant species selection should avoid plants with allergenic or toxic properties. At the same time, due to their proximity to living spaces, the plants used should match the inhabitants' aesthetic preferences (Priya and Senthil 2025). From an ecological conservation standpoint, it is fundamental to avoid invasive species that could compromise the ecological value of other urban green spaces (Priya and Senthil 2025).

9.4.3 Specific Benefits of Bioarchitecture, Street Furniture, and Indoor Greening

One of the most consistently documented benefits of terrace gardens is their ability to regulate microclimates. Vegetated balconies and terraces reduce thermal stress through shading and evapotranspiration, mitigating heat accumulation on façades and in semi-outdoor living spaces (Visvanathan et al. 2024). Terrace gardens are particularly effective due to their larger surface area and deeper substrates, which enhance water retention and evaporative cooling. At street level, vegetated street furniture provides localised shading and cooling in highly sealed environments, thereby improving pedestrian thermal comfort during heatwaves (Ordóñez Barona and Trammell 2022).

Indoor greening extends these benefits into enclosed environments, considering that urban populations spend most of their time indoors. Indoor plants, living walls, and hydroponic systems can improve indoor environmental quality by affecting air composition, humidity level, and perceived comfort (Zorić et al. 2019). Such plants cannot act as a complementary biological component within integrated indoor environmental management strategies, enhancing biophilic experience, visual comfort, and psychological well-being (Wijesooriya et al. 2023).

Another key benefit is water regulation, particularly relevant for terrace and balcony systems, which intercept rainfall, retain water, and delay runoff, thereby reducing peak flows at the building scale. Although the retention capacity of individual systems is limited, their distributed implementation supports decentralised stormwater management in highly sealed urban areas (Aronson et al. 2017). Indoor greening systems, especially hydroponic installations, indirectly promote water efficiency by enabling precise control of irrigation and nutrient delivery, thereby reducing water waste (Liu et al. 2022a).

Despite their small size, bioarchitectural systems and vegetated street furniture can contribute to urban biodiversity by providing microhabitats and functioning as ecological stepping stones within dense urban areas (Cabanek et al. 2020; Fineschi and Loreto 2020). Flowering plants on balconies, terraces, and street furniture provide insects with nectar and pollen resources, while structurally diverse plant assemblages offer shelter and nesting opportunities (Bowler et al. 2025). Evidence from urban ecology indicates that increasing vegetation diversity and incorporating native or functionally appropriate *taxa* enhances biodiversity outcomes, particularly for pollinators (Aronson et al. 2017).

Finally, exposure to nature and everyday interaction with plants are closely linked to psychological and social benefits. Vegetated balconies and terraces provide accessible restorative spaces within residential environments, supporting stress reduction and engagement with seasonal change (WHO 2016; Wijesooriya et al. 2023; Nazif and Yousefi 2024). Indoor greening improves perceived environmental quality and well-being in workplaces and homes, while vegetated street furniture enhances the attractiveness and usability of public spaces. This encourages social interaction and prolonged outdoor activity, particularly in warm climates (WHO 2016). Collectively, these benefits contribute to the multifunctionality and resilience of urban environments when bioarchitecture, street furniture, and indoor greening are implemented as complementary components of urban green infrastructure (Wang et al. 2024).

9.4.4 Maintenance and Monitoring

Maintenance requirements vary by bioarchitectural typology and are a decisive factor in long-term performance (Priya and Senthil 2025). Balcony systems require regular irrigation, seasonal pruning, and pest control, while terrace gardens demand more complex management, including scheduling irrigation, monitoring nutrients, and periodic structural inspections. Indoor greening systems require careful regulation of lighting, watering, and, in the case of hydroponic installations, nutrient concentrations to maintain plant health and functional ecosystem service delivery under indoor environmental constraints (Liu et al. 2022a, b).

Monitoring plays an increasingly important role in ensuring system resilience to climate variability, and long-term performance evaluation is essential for tracking the effectiveness of nature-based solutions in urban green spaces (Cook et al. 2025). Sensor-based approaches, including proximal and remote sensing techniques, enable the comprehensive monitoring of vegetation health and environmental conditions across urban landscapes, thus supporting adaptive management and evidence-based decision-making (Bonifazi et al. 2023). For indoor systems, monitoring of air quality parameters can inform the assessment of ecosystem service delivery and guide maintenance decisions (Zorić et al. 2019). Integrating monitoring technologies with maintenance strategies transforms bioarchitecture from a static design feature into a dynamic system that adapts to environmental change.

9.4.5 *Challenges and Limitations*

Despite their potential, bioarchitectural systems face significant constraints. Limited rooting volume, substrate weight restrictions, and exposure to microclimatic extremes can compromise plant performance, particularly on balconies and terraces (Nazif and Yousefi 2024). Plants within indoor systems are limited by the availability of artificial or low-natural light and by humidity fluctuations associated with indoor ventilation (Liu et al. 2022a, b). Poorly designed plant layouts may also alter local airflow patterns or increase allergenic exposure, necessitating careful species selection and canopy-density control.

Maintenance failures pose a major risk; common causes of system degradation include inadequate irrigation, nutrient imbalance, and pest outbreaks. While automation and monitoring can improve performance, they also introduce additional costs and technical complexity (Bonifazi et al. 2023). Economic barriers, governance fragmentation, and uneven regulatory support further limit widespread adoption of bioarchitectural systems, despite enabling policy frameworks at the EU level (EC 2021; EEA 2021).

Ecological benefits also depend on plant selection. Simplified plantings or heavy reliance on exotic ornamental plants may reduce habitat value, whereas evidence from urban biodiversity research highlights the importance of native and structurally diverse *taxa* for supporting a wider range of organisms (Aronson et al. 2017).

9.4.6 *Case Studies*

9.4.6.1 Bosco Verticale Apartments (Milan, Italy)

Bosco Verticale, located in the Porta Nuova district of Milan, is an example of bioarchitecture integrating vegetation into high-rise residential buildings. Completed in 2014, the project consists of two residential towers (approximately 80 m and 112 m high) designed to host a vertically distributed vegetation system integrated into cantilevered balconies and façade elements (Giacomello and Valagussa 2015). The buildings are fully accessible and observable from the public realm.

From a botanical perspective, Bosco Verticale is characterised by the integration of approximately 13,000 plants, belonging to around 90–100 different plant *taxa*,

Fig. 9.10 Green balconies of Bosco Verticale (**a**) and green park surroundings (**b**), Milan, Italy. (Photos by E. Roccotiello)

both native and exotic. The vegetation comprises trees, shrubs, and perennial *taxa* intended to introduce a vertically layered arboreal and shrubby component into the built environment, reflecting the conceptual idea that urban density can coexist with the presence of a "forest", even in vertical form (Giacomello and Valagussa 2015).

Evergreen and deciduous tree *taxa* are combined with shrubs and perennial understorey plants, providing structural continuity, seasonal variation, and functional diversity. Plant distribution varies according to façade orientation, height, and microclimatic exposure. Evergreen *taxa* are predominantly located on north-facing façades to provide year-round cover, while deciduous *taxa* on south- and west-facing exposures contribute to seasonal shading and solar modulation (Giacomello and Valagussa 2015).

While integrating potted vegetation into façades can provide specific ecosystem services, a comprehensive assessment of such systems would require consideration of management costs, maintenance requirements, and long-term performance, which are beyond the scope of this chapter (Fig. 9.10).

9.4.7 Green Living Terrace Roofs Initiative (Barcelona, Spain)

Barcelona illustrates how bioarchitecture can be implemented at the city scale through distributed terrace and rooftop greening initiatives embedded within municipal climate adaptation policies. Rather than focusing on a single iconic building, this case reflects a policy-driven approach that promotes the gradual transformation of existing rooftops and terraces into green spaces, many of which are publicly or semi-publicly accessible and therefore directly observable and visitable.

As part of its urban resilience and climate adaptation strategies, the City Council of Barcelona has promoted green roofs and terraces through incentive programmes, technical guidelines, and pilot actions aimed at increasing vegetation cover in a compact Mediterranean city with high solar radiation, summer drought, and sporadic intense rainfall (Ajuntament de Barcelona 2013, 2015). These initiatives explicitly recognise terraces and rooftops as underutilised spaces with high potential for climate regulation and the provision of ecosystem services.

From a botanical standpoint, the municipal guidelines emphasise the use of Mediterranean and Mediterranean-adapted *taxa* selected for their drought tolerance, high resistance to high solar exposure, and low maintenance requirements. Although individual project documentation does not always provide exhaustive species lists, official technical documents consistently recommend stress-tolerant shrubs, perennial herbs, and groundcovers capable of surviving shallow to moderate substrate depths and irregular water availability (Ajuntament de Barcelona 2015).

Preferred plant palettes include evergreen and semi-evergreen shrubs, flowering perennials, and aromatic herbs that contribute to thermal buffering, rainfall interception, and biodiversity support. Using flowering *taxa* increases the availability of nectar and pollen resources for urban pollinators, while structurally diverse plant assemblages improve habitat value compared to non-vegetated roofs. Where structural conditions permit, deep-substrate terraces allow the inclusion of small woody *taxa*, further increasing ecological complexity and shading capacity.

Monitoring reports indicate measurable reductions in roof surface temperatures and improved stormwater management performance in vegetated systems compared with conventional roofs, particularly during summer heatwaves (Ajuntament de Barcelona 2013).

References

Adilkhanova I, Santamouris M, Yun GY (2024) Green roofs save energy in cities and fight regional climate change. Nat Cities. https://doi.org/10.1038/s44284-024-00035-7

Ajuntament de Barcelona (2013) Barcelona green infrastructure and biodiversity plan 2020. Urban Ecology Department, Barcelona City Council, Barcelona. https://ajuntament.barcelona.cat/ecologiaurbana/sites/default/files/Barcelona%20green%20infrastructure%20and%20biodiversity%20plan%202020.pdf

Ajuntament de Barcelona (2015) Guide to living terrace roofs and green roofs. Barcelona City Council, Barcelona. https://bcnroc.ajuntament.barcelona.cat/jspui/bitstream/11703/98795/5/Guia%20de%20terrats%20vius%20i%20cobertes%20verdes%20angl%C3%A8s.pdf

Akhie AA, Joksimovic D (2023) Monitoring of a productive blue-green roof using low-cost sensors. Sensors 23(24):9788. https://doi.org/10.3390/s23249788

Aronson MFJ, Lepczyk CA, Evans KL, Goddard MA, Lerman SB, MacIvor JS (2017) Biodiversity in the city: key challenges for urban green space management. Front Ecol Environ 15(4):189–196. https://doi.org/10.1002/fee.1480

Bakhshoodeh R, Ocampo C, Oldham C (2022) Evapotranspiration rates and evapotranspirative cooling of green façades under different irrigation scenarios. Energ Buildings 270:112223. https://doi.org/10.1016/j.enbuild.2022.112223

Bao S, Zou S, Zhao M, Chen Q, Li B (2022) Experimental study on modular vertical greening shading in summer. Int J Environ Res Public Health 19:11648. https://doi.org/10.3390/ijerph191811648

Batista T, Barros R, Domínguez JF, Cuevas RM, Ruiz JS, Casado ST (2022a) Climate change adaptation of buildings using nature-based solutions: application in Alentejo Central – LIFE myBUILDINGisGREEN. In: Calabrò F et al (eds) New metropolitan perspectives. LNNS 482. Springer, Cham. https://doi.org/10.1007/978-3-031-06825-6_159

Batista T, Torre S, Fermoso J, Morais N, Avila R, Pinto Gomes C, Castanho RA, Cabezas Fernández J (2022b) Nature-based solutions for climate change adaptation in scholar and social buildings. In: Castanho RA, Cabezas Fernández J (eds) Urban green spaces. IntechOpen. https://doi.org/10.5772/intechopen.105701

Bellini A, Bartoli F, Caneva G (2024a) Extensive Green Roofs (EGRs) and the five Ws: a quantitative analysis on the origin and evolution, aims, approaches, and botanical views. Sustainability 16(3):1033

Bellini A, Bartoli F, Kumbaric A, Caneva G, D'Amato L (2024b) Evaluation of Mediterranean perennials for extensive green roofs in water-limited regions: a two-year experiment. Ecol Eng 209:107399. https://doi.org/10.1016/j.ecoleng.2024.107399

Bellini A, Bartoli F, D'Amato L, Casalini R, Caneva G (2025a) Enhancing biodiversity and functionality of extensive green roof: a comparative study of five native Mediterranean perennial species in Rome. Build Environ 282:113285. https://doi.org/10.1016/j.buildenv.2025.113285

Bellini A, Savo V, Caneva G, D'Amico E, Casalini R, Bartoli F (2025b) How a green roof becomes biodiverse: vegetation analysis on a green roof with no maintenance in Rome (Italy). Plants 14:3180. https://doi.org/10.3390/plants14203180

Bonifazi G, Gasbarrone R, Serranti S (2023) Assessing the environmental performances of nature-based solutions implementation in urban environments through visible and near-infrared spectroscopy: a combined approach of proximal and remote sensing for monitoring and evaluation. Sustainability 15:16076. https://doi.org/10.3390/su152216076

Bosio M, Mazzei F, Brunoldi M, Massabó D, Vernocchi V, Parodi F, Prati P, Roccotiello E (2025) Experimental investigation of plants interactions with airborne pollutants in an atmospheric simulation chamber. Atmos Environ X 27:100355. https://doi.org/10.1016/j.aeaoa.2025.100355

Bowler DE, Callaghan CT, Felappi JF, Mason BM, Hutchinson R, Kumar P, Jones L (2025) Evidence-base for urban green-blue infrastructure to support insect diversity. Urban Ecosyst 28:54. https://doi.org/10.1007/s11252-024-01649-4

Burlando P, Ferrante A, Vagge I (2025) Multiscale and multidisciplinary method for plant selection to design green urban and peri-urban areas. Front Hortic 4:1354764. https://doi.org/10.3389/fhort.2025.1354764

Cabanek A, Zingoni de Baro ME, Newman P (2020) Biophilic streets: a design framework for creating multiple urban benefits. Sustain Earth 3:7. https://doi.org/10.1186/s42055-020-00027-0

Caddeo A, Cogoni A, Iiriti G, Loi MC, Marignani M, Podda L, Brundu G, Genovesi P, Carnevali L, Lazzaro L, Ardenghi NMG, Assini S, Arduini I, Viegi L, Bolpagni R, Stinca A, Bonini I, Buono S, Carranza ML, Cianfaglione K, Domina G, Guarino R, Lazzeri V, Magrini S, Marinangeli F, Giovino A, Montagnani C, Musarella CM, Spampinato G, Ranfa A, Proietti C, Salerno G, Villani MC (2020) Dai balconi ai parchi urbani: buone pratiche per un giardinaggio consapevole. Life ASAP (LIFE15 GIE/IT/001039 "Alien Species Awareness Program" – ASAP). ISBN 9788894354423

Caneva G, Kumbaric A, Savo V, Casalini R (2015) Ecological approach in selecting extensive green roof plants: a data-set of Mediterranean plants. Plant Biosyst 149(2):374–383

Capotorti G, Bonacquisti S, Abis L, Aloisi I, Attorre F, Bacaro G et al (2020) More nature in the city. Plant Biosyst 154(6):1003–1006

Cook EM, Kim Y, Grimm NB, McPhearson T, Anderson P, Bulkeley H, Collier MJ, Diep L, Morató J, Zhou W (2025) Nature-based solutions for urban sustainability. Proc Natl Acad Sci USA 122:e2315909122. https://doi.org/10.1073/pnas.2315909122

De Cristo M, Rossi F, Bianchi L, Esposito V, Marino A (2025) A systematic review of green roofs' thermal and energy performance in the Mediterranean region. Sustain Cities Soc 89:102517. https://doi.org/10.3390/en18102517

Dunnett N, Kingsbury N (2008) Planting green roofs and living walls. Timber Press, Portland

European Commission (2021) Communication from the Commission to the European Parliament, the Council, the European Economic and Social Committee and the Committee of the Regions: Forging a climate-resilient Europe – the new EU Strategy on Adaptation to Climate Change. COM(2021) 82 final. Brussels, 24 February 2021. https://eur-lex.europa.eu/legal-content/EN/TXT/?uri=CELEX:52021DC0082

European Environment Agency (2021) Nature-based solutions in Europe: policy, knowledge and practice for climate change adaptation and disaster risk reduction, EEA Report No 01/2021. Publications Office of the European Union, Luxembourg. https://doi.org/10.2800/919315

European Environment Agency (2024) Trends and projections in Europe 2024, EEA Report 11/2024. Publications Office of the European Union, Luxembourg. https://doi.org/10.2800/7574066

Farrell C, Livesley SJ, Arndt SK, Beaumont L, Burley H, Ellsworth D, Esperon-Rodriguez M, Fletcher TD, Gallagher R, Ossola A, Power SA, Marchin R, Rayner JP, Rymer PD, Staas L, Szota C, Williams NSG, Leishman M (2022) Can we integrate ecological approaches to improve plant selection for green infrastructure? Urban For Urban Green 76:127732. https://doi.org/10.1016/j.ufug.2022.127732

Fineschi S, Loreto F (2020) A survey of multiple interactions between plants and the urban environment. Front For Glob Change 3:30. https://doi.org/10.3389/ffgc.2020.00030

Fonseca F, Paschoalino M, Silva L (2023) Health and well-being benefits of outdoor and indoor vertical greening systems: a review. Sustainability 15(5):4107. https://doi.org/10.3390/su15054107

Gaffin SR, Khanbilvardi R, Rosenzweig C (2009) Development of a green roof environmental monitoring and meteorological network in New York City. Sensors 9:2647–2660. https://doi.org/10.3390/s90402647

Getter KL, Rowe DB (2006) The role of extensive green roofs in sustainable development. HortScience 41(5):1276–1285. https://doi.org/10.21273/HORTSCI.41.5.1276

Giacomello E (2021) Green roofs, façades and vegetative systems. Butterworth-Heinemann/Elsevier, Oxford. https://doi.org/10.1016/C2017-0-01917-1. ISBN 978-0-12-817694-8

Giacomello E, Valagussa M (2015) Vertical greenery: evaluating the high-rise vegetation of the Bosco Verticale, Milan. Council on Tall Buildings and Urban Habitat, Chicago. https://cloud.ctbuh.org/research/seed-funding/past-projects/BoscoResearchReport.pdf

Green Roof Organisation (2021) The GRO green roof code of best practice: incorporating blue roofs and biosolar applications, Anniversary edition 2021. Green Roof Organisation, London. https://www.greenrooforganisation.org

Hodgson JG, Wilson PJ, Hunt R, Grime JP, Thompson K (1999) Allocating C-S-R plant functional types: a soft approach to a hard problem. Oikos 85:282–294. https://doi.org/10.2307/3546494

Hoffmann KA, Šuklje T, Kozamernik J, Nehls T (2021) Modelling the cooling energy saving potential of facade greening in summer for a set of building typologies in mid-latitudes. Energ Buildings 238:110816. https://doi.org/10.1016/j.enbuild.2021.110816

IPCC (2023) AR6 synthesis report: climate change 2023. Intergovernmental Panel on Climate Change, Geneva

ISO (2024) ISO 16813:2024 Building environment design – indoor environment – general principles. International Organization for Standardization, Geneva

Kim ES, Yun SH, Lee DK, Kim NY, Piao ZG, Kim SH, Park S (2023) Quantifying outdoor cooling effects of vertical greening system on mean radiant temperature. Dev Built Environ 15:100211. https://doi.org/10.1016/j.dibe.2023.100211

Köhler M (2008) Green facades—a view back and some visions. Urban Ecosyst 11:423–436. https://doi.org/10.1007/s11252-008-0063-x

Laface VLA, Musarella CM, Tavilla G, Cambria S, Maruca G, Giusso del Galdo G, Spampinato G (2023a) Carpological analysis of two endemic Italian species: *Pimpinella anisoides* and *Pimpinella gussonei* (Apiaceae). Plants 12:1083. https://doi.org/10.3390/plants12051083

Laface VLA, Sicari V, Musarella CM, Spampinato G, Maruca G, Giuffrè AM (2023b) Morphological and phytochemical features in fruits of *Pimpinella anisoides* V.Brig. (Apiaceae). Plant Biosyst 157:210–220. https://doi.org/10.1080/11263504.2022.2160025

Laface VLA, Cavallini M, Di Iorio A, Lombardo G, Binelli G, Sorgonà A, Musarella CM, Spampinato G (2024) Genetic structure of populations of *Salvia ceratophylloides* endemic to southern Calabria (southern Italy). Heliyon 10:e35875. https://doi.org/10.1016/j.heliyon.2024.e35875

Landscape Development & Landscaping Research Society e.V. (FLL) (2018) Green roof guidelines: guidelines for the planning, construction, and maintenance of green roofs, 2018th edn. FLL, Bonn. https://www.fll.de

Lima E, Ferreira H, Mateus L, Arruda A (2025) Guidelines for the convergence of bio-architecture and neuroarchitecture based on the WELL building standard. Energ Buildings 346:116141. https://doi.org/10.1016/j.enbuild.2025.116141

Lin C, Zhang Y, Wang H, Liu J, Chen X (2024) Impact of green roofs and walls on the thermal environment of pedestrian heights in urban villages. Buildings 14:4063. https://doi.org/10.3390/buildings14124063

Liu F, Yan L, Meng X, Zhang C (2022a) A review on indoor green plants employed to improve indoor environment. J Build Eng 53:104542. https://doi.org/10.1016/j.jobe.2022.104542

Liu J, Wang X, Chen Y (2022b) Light tolerance and physiological responses of indoor ornamental plants. Urban For Urban Green 68:127484. https://doi.org/10.1016/j.ufug.2022.127484

Lundholm JT, Walker EA (2018) Evaluating the habitat template approach applied to green roofs. Urban Nat 1(1):39–55

Mohammed AB (2022) Employing systems of green walls to improve performance and rationalize energy in buildings. J Eng Appl Sci 69:99. https://doi.org/10.1186/s44147-022-00154-9

Moroni M, Porti M, Piro P (2019) Design of a remote-controlled platform for green roof plants monitoring via hyperspectral sensors. Water 11(7):1368. https://doi.org/10.3390/w11071368

Nash C, Clough J, Gedge D, Lindsay R, Newport D, Ciupala MA, Connop S (2016) Initial insights on the biodiversity potential of biosolar roofs: a London Olympic Park green roof case study. Isr J Ecol Evol 62(1–2):74–87. https://doi.org/10.1080/15659801.2015.1045791

Nash C, Ciupala MA, Gedge D, Lindsay R, Connop S (2019) An ecomimicry design approach for extensive green roofs. J Living Archit 6(1):62–81. https://doi.org/10.46534/jliv.2020.06.01.062

Nazif H, Yousefi NF (2024) The effects of greenery in balconies of apartments on people's well-being: using virtual reality. Qeios 10(1):juae005. https://doi.org/10.32388/XZAMMG

Oberndorfer E, Lundholm J, Bass B, Coffman RR, Doshi H, Dunnett N, Gaffin S, Köhler M, Liu KKY, Rowe B (2007) Green roofs as urban ecosystems. Bioscience 57(10):823–833. https://doi.org/10.1641/B571005

Ordóñez Barona C, Trammell TLE (2022) Editorial: urban trees in a changing climate: science and practice to enhance resilience. Front Ecol Evol 10:882510. https://doi.org/10.3389/fevo.2022.882510

Panuccio MR, Fazio A, Musarella CM, Mendoza-Fernández AJ, Mota JF, Spampinato G (2018) Seed germination and antioxidant pattern in *Lavandula multifida* (Lamiaceae): a comparison between core and peripheral populations. Plant Biosyst 152:398–406. https://doi.org/10.1080/11263504.2017.1297333

Patti M (2024) Conservation and valorization of two endemic Calabrian species used as food within the Graecanic Area (Southern Italy) through domestication trials. In: Calabrò F, Madureira L, Morabito FC, Piñeira Mantiñán MJ (eds) Networks, Markets & People, Lecture notes in networks and systems, vol 1185. Springer, Cham. https://doi.org/10.1007/978-3-031-74672-7_39

Patti M, Musarella CM, Spampinato G (2025) A habitat-template approach to green wall design in Mediterranean cities. Buildings 15:2557. https://doi.org/10.3390/buildings15142557

Perini K, Roccotiello E (2018) Vertical greening systems for pollutants reduction. In: Pérez G, Perini K (eds) Nature-based strategies for urban and building sustainability. Butterworth-Heinemann, Oxford. https://doi.org/10.1016/B978-0-12-812150-4.00012-4

Perini K, Ottelé M, Haas EM, Raiteri R (2013) Vertical greening systems, a process tree for green façades and living walls. Urban Ecosyst 16:265–277. https://doi.org/10.1007/s11252-012-0262-3

Perini K, Ottelé M, Giulini S, Magliocco A, Roccotiello E (2017) Quantification of fine dust deposition on different plant species in a vertical greening system. Ecol Eng 100:268–276. https://doi.org/10.1016/j.ecoleng.2016.12.032

Pianta M, Calbi M, Cerabolini BEL, Poschlod P, Schloter M, Roccotiello E (2025) How substrate depth shapes spontaneous plant colonisation on Mediterranean green roofs: a case study. Urban For Urban Green 114:129130. https://doi.org/10.1016/j.ufug.2025.129130

Piro P, Porti M, Veltri S, Lupo E, Moroni M (2017) Hyperspectral monitoring of green roof vegetation health state in sub-Mediterranean climate: preliminary results. Sensors 17(4):662. https://doi.org/10.3390/s17040662

Priya UK, Senthil R (2025) Influence of balcony greenery on indoor temperature reduction in tropical urban residential buildings. Energ Buildings 343:115915. https://doi.org/10.1016/j.enbuild.2025.115915

Raimondi A, Becciu G (2021) Performance of green roofs for rainwater control. Water Resour Manag 35:99–111. https://doi.org/10.1007/s11269-020-02712-3

Regulation (EU) 2024/1991 on nature restoration and amending Regulation (EU) 2022/869. Official Journal of the European Union, 24 June 2024. https://eur-lex.europa.eu/legal-content/EN/TXT/PDF/?uri=OJ:L_202401991

Reyes R, Bustamante W, Gironas J, Pastén PA, Rojas V, Suárez F, Vera S, Victorero F, Bonilla CA (2016) Effect of substrate depth and roof layers on green roof temperature and water requirements in a semi-arid climate. Ecol Eng 97:624–632. https://doi.org/10.1016/j.ecoleng.2016.10.025

Safikhani T, Abdullah AM, Ossen DR, Baharvand M (2014) A review of energy characteristic of vertical greenery systems. Renew Sust Energ Rev 40:450–462. https://doi.org/10.1016/j.rser.2014.07.166

Soni L, Farrell C, Fletcher TD, Szota C (2025) Climate is more important than green roof design for rainfall retention and plant drought stress in temperate and semi-arid climates. Urban Water J 22(7). https://doi.org/10.1080/1573062X.2025.2509071

Spampinato G, Laface VLA, Posillipo G, Ortiz AC, Canas RC, Musarella CM (2022) Alien flora in Calabria (Southern Italy): an updated checklist. Biol Invasions 24:2323–2334. https://doi.org/10.1007/s10530-022-02800-y

Todeschini CC, Fett-Neto AG (2025) Life at the top: extensive green roof plant species and their traits for urban use. Plants 14(5):735. https://doi.org/10.3390/plants14050735

Treder M, Joedecke V, Schweikert K, Rosenkranz P, Ruttensperger U, Traynor K (2024) Vertical greening systems serve as effective means to promote pollinators: experimental comparison of vertical and horizontal plantings. Landsc Urban Plan 243:104951. https://doi.org/10.1016/j.landurbplan.2023.104951

UNEP (United Nations Environment Programme) (2022) Nature-based solutions: opportunities and challenges for scaling up. United Nations Environment Programme, Nairobi

UNI (2015) UNI 11235:2015. Istruzioni per la progettazione, esecuzione, controllo e manutenzione di coperture continue a verde. Ente Nazionale Italiano di Unificazione (UNI), Milano. https://store.uni.com/uni-11235-2015

Van Mechelen C, Dutoit T, Kattge J, Hermy M (2014) Plant trait analysis delivers an extensive list of potential green roof species for Mediterranean France. Ecol Eng 67:48–59. https://doi.org/10.1016/j.ecoleng.2014.03.043

Visvanathan G, Patil K, Suryawanshi Y, Meshram V, Jadhav S (2024) Mitigating urban heat Island and enhancing indoor thermal comfort using terrace garden. Sci Rep 14:9697. https://doi.org/10.1038/s41598-024-60546-0
Wang D, Xu ABW, Guo QPPY (2024) Urban green infrastructure: bridging biodiversity conservation and sustainable urban development through adaptive management approach. Front Ecol Evol 12. https://doi.org/10.3389/fevo.2024.1440477
Wijesooriya N, Brambilla A, Markauskaite L (2023) Biophilic design frameworks: a review of structure, development techniques, and their compatibility with LEED sustainable design criteria. Clean Prod Lett 4:100033. https://doi.org/10.1016/j.clpl.2023.100033
Wilson EO (1984) Biophilia: the human bond with other species. Harvard University Press, Cambridge, MA
World Green Infrastructure Network (2021) CaixaForum Museum Vertical Garden. Project profile. https://worldgreeninfrastructurenetwork.org/caixa-forum-museum-vertical-garden/
World Health Organization – WHO (2016) Urban green spaces and health: a review of evidence. WHO Regional Office for Europe, Copenhagen
Zorić M, Simic M, Orlović S, Miladenovic E (2019) Indoor ecosystem services: impacts of plants on air quality. Contemp Agric 68:12–16. https://doi.org/10.2478/contagri-2019-0003

FSC
www.fsc.org
MIX
Papier | Fördert gute Waldnutzung
FSC® C083411

Kolibri 360 GmbH
Ferdinand-Jühlke-Straße 7
99095 Erfurt, Deutschland
produktsicherheit@kolibri360.de